Praise for

The Story is in Our Bon

These pages summon from our bones our commitment to defend this living Earth. I bow to Osprey in deepest respect and gratitude for her years of inspired activism and this brilliant book.

—Joanna Macy, environmental activist, scholar, Buddhism, general systems theory, and deep ecology, author, *Coming Back to Life* and *Active Hope,* and featured, *A Wild Love for the World: Joanna Macy and the Work of Our Time*

Osprey Orielle Lake has given us a magnificent book loaded with knowledge, wisdom, and fine storytelling. In it she lays out a tapestry of multiple pathways that unite to demand humility in our relationship with Mother Earth. The book exposes colonialism, imperialism, racism, capitalism, and patriarchal systems as the underlying factors that have fostered an extractivist, ecologically degrading mindset that drives the current polycrisis. With lavish examples of traditional ecological knowledge, reciprocal economic and governance frameworks, and new narratives, *The Story is in Our Bones* does not leave the reader exasperated and helpless—it is an empowering call for action.

—Nnimmo Bassey, author, *To Cook a Continent: Destructive Extraction and the Climate Crisis in Africa,* Right Livelihood Award winner

As a young Indigenous woman, it is important to me that we consider all the complex intersections of colonialism, racism, patriarchy, capitalism, and ecocide while building a better world. This incredibly important and timely book includes the memory and knowledge of how we can live in balance with nature, which still lives on in Indigenous communities and is crucial to solving the multiple crises we are facing!

—Helena Gualinga (Kichwa from Sarayaku), Indigenous youth climate leader, Ecuadorian Amazon

The Story is in Our Bones is a remarkable achievement, a rich read, and one surely not to miss. For anyone who wonders—as I often do—how on Earth we're going to navigate the seemingly intractable confluence of crises, this extraordinary book offers a very potent recipe, spanning culture, global systemic change, sense-making, and remembrance of our Earth legacy. The book resonates from mind to belly to bones.

—Nina Simons, co-founder, chief relationship officer, Bioneers

Osprey Orielle Lake guides us on a majestic journey of sense making for the 21st century as we attempt to emerge from emergency. She leads us through the importance of adopting a systems approach that fosters new economic models and the need to value nature and climate justice. The resounding message throughout this book is to act with urgency and purpose in these times of interlocking crises.

—Sandrine Dixson-Declève, co-president, The Club of Rome,
co-author, *Earth for All*

In this beautifully written book, *The Story is in Our Bones* offers a frank acknowledgment of the Anthropocene that serves as a vital, yet sober grounding in what we should already know but many are in denial to fully admit. At the same time, Osprey skillfully weaves history, mythology, anthropology, climate and earth science, sociology, and spirituality to illustrate the central message. Capitalism and colonialism have gotten us on this path of catastrophic climate change, but as she says, they can be transformed. Whether it is learning from ancestors from Ukraine, movements like *Via Campesina,* or women foresters from the Democratic Republic of Congo, the path to a Just Transition and healthy ways of living like *Buen Vivir,* are rooted in interconnection and in learning from the story. If we do this, we all thrive, nestled in the bosom of Mother Earth.

—Jacqui Patterson, founder, executive director,
The Chisholm Legacy Project

This book traces luminous threads of possibility away from extractive collapse, coalescing back into reciprocity with sovereign living processes. Osprey Orielle Lake reminds us of the ancient lineage of regeneration, alive in our cells, awakening now in sacred form and practical action, in just the right places and forms to bring down the planetary fever.

—Stuart Cowan, executive director, Buckminster Fuller Institute,
co-author, *Ecological Design*

In this landmark offering, Lake, a tireless campaigner for a just and vibrant world, gives voice to those who have long been marginalized by the dominant culture: Indigenous and Black women from around the world along with the multitudes of our nonhuman relatives. At its core, this marvelous wide-ranging book takes us on a deep dive into root causes of our polycrisis and with flair and scholarship delivers a roadmap toward cultural transformation.

—Jeremy Lent, author, *The Web of Meaning*
and *The Patterning Instinct*

Reading this book in these dark times of increasing ecological destruction, is like being a salmon in the depths who scents the stream of origin that will guide it home. Osprey Orielle Lake speaks with great wisdom and scholarship—interweaving her exquisite sensitivity for the voice of the wild with her vast experience as a movement leader, and the knowledge of the many frontline communities she stands with. Reassuring us that the wisdom of our Earth-loving ancestors is still within us, Osprey shows how people all over the world are rising to defend Earth and bring more just and ecologically benign societies into being.

—Cormac Cullinan, author, *Wild Law: A Manifesto for Earth Justice,*
director, Wild Law Institute

Osprey Orielle Lake, in her magnificent *The Story is in our Bones,* offers us a new cosmology and a new lens with which to see reality. By combining the wisdom in Indigenous origin stories from around the planet with modern ecological knowledge, her work awakens a radical imagination capable of ushering forth a vibrant Earth Community. If you read her book and dwell in its wisdom, you will soon find yourself in the next era of your creative life.

—Brian Thomas Swimme, author, *Cosmogenesis: An Unveiling of the Expanding Universe,* director, Human Energy

This is a very valuable book. It delves deep into what we can and must learn from both Indigenous worldviews and the natural world that has helped inform them, and it does so without sentimentality or rancor; in so doing, it opens a number of paths for everyone trying to think more wisely about how we can inhabit a planet in fundamental crisis. It would best be read not as an intellectual exercise but as a guidebook to real change.

—Bill McKibben, author, *The End of Nature,* founder, Third Act

This is a profound and much-needed book. I am grateful to Osprey Orielle Lake for presenting an in-depth analysis of the root causes of the ecological crises we face and for paths forward to secure the future of humanity in harmony with nature. With gorgeous poetics and precise logic, the chapters show us how to build a thriving future informed by radical imagination, science, Indigenous People's wisdom, and principles of climate justice. Simply stunning.

—Farhana Yamin, lawyer, climate activist, Honorary Fellow, Somerville College, Oxford University

THE STORY IS IN OUR BONES

How Worldviews and Climate Justice Can Remake a World in Crisis

OSPREY ORIELLE LAKE

Foreword by Casey Camp-Horinek, Ponca Nation

Cover design by Diane McIntosh.

Cover art by Christi Belcourt, a Métis visual artist and author living and working in Canada. The title of the painting is *Offerings and Prayers for Genebek Ziibiing.* As Christi explains, "Between 1955 to 1978, there were over 30 tailings dumps and spills from uranium mines at Elliot Lake into 10 lakes and Serpent River (Ontario, Canada). The radiation from uranium dumps completely killed the life in the waters, and the people of Genaabaajing are still living with the devastating environmental effects today. Water is the very lifeblood of Mother Earth. Water, and our connection to everything in the spirit world, depends upon us keeping everything in balance. Every time we make our offerings and say words to the waters, it's helping to restore the balance the Earth needs. I send my love to this river, and all waters. This painting was inspired by the stories and teachings of Isaac Murdoch. Miigwetch Issac for continuing to teach the people how to live in balance." https://christibelcourt.ca/

Printed in Canada. First printing January, 2024.

Any other inquiries can be directed by mail to:

New Society Publishers, P.O. Box 189, Gabriola Island, BC, V0R 1X0, Canada
(250) 247-9737

Library and Archives Canada Cataloguing in Publication

Title: The story is in our bones : how worldviews and climate justice can remake a world in crisis / Osprey Orielle Lake ; foreword by Casey Camp-Horinek.
Names: Lake, Osprey Orielle, author.
Description: Includes bibliographical references and index.
Identifiers: Canadiana (print) 20230562604 | Canadiana (ebook) 20230562671 | ISBN 9780865719941 (softcover) | ISBN 9781550927870 (PDF) | ISBN 9781771423830 (EPUB)
Subjects: LCSH: Traditional ecological knowledge. | LCSH: Human ecology. | LCSH: Climate justice. | LCSH: Sustainability.
Classification: LCC GN476.7 .L35 2024 | DDC 304.2—dc23

Funded by the Government of Canada | Financé par le gouvernement du Canada | Canada

New Society Publishers' mission is to publish books that contribute in fundamental ways to building an ecologically sustainable and just society, and to do so with the least possible impact on the environment, in a manner that models this vision.

For the Earth and All Generations

You have to act as if it were possible to radically transform the world. And you have to do it all the time.

—Angela Davis

The story is not told to lift you up, to make you feel better, or to entertain you, although all those things can be true. The story is meant to take the spirit into a descent to find something that is lost or missing and to bring it back to consciousness again.

—Clarissa Pinkola Estés

Contents

Part IV

Part V

Author's Note

On this misty morning, I am looking out at the coastal redwood trees near my home in Northern California. The tallest trees in the world, their massive columns lift up the heavens, reminding me of the tradition of the World Tree. This ancient mythology encapsulates the whole cosmos: the vast skies in the branches, the Earth in the trunk, and the life-nourishing underworld in the roots.

Cultures around the globe each have unique World Tree stories and teachings specific to their home ecologies, and yet in each case, the World Tree is an expression of the axis mundi—the world pillar and center of the cosmos, which represents the universe in balance. The World Tree traditions convey a respect for nature and our responsibility to maintain a healthy relationship with the Earth and all the realms. Human systems have now catastrophically disrupted this interconnected living ecology. The subsequent dangerous imbalance, and all that it means for the Earth community's future, has compelled me to write this book.

Along with many people globally, I have committed my life's work to addressing the dire circumstances we face, and to envisioning and building the healthy and just world we know is possible. This moment calls for deep systemic change entailing a metamorphic shift in worldview: literally, how we understand the world and our relationship and responsibility to the web of life and each other. Worldview is a vast topic that is as crucial and marvelous as it is mammoth, with dramatic differences depending on culture, environment, ancestry, mythology, epistemology, and so much more. Thus, the chapters ahead, though they cover a range of geographies, are just one offering to a collective conversation that is sprouting in varied fields of knowledge and around the world in diverse contexts.

Many of us are in the midst of the great task of changing who we are and how we interact with each other and the land, which means fundamentally addressing how we see the world. I invite you in this book to bring our collective learning together so we can coherently see the world in balance once again within the many unique expressions of the ancient World Tree—meaning, here, restoring and generating flourishing and equitable relationships with people and the web of life.

Because I have experienced Indigenous worldviews and knowledge as central to our path forward, I have featured Indigenous leaders throughout numerous chapters. Where I have quoted these leaders or shared their stories, I have either asked their permission directly or have quoted previously published materials with citations. Additionally, in some chapters, I share longer block quotes from frequently marginalized voices that, due to our current unjust social constructs, are not heard. Yet these are the voices we need to listen to, and I kindly welcome you to sit with their words.

Thank you for picking up this book and engaging with me in this vital conversation as we explore and take action for a different future in urgent times. The World Tree may be withering under current assault, but together and with future generations, we have the possibility to bring her magnificently back to health.

Foreword

By Casey Camp-Horinek, Ponca Nation—Environmental Ambassador,
Hereditary Drum Keeper of the Ponca Nation

THIS IS A CRITICAL TIME IN THE HISTORY of our species on Earth. Our spiritual leaders call it the Great Purification, when the land rises up to counter the effects of humans tampering with the sacred systems of life. To heal herself, Mother Earth's fires, strong winds, ice storms, floods, and hurricanes are clearing the landscape. Mother Earth and her unbreakable laws will endure—yet will our species continue to exist?

I am called Casey Camp-Horinek—Zhuthi is my true name. My people are the Ponca, meaning "Sacred Head," and Wakonda (Creator) placed us in the northern plains of Turtle Island (North America). We were caretakers of over 2.4 million acres in what is now called Nebraska and South Dakota. Nebraska is derived from a Ponca word, *Ni'bthaska,* meaning "water that is flat" because it lies atop an enormous aquifer, the Ogallala. We hold much knowledge in our stories of the water, land, and sky.

After the federal government forcibly removed the Ponca from our land under the Doctrine of Discovery, we, like most other Indigenous Nations, walked a trail of tears, having to leave behind our seeds, our hunting and fishing, our homes, and our relationship with our lands. Many people died on the long journey—and those who survived were forced to model their Nations and families after the colonizers, who broke treaty after treaty.

As a Ponca woman living on the occupied land now called Oklahoma, I am intimate with what happens when people attempt to live outside the laws of nature. This inclination comes from a worldview dominated by ongoing extraction from the Earth and a willingness to take without any consideration of the devastating consequences.

Years after the Ponca were removed from our homelands, we're still experiencing genocide—now by fossil fuel companies through fracking, injection wells, and the male workers who set up camps on Tribal Lands and assault Native women. The fossil fuel corporations, along with industrial farms, defile our Mother Earth by pouring toxins into the land, air, and water. Our small community conducts funerals almost every week due to illnesses directly caused by fossil fuel pollution—all reflections of the harms Mother Earth herself is feeling.

Long ago, humans were part of the sacred circle of life that existed in perpetual reciprocity. In order to join that community, humans entered into an agreement to live in harmony with all other beings. My ancestors understood the importance of following the Original Instructions for interacting with the web of life—and I believe your ancestors made that promise too, wherever they are from. It is now time to bring ourselves back into balance with our sacred agreements.

In my capacity as the Environmental Ambassador for the Ponca Nation of Oklahoma, I have been honored to educate Indigenous and non-Indigenous people about environmental and human rights, including Rights of Nature. Within our own sovereign laws, the Ponca are re-establishing the legal Rights of our Mother Earth in order to protect the area in which we live, which includes not only our own community but also our relatives the four-legs, our relatives with fins, our relatives with wings, and even the fossil fuel workers and industrial farmers. In 2022, we legally recognized the sacred rights of the two rivers that surround us, protecting them from fracking, which kills the fish and causes devastating earthquakes.

Our collective way forward means realigning our human laws and ways of living with the natural laws. After all, what are we if not Nature herself? As a living entity, you must eat, drink, breathe, and be warm. All of the food we eat comes from the source of life, which is our Mother Earth who grows the plants and nourishes the animals. We breathe the sacred breath, the trees' gift of oxygen that travels on the winds from all directions, passing through the deer and flying on the wings of an eagle. Mother Earth nurtures us with unconditional love, but we act like we own everything. For humans to imagine themselves as separate from and above nature is a huge mistake. By removing ourselves from the natural systems through our arrogance, we are likely to remove ourselves from the face of the Earth.

Remember, if you ate this morning, if you breathed air, if you drank water, then it is time for you to acknowledge your responsibility to and relationship with Mother Earth and Father Sky. We cannot wait any longer. This movement away from fossil fuels and other extractive industries will require a monumental shift in the way society thinks, what we value, and how we make decisions.

In our community, matriarchs shape and share ways of living and understanding the world in balance with masculine energies. As a Ponca, I act as an Elder and traditional Drum Keeper in the Women's Society. Women have an immense capacity for leadership, and now is the time for us to use that ability to stand up for our Mother Earth. Around the world, from South America to Africa, to Asia, to Europe, to the Island Nations, and to North America, many of Mother Earth's strongest protectors are women. We need to gather together—women and allies of all genders—to discuss what must be done for the next seven generations.

In this defining moment in time, I am honored to serve as a board member of the Women's Earth and Climate Action Network (WECAN), an organization founded by Osprey Orielle Lake to mobilize women in all of their diversity to protect and defend Mother Earth. I have known Osprey for over a decade, and she is my adopted niece. I can confirm that she stands for the same values I do. And far beyond writing about these issues, she stands on the front lines with us in our struggles and in our organizing to bring forth a future of hope and beauty.

I was so pleased when Osprey told me she was writing this book because the deep discussions she presents regarding understanding and changing our worldviews are crucial to the discourse humanity needs and actions we need to take to imagine our future. She has laid out a clear road map to move through this time of crisis, to help shift the destructive mindset of our societies, and to remember the sacred agreement our species made with the Creator, the Earth Mother. This book expresses the hopes, ideas, and aspirations of many who love the Earth, who stand with Her as protectors and healers, with no way back—only forward. I hope these visionary and transformative words are shared far and wide so that more and more of us can gather around the table, discuss our human legacy, and take action together.

You have the spirit living within you that is connected to all of life. You have the capacity to enact changes that honor all of Creation. Mother Earth needs all of our voices and energy in these precious years ahead, so let this truly vital and vibrant work offered by Osprey deeply nourish you and carry you onward.

Part I

Entering the Terrain of Worldviews and Climate Justice

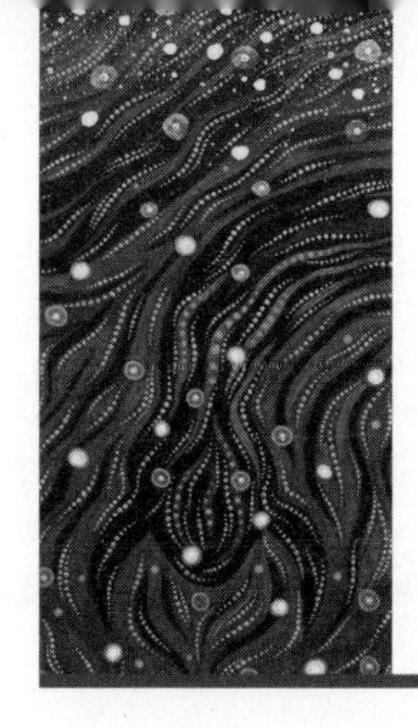

Chapter 1:

Worldviews Are a Portal

Tutored by the Stars

NIGHTFALL ARRIVES IN TEXTURED INDIGO and sable-black banners as I make my way along a narrow trail to the brim of the headlands and breathe in nontoxic air for the first time in many weeks. Another climate crisis-induced fire season lingers on this Northern California coast, and I have come to stand on the ocean cliffs to embrace an exquisite fresh breeze. The dense smoky haze has finally dispersed after seemingly endless fires and hazardous air quality, which for several days measured the worst in the world.

Breathing in the smoke that blanketed California for those weeks was to be besieged with a visceral experience of devastating sorrow. The smoke and ash were the aching remnants of wondrous pine and oak trees, deer, foxes, squirrels, bobcats, and many other creatures who lost their lives in the fire. The smoke mixed with particulates of destroyed homes and entire towns burned to the ground, toxins from plastics and chemicals—and yes, the airborne remains of family members and beloveds who died trying to escape the unrelenting flames. This is what climate chaos looks like and feels like, up close and personal—in our bodies.

This fire, the Camp Fire of November 2018, marks one of many distinct climate tipping points, a grisly augury of California's intensifying fire seasons. Each year, infernos blaze fast and furious across the state, with valiant firefighters risking their lives to quench the flames, and residents poised to flee at a moment's notice. The Camp Fire killed 85 people and scorched 153,336 acres.[1] Since then, even larger fires have raged in the Amazon, Africa, and Australia. It seems whole continents are ablaze. Errant human activity has created the precise conditions for increased fires—and Nature is responding resolutely, in the language of infernos.[2]

Standing on the bluffs this clear night, I feel some blessed relief, I breathe deeply, trying to clear my mind, and to assemble some sense-making of this moment in the human journey. I stand not only at the edge of night and sea but at the edge of a world.

The crises around us are cascading and interlocking—climate chaos, colonization, racism, patriarchy, ecological systems in collapse, an economic system based on endless extractivism. It's clear that we cannot continue as we are. In fact, it is impossibly arrogant to parley with the natural laws of the Earth, and unacceptable that egregious, long-standing human injustices persist. Our age cries out for a multitude of reparations, rememberings, reimaginings, and restorations—both within our personal lives and the public sphere.

As I see it, while one world is burning and unraveling, another world is being called forth by forces stirring deep within the spirit of our collective psyches.

As the full darkness of the night settles around me, my attention falls upward into the vast bejeweled landscape of the starry firmament. I have entered the darkness to listen to the stories and ancestors in the sky. Not to seek transcendence away from Earth—no, not at all—but to seek guidance and perspective from the great wheel of time and from the old ones who peer downward with their glimmering eyes. I ask how humanity, and I personally, can live responsibly and vibrantly in this perilous, uncertain moment. And in this quest, how can we restore all that we can of life-sustaining ecosystems while becoming good ancestors for future generations—here, now, in the liminal time of the Anthropocene.*

Can we undo the doing?

I passionately explore this realm of inquiry as part of my work at the Women's Earth and Climate Action Network (WECAN), which I had the honor of founding in 2009. Women in all their diversity are critical to implementing successful solutions to the climate crisis and environmental degradation. Globally, women are responsible for the majority of the world's food production, and in most Global South countries, women

* The Anthropocene is the current geological epoch, defined by the domination of the Earth by human activity, which has affected Mother Earth's natural processes and systems. I note that this term is not quite sufficient on its own because it flattens responsibility across all of humanity, and names that the defining problem is human activity, not particular systems and structures.

produce between 40 and 80 percent of food, and are central stewards of seeds and agricultural biodiversity.[3] A detailed study found that a one-unit increase in a country's score on the Women's Political Empowerment Index leads to an 11.51 percent decrease in the country's carbon emissions.[4]

Yet, women are at the same time disproportionately impacted by ecological crises because of global gender inequality.[5] Women more than men engage in subsistence farming and fuel and water collection. As droughts, floods, and other erratic weather events increase, there is an increased burden on women holding responsibility for their family's food, water, and energy needs.[6]

Globally, women's basic rights continue to be denied in varying forms and intensities, from lack of education opportunities to gender-based violence—and we cannot discuss gender inequality without addressing its inextricable relationship to racism and the additional disproportionate impacts of extractive industries and socio-ecological harms to Indigenous, Black, and Brown women. Because of these interlocking injustices, there's a tacit understanding in our global advocacy networks of the need to stand together, including with many of our brothers in solidarity, as we fight interconnected systems of oppression in the "doings" of these times.

Along with a wide range of topics in the chapters ahead, you will hear from many of the truly remarkable women and feminist leaders across the gender spectrum whom I have had the honor of learning from and working with in regions around the world. Through their stories, advocacy, and campaigns, we can learn a great deal about undoing harms to our communities and the Earth—and, importantly, about forging new ways of thinking, healing, and being.

Collectively, many of us who are working to avert the worst of the climate crisis are endeavoring to protect forests, waterways, land, oceans, biodiversity, and frontline communities, while also creating thriving spaces of decolonized discourse, sites of safe harbor, and lands of liberation. For my part, the simple fact is that I am wildly in love with our beautiful planet, and I will never stop doing all that I can to defend and protect Mother Earth, our communities, and future generations. Through this work, I often find my deepest hope in the efforts of everyday people, civil society, and global movements. We can be found in the streets, in the fields, in the boardrooms, in educational institutions, and in the halls of power. At critical times, we are putting our bodies on the line, standing up for the last of the clean water, the last of the forests, for what we hold sacred—for the last chance for our climate and a livable future.

To state it plainly, what is happening in our world right now is not only unjust, it is insane—our very home is being defiled. And within this reality, many of us are not willing to sacrifice current and future generations for reasons of misbegotten fears and belief systems, greed, political power, and hubris. Rather, we are willing to use our imaginations, to be alive, to thrive, to be courageous, to be uncomfortable in order to enact positive changes, and to strive for a future that prioritizes social, racial, and economic justice.

In this endeavor, so close to my heart, I have stood with my sisters as we have been criminalized and arrested while protecting the land, water, and air from climate-wrecking oil pipelines in peaceful, nonviolent acts of civil disobedience. With Indigenous, Black, and Brown sisters, we have directly engaged with the largest banks and asset management firms in the world that finance deforestation and fossil fuel companies that drive the climate crisis. We have advocated for financial institutions to divest from harmful projects in marginalized communities and instead invest in community-led solutions and uphold human rights. Alongside many civil society groups, our organization has advocated at the annual United Nations climate conferences for over a decade demanding no more than 1.5 degrees Celsius in global temperature rise, phasing out all fossil fuels, a Just Transition, gender-responsive climate policies, and implementation of Indigenous rights and Rights of Nature.

Standing with sisters in sacrifice zones, we have held each other in outrage and heartbreak as Black and Latina women miscarry babies and experience cancer at alarming rates due to the pollution forced into their communities. Some of my Indigenous sisters have been attacked and even murdered in their fierce dedication to land defense. Patricia Gualinga, who is a Kichwa leader of the Amazonian Women Defenders of the Jungle in Ecuador, and a dear friend, says, "We have been criminalized, we have been persecuted, many times threatened and sometimes murdered. This has to stop because our country and the world needs to be conscious that our fight is not an isolated fight of the environmental defenders or the Indigenous Peoples. It is a fight that allows the world to survive."[7]

Within these struggles, I also stand in my white privilege and question how I think and act, mistakes I've made, and what I can do in my never-ending quest to be a wiser person and open my heart and mind ever more. It never stops, nor should it—this is how we learn and change. I mention these things not to call attention to myself, as there are colleagues, movement leaders, and dear friends whose credentials and experiences are far more expansive, but rather to describe the ground upon which I stand. These experiences form the lens of my reflections, and they also are an embodied experience of the very stark reality of what is deemed valuable or not in the dominant culture.*

Importantly, my sisters and I have also gathered in moments of significant victories with campaigns and programs that have protected millions of acres of old-growth forests, stopped oil pipelines, changed policies and laws, created food sovereignty networks, replanted forests on clear-cut lands, practiced healthy and mystical relationships with nature, generated serious discourse on care economies and feminist futures, dug deep into complex and uncomfortable conversations on racism and colonization, centered Indigenous, Black, and Brown women's knowledge and leadership, witnessed rivers being acknowledged as living relatives protected under new laws, and more. From these experiences, I know that transformational change is possible—and this kind of deep societal metamorphosis is a central thrust of the book you now hold in your hands.

Under the star-filled sky, I ask who we need to become in the Anthropocene in order to heal ourselves, our communities, and Mother Earth—because clearly the harmful actions of the dominant culture have not stopped.

To address any great dilemma, we must examine how we perceive it and approach it. After decades of working in social and environmental movements, I continually circle back to a core entry point of transformation: the portal of worldviews. We have arrived at a juncture where the

* Throughout, there will be times I use the term "dominant culture" rather than the proverbial "we" that often denotes collective society in general. It is important to acknowledge that Indigenous Peoples, Black and Brown people, and land-based communities have different cultural frames, knowledges, and lived experiences than what falls under the umbrella of Western culture or white mainstream society. Thus, by using the term "dominant culture," there is clarity about the specific part of society I am referencing.

collective efforts of countless people to protect our Earth and all generations are invoking a worldview analysis and approach at the precise moment in which we face a collective catastrophic event in the human experience.Without examining worldviews, we will not conjure and enact the level of imagination, relationships, and ethics now needed. Worldviews express the fundamental nature of a particular culture—how it operates and constructs perceptions and relationships. They provide an all-embracing assemblage of the world around us, including the greater cosmos, and how the world came into being. Worldviews influence who we are, how we behave, our dreams, imaginings, and our relationship to the very web of life. A worldview is a composition of principles, narratives, codes of conduct, and foundational assumptions about reality, which inform our scientific discourse, cultural norms, political and social ideologies, and all aspects of our lived experiences.

Worldviews shape our world and influence the course of history. When threatened by law enforcement at an Indigenous-led peaceful prayer vigil to protect water from the construction of an oil pipeline, a vast differentiation in worldviews was not abstract. Why were we being criminalized for protecting the very source of life? Reality seemed upside-down. We cannot live without precious water, and Indigenous leaders have been at the forefront of voicing this awareness and leading struggles to protect water for decades. Worldviews clashed in that pipeline struggle—one bent on financial profit at any cost, entitlement, and destruction; the other fighting for life, rights, and justice.

We are at a point where we must ask, *How did the dominant culture arrive at this paradoxical moment, when we have healthy and equitable pathways to address our greatest problems, but the central worldview is fixated on cataclysmic exploits?*

Some of the most damaging and pervasive worldviews are human dominion over nature, separation from a living Earth, and structural patriarchy and white supremacy. These worldviews, primarily housed in what we might call Western worldviews, perpetuate deleterious fears, ignorance, personal traumas, avarice, violence, and dangerous cultural norms—all of which have led to a huge deficit in political will and necessary action.

What I have come to learn is that the current dominant social mindsets and systems are incapable of addressing the crises we face or providing a flourishing way ahead.[8] Consequently, in this tenuous moment, understanding worldviews is vital to entering new thresholds of living—some worldviews need radical dismantling and reimagining, and others need to be thoroughly bolstered, remembered, or rescued. What I hold to be true is that another world is possible, and the portal exists within this one to build the world we seek.

Networks of change-makers, systems thinkers, activists, writers, and artists from every background have recognized the need for systemic change because we know that while installing solar panels, deploying wind turbines, recycling, and implementing energy efficiency are real solutions and absolutely paramount, these alone cannot address the deep layers of societal evolution that this moment demands. We are striving in our work to transform oppressive systems and processes into radically beneficial and thriving cultural configurations. The emergent societal scaffolding and cultural story we are engaged in seeks to disentangle itself from a violent colonial and patriarchal stronghold, and in that ever-burgeoning effort, there can be no throwaway people, lands, or species. Instead, we are weaving life-enhancing worldviews and new ways of being and thinking into our projects, narratives, and programs. The core focus is to transition from an extractivist paradigm of exploitation, hyper-individualism, and supremacy to a relational Earth-conscious understanding of respect, reciprocity, and restoration.

To approach this prodigious and epic topic of worldviews, I need tangible spaciousness. So, I have come this night to these oceanside cliffs to be tutored by the stars. These ancient celestial lights offer a unique perspective, allowing me to pause and consider, with breathtaking wonder, our very existence in a greater universe—one so unimaginably immense, we can barely comprehend it. I situate myself in the profound reality that I am spinning on a living planet in a boundless continuum of space—and astonishingly, like most of us, I almost never think about it.

In my contemplation, it comes to me that there are stories and navigational maps in the night sky. These are held within the stellar configurations identified by our forebears from many traditions and lands. It is here that I enter the realm of wayfinding with worldviews that intertwine memory, time, language, history, diverse cultures, different ways of knowing, and relationships with land and each other.

I invite you to join me in a mapping of sorts. Together we will journey from where we are now in the dominant culture, a worldview with detrimental cultural norms in our personal lives and societal experiences, to other possible futures and worldviews that center justice, holistic and other ways of knowing, living landscapes, ancestral remembrances, and thriving Earth communities. This map does not strive to provide tidy answers to wildly complex conditions, but rather to open questions and invitations along the way on a nonlinear course.

This worldview cartography includes a story about the elaborate contours of time. As an example, nocturnal time can influence our perceptions. When the light of our home star, the Sun, ebbs at dusk after shining brightly all the day long, a great darkness appears, exposing a limitless time and space realm, interpreted in barely conceivable light-year measurements and galaxy-forming fractals.

When we peer into the starlit heavens, we look into the past—and witness events from multiple points in history simultaneously. The rays of our Sun, approximately 93 million miles away, travel at the speed of light to arrive on Earth in only eight minutes, but the light from other stars must journey many light-years to reach us through the great expanse of the universe. Even within a particular constellation, what we see is a collection of various points in time. Take Orion, the hunter. The brilliant blue supergiant Alnilam (Arabic for "sapphire"), which forms the center of Orion's belt, is nearly 2000 light-years away. But Gamma Orionis, forming one of the hunter's shoulders, is only 244.6 light-years away, and the reddish supergiant Betelgeuse, Orion's other shoulder, is 548 light-years away from us. To gaze at Orion is to stand at the nexus of multiple timelines, and this remarkable convergence causes me to expand my awareness to multiple ways of experiencing this present moment and a myriad of differing perceptions. This is one way that I enter the portal of worldview transformation.

This constellation of varied timelines presents an aggregate of story threads happening all at once, in which seemingly contradictory realities can overlap: we are witnessing and experiencing the unraveling, grief, and dying of the world as we know it; meanwhile, we are also actively building spaces of liberating possibilities and the emergence of a healthier world we have been envisioning and summoning for many years.

I know that time is not linear—it dissolves and bends in space—so I sense that its multivalent nature is critical to understanding and coping with this moment and the metamorphosis of our worldviews. The colossal time measurement, imparted by the ancient stars, can remind us of our ancestral lineages, our human arrival in the unfolding universe, and our responsibilities to generations to come. Past, present, and future bend together in overlapping realities that can inform and guide our actions. We can see in the time curvature that the future, in great part, is being chosen in the present. What we do right now will be what our children and all the beings of the Earth will experience for generations. In this most urgent moment, our every course of thought and action matters. We are engaged in intergenerational work, and we must hold future generations as a guiding light.

The gemlike constellations watching us will not only witness this present instant but have already observed eons upon eons of life on Earth, and are reflecting our stories back to us from the ages. We need to learn from these ancient, echoed stories.

Is there time to undo the doing?

As I search time cycles in the night, I remember something I learned from Pandora Thomas, the brilliant African American founder of EARTHseed Permaculture Center and Farm.[9] Pandora taught me about Sankofa, which is an African concept from the Akan People of Ghana. The symbol of Sankofa is a great mythical bird with her feet firmly planted forward and her head turned backward. And, as I understood from Pandora and further research, Sankofa conveys the wisdom of learning from the past to ensure a strong future. Sankofa is an acknowledgement that critical examination is vital in gathering the knowledge we need, and that remembering the course of the past provides irreplaceable guidance in charting a wise and strategic path ahead. This approach and worldview keenly resonate with me because if we wish to move forward with the skills to shape a healthy and just world, it is vital that we comprehend and speak to the origins and histories of this quite epic choice-point in the human enterprise. An ahistorical deciphering will lack the thick strata of wisdom and understanding to garner a worldview and perspective that is required for proper reconciliations and restorations, and for creating resilient, thriving communities for all.

The compounded tsunami of societal dysfunction and ecological devastation did not appear suddenly on the global stage—these behaviors and consequences are rooted in a long-standing disrespect for neighbor and nature, fueled by the dominant culture. To add insult to injury, we are often told that we, as a global community, are "all in this together," but the fact is, we experience these struggles neither together nor equally. Thus, we cannot proceed to effectively repair in a new way and heal without addressing the root causes of these crises and the ongoing inequities and injustices.

As I ponder these struggles in the starlit night, I return to the inquiry and contemplation of time. Humanity is in quite a peculiar relationship with time. The Intergovernmental Panel on Climate Change (IPCC) report on the impacts of global warming of 1.5 degrees Celsius above preindustrial levels, and subsequent IPCC reports, tell us that we have a frighteningly short window to avert the worst effects of the climate crisis.[10]

Simultaneously, we cannot allow the powers that be, including governments, corporations, and financial institutions, to rush to so-called solutions that are born from the same deleterious paradigms and ideologies that produced the problem itself—advancing greenwashed, false, narrow-minded, and unjust solutions under the cover of this harrowing crisis.

It is in the present moment, under intense pressure to act precipitously, that we need to break open fresh space and collectively draw upon the multivalent nature of time. We are in a massive emergency that requires haste, and yet we also need to slow down. But with a special kind of slowness. This slowness notices the well-being of every pollinator relative, like the bees and butterflies that our entire food ecosystems depend upon; it helps us untangle ourselves from the grinding pace of the over-production economy in our daily lives; it helps us grow our own food, patiently and joyously; it helps us delve into the innermost layers of ourselves—our beliefs, fears, and societal constructs—in order to learn, heal, and grow, and bring forth flourishing communities. And yet, we must also move quickly, like at the speed of light, to meet the demands of the laws of physics and the vital cries of "climate emergency!" from youth leaders globally.

I do not have a definitive answer to this precarious time riddle, but I recognize that it calls for other ways of knowing, mindsets, and approaches to the manner in which we do or undo every action and thought in our lives. Here we can ask not only *what* we are doing but *how* and *why*, and in this way the route we chart might lead us to new perspectives, perceptions, and worldviews concerning the terrain ahead. And while indeed dire conditions are here, we must not abandon hope in the effort. There is real reason not to give up. While scientists are sounding the alarm, as they should, they are also telling us that though serious climate disruptions will continue, we can avert the worst impacts of the climate crisis and re-establish ecosystem stability—but we must mobilize and act now.[11]

Wrapped in the shawl of night and this choice-point in time, the stars elicit in me the need to contemplate and journey to more expansive coordinates. Here we can draw upon the power of radical imagination that resides within us and recall ancient alchemies—these are primordial coordinates inside of us that remember, as our ancestors did, that we humans are inseparably a part of Nature in an animate cosmology that is ever unfurling.

The shining celestial fires, and their stories, speak to me in mellifluous tones. I recognize and say out loud the names of several constellations, finding great comfort in the familiar groupings. There is something inwardly quieting about this star map that I have viewed all my life from the Northern Hemisphere. The massive skyscape brings human toiling into perspective as it lifts my heart and mind into broader dimensions.

Tonight, I am drawn to Ursa Major, the Great Bear, and find joy in her easy identification. She is widely known as the Big Dipper for her unmistakable ladle-like shape. The star at the lip of the dipper, Dubhe (from the Arabic word for "bear"), works with Merak, just below her, to form a line that guides wayfinders toward Polaris, the not-so-brightly lit but profoundly important North Star, who remains almost completely still above the North Pole, like a maypole around which the other stars revolve, dancing through the nights and seasons.

The North Star is actually a system of three stars, and the trio is the center of a cosmic clock, as the Great Bear revolves around them once per day. All at once, I imagine the celestial lights wheeling around Polaris, the center of the heavens, and envision this North Star as the belly button of the sky, in the way the navel is the center of our bodies. As our belly buttons mark the umbilicus that bound us to our mothers, the North Star fastens the sky to the mother of our solar system, to the mother of the cosmos.

Through my own navel, I know I am wondrously and forever tethered to my mother and to her mother and her mothers, mothers, mothers, through the life-giving umbilicus. No matter how alone we might feel, our navel plainly announces that we are all here by virtue of others, through our ancestral lineages. Our belly buttons are physical memories of other bodies that carried us through generations, in a universal cordage of belonging.

I ask the Polaris-navel, the lodestar at the center of the sky, at the fulcrum of our epoch, how our dysfunctional, distorted, and disconnected dominant culture can ever retrieve the full memory of our umbilicus. Through our umbilicus, we are primordially connected to the greater universe as a consequence of the evolution of our planet because, in fact, we are created from the elements. Our literal ancestral origins are in the sky—most of the elements constituting our bodies, including oxygen, hydrogen, carbon, nitrogen, and phosphorus, were forged in the stars over billions of years. The North Star tells me a story, shimmering an often-forsaken message that all of us belong here, we are part and particle of nature, even in our mammoth and dangerous amnesia. This guiding star is teaching me that it is quintessential we regenerate a worldview of ancestral belonging.

Tonight, I feel at home on the familial Earth because the stars beseech me to widen my perceptions. I remember the umbilical points around the planet, marked by many Indigenous and place-based peoples with omphalos stones—*omphalos* being the Greek word for "navel".[12] These are often human-carved stones marking sacred places of origin and recognizing forces of the natural landscape. To further identify these sites as sources of life and the loci of origin for various peoples, most omphalos stones were formed to resemble eggs, a universal symbol of birth and new life.

According to Greek legend, Zeus's umbilical cord fell to Earth after his birth and became the island of Crete, where an oft-visited omphalos stone still stands. Many more omphalos stones exist throughout Europe (some worked by human hands and others not), as different peoples recognized their unique places of origin. The Stone of Divisions, located at the center of Ireland, marks the spot where the archdruid Midhe would light a fire at the beginning of each year. This fire would then be carried from the navel site to light all of the hearth fires of the community. The spiritual center of the Hopi People is not a stone, but an egg-shaped geyser called the Sipapu. Spouting from the bottom of the Grand Canyon in what is now Arizona, the Sipapu is where the Hopi emerged from the last world into this one.

Tragically, precious origin histories like these have been diminished, violated, lost, or consciously eradicated by the dominant culture. Because they help to root us in the living Earth, it is vital that we return these understandings, and other old-time, land-based stories, to our full awareness again. We need to revive these stories and knowledge systems, and then renew them in our current cultural and ecological contexts to heal from the false narrative and lived experience of orphanage from a living Earth. Through this reclaiming, we can regenerate a sense of belonging to place and to an animate landscape that is our living Relative. This shift in worldview to an animate cosmology, in which we are inextricably and wondrously entangled, is an essential transformation and responsibility as we strive to become more human in a frenzied, colonized, modern world—a world that tries relentlessly to separate us from each other, Mother Earth, and the greater cosmos.

The process of shifting our worldviews in mainstream society is likely to be disorienting and challenging as we encounter and invite different concepts and realties. Yet, if we slow down on the trail, we might allow ourselves to see how some of our life experiences are already embroidered with other ways of knowing and seeing—they have only been veiled or devalued by cultural constructs.

In my contemplation of the umbilicus of the sky, I unexpectedly recall a memory of myself as a three-year-old. I was traveling by car across the country with my family to our new home in California. We were returning from Germany, where our family lived for two years due to my father being stationed in Augsburg as a doctor in the US Army. My parents were interested in learning about diverse cultures, and this led us to stop in the American Southwest on our homeward travel. Once there, we participated in a public Indigenous ceremonial dance that was open to outsiders and tourists. I remember being completely enthralled by the singing and the incredible colors and beauty of the traditional regalia of the dancers and the movement of their bodies. Some part of me was swept up into elliptical rhythms and sensations I had never encountered before.

At one point, a dancer bent on one knee and shot a thunder arrow from his bow into the clouds, accompanied by the sung prayers of the people to bring rain. I distinctly remember "seeing" the arrow flying high, high up to the stars and then coming back to Earth, flying directly into my belly button, and resting there. This experience was exhilarating rather than scary, and when the ceremonial dancing stopped, I recall telling my parents that the arrow was in my stomach and that I wanted to stay with the people in that place and help the rain to come. My parents, perplexed, explained to me there was no arrow released from the bow, that the archer had simulated the action of shooting an arrow as part of the special dance. I remember my frustration with my parents as I insisted that the arrow was in my belly button.

When it was time to leave, I was still trying to explain what had transpired. My parents tried to reason with me, and I simply became silent. Before our departure, a family photo was taken, which I still have; in it, my sister and I, in matching summer outfits, are standing next to our parents, and everyone is smiling except for me. I am not unhappy, but rather staring into the distance, intent on my own world.

Finally, an older man from the community approached me—I'm not sure if my parents had asked him to do so—and bent down and spoke with me. I don't recall many of his words, except the part where he said that, yes, the arrow was in my belly, and he gave me a big, warm smile. This calmed me, and I left with my parents and sister without further issue, and for years I assumed I had a thunder arrow in my belly button. It seemed perfectly normal to me, though I did not speak of this any further with my family until many years later when I inquired about some of the details from my mother.

As a child, this was a profound experience of a different worldview than that in which I was raised. Though I lacked the words to express myself at the time, I knew that something had shifted at the core of my being, forming an experience that has shaped me.

I do not believe I am alone in this—it seems most everyone has had an experience or moment that enlightens us to the fact that the worldview in which we are operating is not the only way of perceiving reality.

When it happens to us as children in the current dominant culture, our voices are often silenced or shunned when we try to speak of our experience. Because most of our parents, teachers, and other adults in our lives are caught in the worldview of mainstream society, they often unwittingly negate dissenting voices and impulses that do not conform to the accepted societal narratives and norms. Yet, despite the demurring voices, we can maintain an unfeigned interior space that authentically senses and feels other ways of knowing and other spheres of experience. It is through this still-alive voice within us that we can keep listening to and learning from the stories the universe is whispering to us beneath the incessant roar of a culture devoid of a living-Earth cosmology and worldview.

In this time of escalating socio-ecological crises, we need to nurture and kindle this aliveness to trace our way out of the soul-crushing business of destroying our sensitivity to deep remembrances and understandings of the world. In essence, we are on a journey, wayfinding to the center of our spiritual umbilicus and our path home.

Without an honest, decolonized historical narrative dwelling in broader multivalent worldviews, we will be unable to build a life-enhancing and equitable future for all beings. Most of us carry unresolved intergenerational traumas in our minds and bodies, each in unique ways depending on gender, ethnicities, sexual orientation, economic standing, culture, and many other factors.[13] Thus, our collective past strata must be tenderly surfaced, reconciled, attended to, and mended, or the powers that be will perpetuate ongoing violence to people and the planet.

My endeavor here is to add to a new cartography of healing and justice based on ancient and present-day cultural maps, decolonial perspectives, climate movement analyses, and forgotten or suppressed memories of our Earth-based lineages. While *The Story is in Our Bones* offers coordinates on the map, we are all participants in the wayfinding and learning together, and I am very grateful to many mentors along the way. The collective project is to contextualize ourselves in the bigger narrative that makes up our modern life. This contextualizing is a way of sense-making and orienting to sanity, care, belonging, and effective action in the midst of dangerous mayhem. We need agency to see the arc of the larger narrative of this time in order to anchor diagnosis and administer relevant, systemic responses. No different from a healer or doctor, we must know the origin and constitution of an illness in order to treat it.

In the course of this mapping, remembering and reciting a panoply of stories and histories is an essential part of the healing journey, and a beginning of the conjuring needed to remake the worldviews of the dominant culture in a healthy manner. The great task I am aiming at is to summon deep memory of who we are in the Earth lineage to bring into being the world we keenly long for, residing ever so close, at the delicate threshold of great peril or great promise.

Here within *The Story is in Our Bones* we can sit under the ancient stars and endeavor to undo the doing and remake our world.

We will learn from the wisdom of diverse peoples globally, while we also look directly into the folly of mainstream society. It will take courage and determination to face and transform the sheer terror and brutality of the dominant culture's history—yet layered within unconscionable abuses, we will also find the chronicles of Earth-honoring peoples who fought (and continue to fight) to preserve the remnants of a sacred, respectful, and reciprocal relationship with the Earth and their human family. The ancestors of the current dominant culture in pre-patriarchal and pre-colonial times held practices of reverence and beauty-making and built hundreds of sacred stone wheels across Old Europe; cultivated healing herbal gardens; practiced Well Dressing by adorning sacred water sources with flowers, fruits, stones, and feathers in prayer for the waters; lit ceremonial fires in the spring to welcome the Sun after the long winter; sang with praise the many names of the Mother Goddesses; held dreaming ceremonies to journey to the other world for healing and to retrieve knowledge; upheld laws that for every tree cut a new tree must be planted; and on and on to celebrate the web of life and maintain a harmonious and mutually enhancing relationship with nature. These ancestral whisperings and remembrances have now reached a crescendo, a thunderous crying out for life. I am inspired in this call to action by the words attributed to Nobel Laureate Ilya Prigogine: "When a system is far from equilibrium, small islands of coherence have the capacity to shift the entire system."[14]

Now we need to muster our collective strength to regenerate our umbilical cord home to build a world of justice, accountability, love, creativity, and liberation—thriving Earth communities for all of our kin, including plants, animals, forests, humans, mountains, oceans, deserts, rivers, and all our relatives—and to do so in record time.

Honoring Indigenous Peoples and Learning from Fire

Before proceeding further, I wish to acknowledge and offer respect to the Coast Miwok, the Original Peoples of the beautiful land where I now live. My worldview has changed over past decades, not only from looking into star-filled skies but also from looking directly down and learning to respect and learn from those upon whose stolen lands I stand.

Marin County in Northern California is named after Chief Marin, who was a leader of the Licatiut Tribe, one branch of the Coast Miwok. Chief Marin, like many other powerful Indigenous leaders, resisted the invasion of European colonizers to protect his people and lands.

The Coast Miwok are known as bird people, as described by Tomales Bay Miwok member and cultural carrier Sky Road Webb. Before colonization, they wore clothes woven from grass or bark fibers in summer, and hides and furs in winter, but the Coast Miwok were particularly known to wear elaborate bird feather sets and abalone necklaces. The Miwok practiced the ceremonial catching of particular birds to collect their feathers. The Miwok reverently fed and cared for their winged guests until the bird's feathers had regrown, then released them during a big ceremony and feast. For about three months after release, some birds regularly revisited their specific caretaker, who continued to offer food until the birds would fully rewild.[15]

This extraordinary relationship with the birds demonstrates profound respect and reciprocity in human interaction with the animal world. Learning from this exchange is a critical part of undoing the damage inflicted by the dominant society. How can we remember and practice again the appropriate relationship to place, and to the other creatures in our communities, as we take what we need to live? It is part of the human condition to need water, food, shelter, clothing, and tools, but *how* to conduct this exchange with the web of life is a fundamental worldview question.

The Miwok also cared for the landscape and were a central part of its ecological well-being. They fed their communities in a sustainable manner with seafood, plants, and wild game such as ducks, while ensuring the areas they hunted and gathered were well-tended by their ecologically knowledgeable hands. The Coast Miwok fashioned baskets from seagrass and ornaments from shells, which they often traded with communities who lived farther from the sea.[16]

I recognize and honor the Coast Miwok because to acknowledge the Indigenous Peoples of the land where we each reside is not only an expression of gratitude, but also one step toward decolonizing and undoing racist narratives as well as toxic conventions of Indigenous erasure. It reminds us that colonization is not over and that the violent attacks against Indigenous Peoples in their own territories and the assault on their lands are ongoing. This recognition, along with offering tangible support for the struggles and solutions of Indigenous Peoples wherever we live, is one part of a new way of living that is intrinsic to our collective survival and transformation of the worldviews of the dominant culture.

Those of us who are not Indigenous to a place need to recognize that we are on occupied, stolen land. To begin the reconciliation, reparation, and healing process, we must learn whose territory we are in. We have a responsibility to learn from and respect the Original Peoples, including their rights, treaties, calls to action for Land Back, sovereignty, knowledge, and sacred sites.

Another question that might arise as we participate in land acknowledgement and respect Indigenous leaders and their requests is, if we are not indigenous to the place we were born, then where are we indigenous to? This is especially important for those of us who are settlers and thus part of the dominant, colonizing culture in the Americas and other regions. Many Americans today are both beneficiaries of colonization and descendants of displaced and conquered peoples—these are often traumatic and challenging lineages to reconcile. It takes courage, as an example, to trace your family tree and learn that your ancestors stole lands from Indigenous Peoples and took part in egregious violations. From another angle, some of these questions about indigeneity can guide us in exploring our older histories—including long-ago pre-patriarchal histories, ancestral homelands, pre-colonized cultural worldviews, and practices prior to our own colonization in lands far and wide. These inquiries are vital because forsaken and unreconciled past traumas often lead to personal or societal violence and a warped amnesia that can cause harm as it searches for belonging and home. I will explore these themes in the chapters ahead.

Is it possible to undo the doing?

To reverse some of the destruction the dominant culture has inflicted upon the planet, we must radically change the way we live on the land we inhabit. We can learn much from those who have lived in balance with the natural world for tens of thousands of years and who hold a non-colonial worldview and reciprocal relationship with nature—Indigenous Peoples.

Which brings us back to the 2018 Camp Fire. This inferno, like many others, happened because of extreme and ongoing drought resulting from human-induced climate disruption; however, some of the fires could have been prevented, or their scope significantly reduced, if those in charge had understood and respected Indigenous methodologies and Traditional Ecological Knowledge (TEK) of forest ecology practices.[17]

Tribes across California regularly set controlled fires, which kept the land from cataclysmic wildfires fueled by dense, accumulated undergrowth. These intentional fires also perpetuated beneficial grasslands by clearing shrubs that proliferate in the absence of fire, and these areas were then seeded with desired plants. Fire was also a tool used by Indigenous Peoples to support many species of native plants that require fire for germination.

Due to persistent efforts by Indigenous Peoples, even in the face of colonial land development, these fire practices continue to this day. In my corner of the world, for thousands of years, the Karuk Tribe has applied practices of ancient technologies in what we now call agroforestry, which integrates preferred plants and animals into native forests and grasslands. Prescribed controlled burning is central to this Karuk forest practice, which nurtures and increases the growth of acorns and medicinal herbs as well as plants used in basket-making. This healthy forest ecology in turn supports the proliferation of salmon, deer, elk, and many other forest animals. Indigenous communities worldwide engage in these forest practices, and this knowledge can teach us a great deal about contributing to the living sacred landscape as life-enhancing, beneficial participants in a thriving ecology.

Given the rapidly heating planet and increased conflagrations, I seek to further understand our historical relationship to fire—in the broadest sense of this element from a small flame to the sun's nuclear fusion and light. I recall again the wisdom of Sankofa in reminding us to look to our past to guide our way forward. For this, I turn to the fire knowledge in stories that have been passed on through the generations, stories that reflect the worldviews and knowledge-sharing practices of Indigenous Peoples and of pre-colonial, pre-patriarchal ancestors of the dominant culture.

Modern humans have been around for some two hundred thousand years, but only in the past two hundred years has a portion of humanity consumed vast amounts of the ancient sunlight stored for millions of years in the form of oil, coal, and natural gas—fossil fuels.[18] I say "a portion of humanity" because, although we all share the same ultimate fate in the unfolding climate crisis, the unjust fact remains that those who have used very little fossil fuels are the communities being impacted first and worst.. In less than one hundred years, this huge energy consumption by wealthier, industrialized countries has led to the disastrous climate disruptions we are now experiencing.

During the ongoing California fire seasons, some of the old-time stories about how fire originally came to people have taken on new meaning for me. I have begun to viscerally understand the universal warnings encoded in these fire tales and teachings passed on through generations.

The essence of fire is transformation, a phenomenon that has captivated humans since we have interacted with this element. In the introduction to his book *The Psychoanalysis of Fire*, French philosopher Gaston Bachelard writes, "If all that changes slowly may be explained by life, all that changes quickly is explained by fire."[19]

Our multifaceted, sacred, and distinctive relationship with fire appears to be fairly universal. When we turn to the many fire origin stories and myths from cultures around the world, we find that often they impart a warning of sorts. Many of them express outright that fire is not freely given—it must be stolen. Significantly, this is not true for the other elements of life: water, air, and earth. In the old mythologies, none of these are purloined. In his book *Myths of the Origin of Fire*, social anthropologist James Frazer relates how fire-bringers are often legendary cultural figures—animal or human—who steal the original flame and offer it to the people.[20]

The Andaman Islanders in the Bay of Bengal tell of how Kingfisher stole fire from a magical creature called a Bilik. When Kingfisher was caught in the act of pilfering the flame to deliver to his people, the Bilik threw a firebrand, hitting Kingfisher on the back of his neck. Bright red feathers on the bird's neck still mark the spot today.

Several Indigenous Peoples from Victoria, Australia, tell of how a Fire-tail Wren or Finch brought the original fire to the people. The small red-tailed bird stole it from the sky or, as is sometimes told, from the Crows.

There are numerous Native American stories in which Coyote cleverly snatches the fire from the Sun or other source, but the thief can also be a trickster in the form of a Turtle, Raven, Frog, or Fox who slyly makes off with the flame.

In South Africa, the native San report that Ostrich held an ember under a wing until Praying Mantis stole it and gave it to the people.

Indigenous Peoples living along the Amazon River basin in Brazil convey a story about a boy rescued by Jaguar. The boy learned how Jaguar cooked food over the fire and then appropriated a hot coal for his people so that they, too, could cook their food.

In Coast Miwok stories, Coyote sent Koo-loo'-pis, the swift and tiny Hummingbird, to steal fire from the Sun and bring it back for humans. Brave Hummingbird returned from a perilous journey with a spark tucked beneath her chin. The ember was then placed in the California buckeye tree by Coyote, where it can be reignited with the spinning of a hand drill.

There is the Greek mythology of Prometheus, the hero who steals fire from the Gods for human use and creativity. Zeus had decided to withhold fire, but Prometheus, in his fondness for humans, stole it and delivered it to them, hidden inside a fennel stalk. The myth goes on to relay that Prometheus, chained to a rock, was to be eternally punished for his deed by an Eagle who eats his liver each day, only for it to regenerate each night; in some accounts, he is eventually freed by Heracles.

It would be impossible and inappropriate for me to interpret the meaning of these stories from the many cultures mentioned, as that should respectfully be left to those who carry this knowledge from their Nations and Peoples. I only wish to point out what we might learn if we reflect on why the fire is consistently stolen rather than given by nature freely. The stories suggest that our ancestors the world over wished us to be particularly mindful and respectful of this sacred element.

Unlike the life-giving offerings of water, air, and earth, fire most often has to be consciously made. There is choice involved in how and when to use the fire technology. While indeed, fire can be "found" when lightning ignites a tree, this is not an occurrence we can rely on for our daily needs.

That the flame is not freely given can also warn us that fire is a double-edged sword: it allows us many comforts while also giving us the power to harm ourselves, others, and the natural world in which we live. Fire enables us to change the world for good and for ill; the way we wield it has a tremendous impact on the natural environment. And here, again, I mean fire in the broadest sense, including all that comes from this element through energy production and the use of energy. The old stories indicate an unusual relationship—fire has to be stolen because the guardians of the flame are not sure that humans should possess it or can control it. If we are to have it, we need to understand that fire is a hard-won prize, and we need to make an adequate exchange for its use; otherwise, it is a theft with serious consequences. Clearly, the dominant culture has not made the proper offering in return, nor learned how to use fire and its innate energy responsibly.

I am not suggesting that we should not have fire and the energy it generates. We are all the better for fire's gifts, and our very humanity and evolution as a species in part depend upon our relationship to fire. Rather, the point is that fire is all at once a costly, beneficent, and dangerous gift, and it requires an enormous amount of reciprocity, maturity, and discipline to use it in a sustainable and balanced way.

Since the beginning of human time until the modern era, we have assembled around the hearth fire to tell our stories and to discuss and deliberate community matters. The hearth fire is where humans have gathered to cook food, warm ourselves, and dream as we look into the enchanting embers and the undulating flames. Now we have great need to return to the central fire of our communities and relearn respect not only for this sacred element, but for our place in the greater cosmos. Here at the center fire and under the stars, we have the possibility to renew our stories and transform our worldviews to help chart us through the formidable passage ahead.

Chapter 2:

The Story is in Our Bones: Origin Stories to Remake Our World

HER SKILLED GOLDEN-BROWN HANDS carried the knowledge of planting ancestral corn and weaving traditional textiles, among many abilities and artistries. They were hands shaped by hundreds of generations of her people working the fertile soil of Mount Imbabura, a volcano rising from the Valley of the Dawn in Otavalo, Ecuador. They were free hands fought for by generations of resistance movements fending off colonization and surviving industrialization, including the enslaving Spanish haciendas of the early and mid-twentieth century.

Now, Blanca Chancoso, an Ecuadorian Indigenous leader of Kichwa nationality and a member of Saramanta Warmikuna's women's collective, opened her hands to welcome us—participants from all over the world who had gathered in her Andean homelands for a Global Rights of Nature Summit. It was a diverse assembly of over sixty scientists, attorneys, economists, Indigenous leaders, authors, cultural leaders, politicians, and activists from sixteen countries and six continents determined to further the work of establishing a new Earth jurisprudence that respects the natural laws of the Earth. Rights of Nature is a legal and cultural framework that recognizes and honors that our ecosystems have rights just as human beings do, and that all life on our planet is to be respected. The global movement for Rights of Nature is rapidly growing, and given its effects on worldview impacts, implications for a systemic response to the climate crisis, and real-time wins for communities and nature, I have devoted Chapter 12 entirely to this topic.

Adorned in a traditional, intricately embroidered white blouse and long skirt of magenta and ocean blue, Blanca delivered one of the opening presentations at the summit. She was requested to do so out of respect for the Indigenous Peoples upon whose lands we had gathered as guests. Her opening remarks were a significant enterprise, as she was tasked not only with orienting participants to the Andes and the story of her homelands but also with providing a transformative framework for our two-day deliberations.

I was honored and elated to be in the room with an Indigenous woman leader guiding us at this critical time of climate disruption, environmental degradation, and social and economic malady and injustice. It is imperative to learn from Indigenous, Black, and Brown women, and women from frontline communities the world over. These brilliant and fierce women emerge with great strength and wisdom (and, yes, often deep wounds) from years of surviving relentless colonial and racist assaults, destruction of their lands, bodily violence, and patriarchy. Their knowledge of resistance and resilience is central to all of our struggles, as they carry invaluable cultural and ecological knowledge with ceaseless determination and love.

Blanca stood silently for a moment. Her eyes scanned the room to register the diverse audience and rested for a moment on the blazing fire in the large stone fireplace, the only source of heat in the chilly high-Andes hacienda meeting hall. She is one of the Daughters of the Corn, or the Saramanta Warmikuna, women who defend and protect life as well as the collective memory of the diversity, multiplicity, and stories of seeds.[1] I felt the ancestors standing with her as she began to share precious traditional knowledge going back thousands of years, now transmitted in her elegant speech.

As she spoke about her Otavalo region, I could sense the powerful presence of Mount Imbabura, not far from our meeting hall. Known by the local people as Taita Imbabura or Papa Imbabura, this volcano rises sixteen thousand feet from the valley floor and faces Mount Cotacachi, the Mother Volcano, across the valley basin. The volcanoes each have their stories of vital spiritual and cultural significance to the Indigenous local people, maintained since the beginning of time, as I was told. Not only is Imbabura abundant in rich soil, but the volcano is also blanketed with flourishing cloud forests, and on the western slope lies a unique area of open earth that distinctly forms the shape of a heart. This "Heart of the Mountain," as local people call it, is known as a place of the spirits. Due to the steepness of the slope, no human has been able to climb to the sacred site. Per many Indigenous Peoples' protocols, sacred areas in nature such as this one are to be respected and recognized as places not meant for humans—a dwelling for the wild spirits to reside undisturbed.

In a striking juxtaposition, Blanca Chancoso stood before us as a strong and proud Indigenous woman, speaking about decolonization and dominant cultural worldviews, within a hacienda that had been a site of the Spanish colonizers. Blanca was reclaiming this space that had been an edifice of oppression and violence for her people. There was a period of Incan incursion in the fifteenth and sixteenth centuries, after which the Spanish colonizers came, conquered the Incas, and built haciendas. The Spanish ruled over the Otavalo and forced them to toil under the blazing sun in gardens and vineyards known as "fields of blood" because so many Indigenous People perished in them.[2] Additionally, the Spanish brought diseases that decimated eighty percent of the Indigenous population. The Otavalo and other Indigenous groups revolted and ended the haciendas in the 1960s and '70s, and they have remained determined and resilient despite ongoing colonial disruptions.

Back in the fire-warmed hacienda meeting hall, Blanca's venerable composure and words brought us to another dimension of thought and energy. She didn't relate any case studies or legal procedures concerning environmental law in Ecuador, although she could easily have addressed these topics. Instead, through stories of her lineage, she commanded us to stretch our minds to touch the still viable but distant call of our own ancestral memories—memories that intimately connect us to the Earth. It was the meeting of two worlds, Western perspectives and an Indigenous worldview. It seemed necessary and fitting that we should open our summit with an important disruption of the dominant cultural system both internally in our way of thinking and being as well as in societal norms. Blanca prepared us to delve into formidable topics such as extractive economic systems, Indigenous rights, property-based legal systems, and environmental racism, but purposely interwoven with teachings of older Earth reciprocity practices and traditional knowledge from her people.

What most caught my attention were Blanca's stories of the origins of her people and their Original Instructions about how humans should live in respect to the land and each other. Through incredible dedication and courage, and often at great risk and cost to many lives due to colonization and genocidal polices, this great wisdom still survives. The Original Instructions, as Blanca explained, have been astutely and caringly developed over millennia of humans living intimately with specific lands, as well as contemplating the cosmos and our human place in the sacred systems of life—consequently, they contain invaluable guidance and perspective.

Not only did her information influence the depth of the discourse at the summit, but it has stayed with me as a transformative sphere of thought ever since. Hearing the Creation Stories and Original Instructions from Indigenous Peoples, which have been handed down for thousands of years through oral tradition, brings forth profound inspiration and insight regarding codes of proper conduct, morality, dignity, wisdom, appreciation, and a deeply embodied relationship to our living Earth.

Just like a colorful shawl being threaded on a backstrap loom from old Otavalo, Blanca's sentences were woven together in elaborate patterns meant to rearrange perception—an ornate textile that opened the imagination. From my notes that day and further clarifying conversations with Blanca, I present just a few of the well-spun threads that she has generously given me permission to convey:

> Our way of thinking and living is based on Pachamama, Mother Earth. The word *Pacha* means Universe, all of Mother Earth, Time, and Space. This word has great meaning in our cosmo-vision of the world. It means a relationship between time and space that celebrates all that exists. Time is not linear but is measured in cycles such as seasons and when we plant our seeds, but these cycles can also be much bigger events in the cosmos. Time is connected to the land and eternity. So in time and place, we are connected to the land with our activities, ancestors, and memory.
>
> And the word *Mama* is the origin, the root, and the very heart of life. That's what we identify when we say Mama; we are talking about the center of our cosmos and communities. Pachamama is the great life-giver on this Earth; she is the womb of our Mother Earth for the people. Most of the problems we have now are because human beings are taking too much from Pachamama. Nature has been violently mistreated due to the extractive economic model. And now, we have the responsibility to help heal and repair the damage to Pachamama.

We need to give back, and this is what our ancestors have told us. We have the Kulla Raymi ceremony to remember our place in Time and Space and to live in reciprocity with the natural laws of Pachamama. We make offerings to the land with Kulla Raymi to remember and to be thankful for all that life gives to us. But, how do we live?

Many people confuse their understanding of agreements and law because they only view these concepts from a modern legal point of view. What is it to have a right? We Indigenous People are trying to interpret from our way of life and worldview what our ways and ideas mean in the industrial world.

When we talk about being able to live, to be able to talk and think freely, to have a home, to have a piece of land, to be able to do, to be able to decide, these kinds of rights are to have power. It is not the power to command others from corporate headquarters, but to live well with Pachamama, Mother Earth. In Kichwa, we say "ushayta charina," the power to have, to even be able to have, to be able to think. This is how we conceive of the idea of a right. When we speak of nature, it is not only the river, the air, the forest we are talking about. We are also talking about us, because we humans are also part of nature.

In a discussion with the Saramanta Warmikuna women, we do not want people to continue talking about natural resources, no, because they are living beings; the word *resources* gives rise to idea that the Earth and all beings are for sale.

The resistance work we do as Saramanta Warmikuna is often translated as opposition. For me it is not only that; for me it is to defend what we believe and oppose what we do not believe. In the case of Ecuador, when talking about pluri-nationality, the government is not complying with the Constitution to respect nature and Indigenous Peoples. When we stand up for the land, the government criminalizes us to make us afraid, specifically persecuting the women. But we continue to resist and defend life.[3]

At this point, Blanca momentarily stopped her presentation. She remained at the front of the room, indicating that she had not given up the floor and was not finished with her comments. There was a long, expanded pause, and in that open space, the living world of nature and Blanca's distinct message embraced the hall as we heard the crackle and snap of the fire. Then suddenly, entering the stillness, a beguiling birdsong echoed loudly in the stone room coming from just outside the old hacienda windows. The hair rose on my arms. Blanca then continued with her closing comments:

> Part of the foundation of our people is Sumak Kawsay; this means to live well, or good living with Pachamama, taking into account the whole of the natural world and human society.
>
> Sumak Kawsay is way of living for our people, thousands of years old, that means to live in harmony within communities, ourselves, and Pachamama. It is a way that respects life and gives back to nature, that gives respect to each unique culture—human and nonhuman cultures—and cares for the next generations by caring for the land.
>
> Sumak Kawsay is not only for Indigenous People because it is in our language, but instead a worldview for everyone in their own way.
>
> And though Sumak Kawsay is an ancient way of living that has helped us survive hundreds of years of colonization, it is not about returning to the past. It is about collectively living in ecological and cultural well-being with each other and Pachamama today. It is a collective life that seeks progress together, and from there we share, but it is a progress not defined by corporate economic structures but rather progress in well-being, creativity, reciprocity, and solidarity—not material expansion.
>
> To move forward now, we have to make a fabric of alliances between non-Indigenous and Indigenous People in our movements. And, our proposals and worldviews as Indigenous Peoples need to be respected in these alliances. For the change we need, we have to sit together as equals, and we have

> to remember Pachamama and what is our place in Time and Space. Let's build the new agenda together![4]

Late that night after Blanca's talk and many presentations by other leaders about the practice of Rights of Nature in different countries, several of us conference participants went for a walk under a bright full moon, the night sky particularly crystalline due to the clear air at high altitude.

I gazed upward and thought about Blanca's presentation, remembering the poet and scholar Drew Dellinger recalling words of his mentor Thomas Berry, the well-known cultural historian and theologian, who said the universe is "the most radical thing that a person can think about."[5]

Being inspired by Berry's and Blanca's words set off a trail of ponderings, which unfurled something like this: The daily thoughts and experiences of the majority of people in our modern culture rarely, if ever, hold a cognition of the astonishing wonder of our planet and the greater cosmos. Most often, we are engrossed in digital devices, workplace and business dealings, personal survival, political and social matters, urban landscapes, and so on. It takes quite a different kind of consciousness to stop and remember the presence of the living universe, which comprises billions of swirling galaxies and baffling phenomena such as black holes, and where each of us exists on an indescribably stunning, but relatively small blue planet in boundless time and space. How might this awareness lead to contemplating the origins of where we as humans come from and how we are literally part and particle of nature? Blanca's description of Pachamama and the Kichwa recognition of Time, Space, Place, Memory, and Cosmos in daily life resonated intensely in my mind.

It seems to me, considering our current state of affairs, that remembering and deeply comprehending our inseparable relationship to nature, each other, and all our nonhuman kin, through reclaiming our collective origins in the greater universe, matters very much.

Given the ferocious urgency of the climate, ecological, and colonization crises, it is often challenging to stop and direct our attention to deeper evaluations such as contemplating the cosmos, which seems abstract and irrelevant in the face of increased emergencies. Yet understanding our place in the universe, while still being aware of the emergency, can offer us perspectives expansive enough for full-spectrum solutions to arise.

In other words, while it is urgent and essential to shut down corrupt and planet-wrecking forces such as the fossil fuel industry that are literally killing us, and to immediately implement renewable, decentralized energy globally for all—we must also commit to resolving the foundations of deadly societal syndromes. For long-term transformation, it is imperative to uproot false and misguided separation-from-nature stories and social injustices. We need to regenerate a different paradigmatic framework based on Earth-honoring knowledge systems, like scientific and cultural Origin Stories that convey our inseparable, equitable connection to each other and the web of life. This worldview shift must happen concurrently with our fight to protect people and the planet from nefarious government and corporate actors. It is now necessary to forge ancient-modern worldviews that ultimately are fiercely in love with, and quintessentially anchored in, a living cosmology that inherently respects all people and all species. Otherwise, the disconnection from the Earth experienced and insisted upon by the dominant culture will further the insatiable hunger to conquer and colonize, reign supreme, and accumulate material wealth in an attempt to create meaning and belonging in the perceived, aching void of separation. It's time to awaken from our long slumber and remember who we are and where we come from.

What is the Origin Story of humanity or of a particular People? There are many layers and approaches to this rather enormous question. But for now, I am narrowing my scope to the dominant culture's basic comprehension of our relationship to the cosmos as understood by science, and will touch on Western religious aspects momentarily.

The evolution of life on our home planet, from the emergence of a single-celled organism that lived between three and four billion years ago to the wildest imaginings of creatures that now swim, fly, walk, dance, run, and sing demonstrates that we are Nature and that we are all relatives in our microbial origins. Yet, this fundamental reality is often missing from conventional histories of civilization: our history books and courses are primarily human-centric, with events divorced from our relationship to the natural world. In the compartmentalized, disembodied worldview of modern thinking, our human history is separated from the natural sciences.

Yet, how can we know our human story detached from nature? In this flawed approach, we sacrifice human well-being to a dominant culture that casts us out from nature to become orphans. We trade relationships of belonging for a notion that we are a superior force that exists above nature with no need to understand our interconnectedness or share in responsibility for the web of life. But the story of our present is forever grounded in a magnificent past that unites us to everything in the greater cosmos.

To help row ourselves a bit forward, here is a piece of the universe's Origin Story. There seems to be consensus among astronomers that the universe was created between thirteen and fourteen billion years ago.[6] Ideas abound regarding this extraordinary moment often referred to as the Big Bang. Like many people, I have wondered: If this emergence occurred suddenly, what physically existed before this moment?

What has helped clarify this quandary for me, at least to some degree, is the theory that the emergence of our universe was the consequence of an earlier collapsed universe, and that potentially there is an ongoing cycle of expanding and collapsing universes. This idea squares well with nature's regenerative systems, much like a forest ecosystem where fallen trees decay and become rich compost for new trees to sprout, creating a cycle. However, the manner in which all of these universe cycles might have begun cannot be addressed within the construct of linear time because, again, we would still require an actual beginning moment for the multiple universes. In order to grasp this idea, we have to suspend our modern view of time—specifically, the notion of linear time.

In the same vein, we cannot travel to the farthest edges of the universe and find a distinct end point. To comprehend both nonlinear time and continuous, boundless space, consider Einstein's theory of relativity, which demonstrates that space, time, and the universe bend, so that space and time are a curvature allowing both endlessness and finiteness. Time was never linear. Many Indigenous sciences have long held knowledge that is older than, and in alignment with, Einstein's theory of relativity as well as quantum physics more generally. Indigenous Creation and Origin Stories and understandings of the workings of the universe mirror these sophisticated concepts, as reflected in languages such as Blackfoot.

As the late physicist F. David Peat points out, Blackfoot traditional science encompasses a living world that is dynamic, interconnected, and in constant flux. Blackfoot relationships to the forest, water, animals, plants, and spirits of a place are all part of their language, which is fundamentally verb-based. The Blackfoot language construct allows natural processes to break free from restrictive modern European language taxonomies and categories to demonstrate the agency and transformation that Western scientists now realize all atoms and subatomic particles possess. The Blackfoot worldview, and consequently their language, allows them to see the world as it is.

In contrast, English is a language based on nouns (that are essentially static), which speakers steer and maneuver using verbs. Most modern European languages developed and evolved to express observations of a visible, tangible, and objective world. Danish physicist Niels Bohr noted that a major reason for the incomprehensibility of quantum theory is the nature of European languages, which are limited to discussing fixed objects and absolutes, when in reality, the world is in constant motion; we the observers are participants in the phenomena we are studying, and therefore unable to be objective.[7] Thus, in European languages, we rely on poetry and metaphor to free language (and our hearts and minds) from the hard confines of a dead-matter reality, which is actually only an illusion—life is far more enchanting and magical!

Another challenge in imagining the truly profound advent of the universe arises in asking ourselves to comprehend not only that it originated from some kind of great nothingness, as cosmologists tell us, but also that this event happened rapidly. In less than a second, the universe was birthed, and in that instance, gravity was produced along with other forces that are, in essence, the governing rules of what we call physics and the beginning of everything else that exists today.

In this first explosive instant, the first stars formed when gases swirled together and began to combust. A process called nucleosynthesis takes place inside a star, and this creates the different elements. The first stars birthed into the universe burned their energy rapidly, and because of this they were only able to create a few elements—in particular hydrogen, helium, and a bit of lithium. However, when those first stars eventually went supernova, which is a massive explosion at the end of a star's life cycle, the elements they had made spread throughout the universe, travelling on to form new entities.

Much like elemental "seeds," these new entities created the next generation of stars and planets, and literally everything in the universe. The new generation of stars burned less quickly, and they were able to create heavier elements such as carbon, silicon, iron, and many others, which comprise the fundamental elements that eventually became Mount Fuji, the Amazon rainforest, ruby-throated hummingbirds, blue morpho butterflies, Ancilla tulips, baobab trees, humpback whales, Golden Delicious apples, Peruvian purple potatoes, the Black Sea, and you, dear reader. As evolutionary cosmologist Brian Swimme deftly summarizes, "It's really simple. Here's the whole story in one line. This is the greatest discovery of the scientific enterprise: You take hydrogen gas, and you leave it alone, and it turns into rosebushes, giraffes, and humans."[8]

Looking at this from another perspective, when a star explodes in a supernova and the elements inside the dying star extend out into space, they are all regenerate, which is known, scientifically, as a galactic chemical evolution. This galactic occurrence is something that really makes my head spin because, as it turns out, the elements in our bodies have been produced in stars over billions of years and through multiple star lifetimes. The elements in our solar system that created planets and formed our Earth and all life on our planet, including us, are, in fact, descendants of many different stars going back for eons. The wild fluke of humans coming into existence from this billion-year-old lineage of celestial bodies reinforces the point of our complete need to respect and be responsible for this tremendous gift of life. It is also quite humbling.

We can learn further of this stellar descendancy in a multitude of origin or creation stories from ancient cultures as well as Indigenous Peoples today, such as the Dogon of Mali, who, without the use of telescopes, cultivated highly advanced astronomical sciences. In Dogon cosmology, the "Egg of the World," also known as the star Sirius B, moves about in infinite space and carries the germ and prototype of all creation. The eight ancestors of the Dogon, called the Nommo, descended to Earth carrying a basket of celestial clay, with which they built their granaries, symbolizing the universe. Within the granaries were scales, representing the female and male forebears of the tribe and the stars. Humankind, our food, our tools, and all life originated from a basket sent from the cosmos.[9] This story, and the worldview it imparts, connects humans directly to a relational, interconnected existence.

Our home solar system is believed to have been created about 4.5 billion years ago via a cloud of supernova dust and gas swirling. Eventually, due to gravitational forces, this cloud began to collapse, creating a huge spiraling disc of solar nebulae. In this incandescent spinning, the nebulae combined into various spiraling arms, and at the center of this dynamic disc, our sun was formed, along with particles that eventually became hot, fiery balls of gas and molten liquid. When these entities began to cool and take on solid form, they turned into planets.

The evolution of life on Planet Earth is the next part of our Origin Story—it is the story of the emergence of the plant people, the tree people, the animal people, and more, who are our collective non-human ancestors with whom we still share much of our DNA; for example, humans share 98.7 percent of our DNA with chimpanzees, 90 percent with Abyssinian cats, and over 50 percent with our plant kin.[10] We are related not only to every single human on the planet, but also to the entire cosmic ecosystem; thus, the hereditary material of our very bodies can awaken an ancient recognition that we are all Relatives. When we gaze into the eyes of our pet cat, run our fingers along a velvety leaf, or search for Orion in the night sky, we can reflect on these embodied connections. Though millennia have passed, the elements in our cells hold an unfailing remembrance of where we come from in the great vault of Time and Space.

Yet, aside from a handful of science classes taken in our early years in school or university, we almost never discuss, contemplate, or celebrate this staggering phenomenon of the origin of the cosmos, the emergence of life on Earth, and how we even came to be human beings on this remarkable, indisputably beautiful planet. Without integrating the perspective of this Origin Story into our lives, we fail to grasp the epic nature and gift of our existence, and instead are confined to a very limited comprehension of life and our meaning and responsibility within it.

How can we possibly live well on Earth without recognizing this Origin Story as we work to reimagine and create healthy and just governments, economies, legal structures, and societal constructs? How can we live in balance with the natural world if we do not first conceive of the universe or hold a real cosmology? If the basic worldview of our dominant culture is not connected to a living universe, how can we possibly envision new systems that respect the natural laws of the Earth? By embracing our Origin Story of the universe on a regular basis through educational, cultural, spiritual, artistic, and personal practices, we have an opportunity to understand that we are kin to all of life in every way possible. Do I think living with this worldview will "solve" all the injustices and environmental crises of our time? Quite simply, no. But I do propose that by regularly integrating science-based and cultural Origin Stories into our lives, there can be no misconception of separation. Instead, we can be guided to understand our undeniable interdependence and responsibilities to the lands where we live and each other, and more deeply ensure the end of white supremacy and inequity at the core of the dominant society's worldview.

For some years, I have been on a quest to further understand cultural Origin Stories, in addition to scientific renderings. This can be challenging if you do not come from an intact Indigenous culture because so many ancient Origin Stories of the modern dominant culture have been long desecrated or co-opted and perverted by thousands of years of colonization and patriarchy.

To rediscover and rekindle the spark of cultural Origin Story knowledge, we must take on the serious business of unlearning internalized narratives that subjugate older nature-based worldviews. One example of this dismantling can be implemented by examining the Bible, which informs us that the correct human-nature relationship is that of dominion: "And God said, Let us make man in our image, after our likeness: and let them have dominion over the fish of the sea, and over the fowl of the air, and over the cattle, and over all the earth, and over every creeping thing that creepeth upon the earth."[11] I do not present this information to judge the Bible, nor am I a biblical scholar attempting to analyze world religions in any depth. Rather, I wish to point out that this hierarchical narrative that consciously or subconsciously influences people—whether they are religious or not—continues to drive the exploitation and extractive economic enterprise of the modern world.

Judaism, Christianity, and Islam are the primary Abrahamic religions and regard the patriarch Abraham as the founder of their traditions. These are monotheistic faiths that conceive God to be a singular, transcendent creator. This God is portrayed as male and has the power of judgment, punishment, and forgiveness, and possesses dominion over nature. This mindset provides the foundation for patriarchal, extractive, and colonial ideologies and worldviews.

Additionally, the biblical story of Adam and Eve, which depicts the origin of humans, portrays a woman who, contrary to the inherent agency and power of woman, is born from the rib of a man. Eve, the first woman, then becomes the source of original sin, as she is the first to eat from the Tree of Knowledge (which seems like a wise thing to do!). Because the culture that has infiltrated much of the world through colonization is Christian, the notion of woman as responsible for the fall of man has affected our worldview, with devastating results. The subjugation of and violence toward women, and people across the gender spectrum, is ongoing and egregious.

The Abrahamic religions were established through the violent downfall of Earth-based traditions that honored a living Earth and revered the Great Mother, with vast expressions of polytheism and animism worldwide[12] (we will delve into this important topic in Chapters 5 through 8). We cannot ignore the deleterious influence of religious beliefs that generate worldviews promoting patriarchy and supremacy over, and separation from, the natural world.

Fortunately, today, many groups and individuals within these religions do call for and powerfully practice an ethical approach to nature and stand strongly for social justice as they embed these values in altruistic maxims found within their faiths. That said, to cultivate the level of necessary ideological change required to meet this crossroads moment, the dominant culture will need to transform its narrative from the very roots. We don't need to throw out world religions, but we do need to bring their central stories and tenets back to the community fire for renewing and retelling. Without this fearless structural approach, we will repeat the same systems of dominance, injustice, racism, and misogyny. The multiple crises we face demand us to excavate all the way to the foundation of our cultural stories, narratives, worldviews, and beliefs.

In addition to Origin Stories acting as the soil from which a culture gathers its nutrients to define itself and grow, Original Instructions are inextricably connected to determining and shaping cultural values, identity, and worldviews.

Indigenous and traditional cultures around the world carry knowledge of Original Instructions, which offer teachings and guidelines about how to live in a healthy relationship with the natural world and to respect the natural laws of the Earth. These instructions, in the form of songs, stories, ceremonies, and dances, are directly gifted to the people from Nature or Creation. These instructions are guidelines and socio-ecological protocols, and must be adhered to or there will be drastic life-altering consequences. For example, some instructions show how and when to collect plants from a region in an ecologically balanced way so as not to overharvest, or how to hunt animals so that a herd is culled in a regenerative manner. Original Instructions can also carry warnings about agreements with nature that humans must maintain to keep balance between the human and non-human world.

It is vital to recognize that nuances of Original Instructions are often specific to place because life-enhancing relationships between nature and humans can primarily be informed by a specific region and ecosystem. Indigenous Peoples globally tell us that their bodies *are* the land itself and that their culture and very existence are bound to the land where their ancestors have lived and are buried. Their bodies and spiritual lives have been nourished and informed by the plants, water, animals, mountains, and spirits of their ancestral homelands since time immemorial. This is also one of the many reasons it is critical to respect, listen to, and learn from Indigenous Peoples in their territories because their life practices and observations are the very expression of how humans can live appropriately in the ecology of a geographical region.

Here it is crucial to recall and respect that eighty percent of the biodiversity left on Earth at this time exists in the territories or lands maintained by Indigenous Peoples, and this is in great part due to the respect and care Indigenous Peoples have for the sacred living systems of life and their commitment to seventh-generation thinking.[13] This is also why it is imperative that we all participate in protecting and defending Indigenous Peoples' rights, sovereignty, and ways of life. To respect their human rights and dignity is morally the right thing to do, but it is also important for all our survival. As we can tragically and readily see, world governments and corporations are leagues away from the notion of Indigenous long-term seventh-generation reasoning and planning, with companies demanding increased quarterly financial returns for activities that are literally desecrating our planet, and with no proper level of government regulation to stop the devastation—demonstrating just how far the dominant culture has strayed from Original Instructions.

This is not to say that modern sciences and technologies are not important, for they too have many critical gifts to offer. Given the current environmental and climate crises, it behooves us to learn from and implement both Indigenous and Western sciences, knowledge, and technologies. That said, given the real need for a moral compass for our endeavors and in-depth knowledge of place, Original Instructions from Indigenous Peoples provide irreplaceable operating frameworks for how we proceed.

I want to highlight an important work called *Original Instructions: Indigenous Teachings for a Sustainable Future,* edited by Melissa K. Nelson, who is Anishinaabe/Métis/Norwegian as well as a member of the Turtle Mountain Band of Chippewa Indians. She is an Indigenous scholar and activist, as well as a dear colleague and friend. In her book, many Indigenous leaders around the world share knowledge from their respective Original Instructions and practices. Nelson tells us in her introduction, "All cultures can learn and be enriched by Native storytelling,"[14] and "Deep within the teachings of the Original Instructions and millennial old oral/verbal art forms are certain story bundles that serve as important warning signs and guideposts."[15]

Rebecca Adamson, who is Cherokee and the founder of First Peoples Worldwide, shares the following in her contribution to Original Instructions, giving us a glimpse into how *Original Instructions* can powerfully inform our disastrous economic system:

> The interdependency of humankind, the relevance for relationship, the sacredness of creation is returning as a fact of life. It is ancient, ancient wisdom. More than any single issue, economic development is the battle line between two competing worldviews. Tribal people's fundamental value was with sustainability and they conducted their livelihoods in ways that sustained resources and limited inequalities in their society. What made

traditional economies so radically different and so very fundamentally dangerous to western economies were the traditional principles of prosperity of creation versus scarcity of resources, of sharing and distribution versus accumulation and greed. Of kinship usage rights versus individual exclusive ownership rights. And of sustainability versus growth.

In the field of economic development, economists like to think western economics is value-neutral, but the fact is, it is not. What does the finance system tell us about function and form, about our very values, when the same system pays a merger acquisitionist millions of dollars and a teacher $40,000? The Cartesian reductionist approach defined success according to production units or monetary worth. The contrast with successful Indigenous development is stark.

For example, because the Northern Cheyenne understand the environment to be a living being, they have opposed coal strip mining on their reservation because it kills the water beings. There are no cost measurements of pollution, production, or other elements that can capture this kind of impact. There is an emerging recognition of a need for a spiritual base, not only in our individual lives, but also in our work and in our communities. Perfect harmony and balance with the laws of the universe means that we all know that the way of life is found by protecting the water beings.[16]

Though all humans can trace their ancestry back to Indigenous roots at some point in time, for many of us those roots have been withered under colonization, religious conquests, patriarchy, and other forms of oppression. Thus, reconnecting with our own Original Instructions is to commence on a long tracking expedition because many of us do not live in the lands of our ancestors or have access to ancestral knowledge directly. We need to engage various ways of knowing through a shifting landscape. The journey can begin by recognizing we each have had, somewhere in our past, Original Instructions, Origin Stories, and homelands. When we realize the importance of this knowledge together, we can collectively embark on research, learning, and remembering. Even if we only find bits and pieces of our ancestral knowledge, every part is filled with layers of meaning, palimpsests holding ancient wisdoms. Through sharing research and scholarship, communicating with people who live where our ancestors once walked, and perhaps, when possible, even traveling to places our ancestors called home, we can learn about our pre-patriarchal, pre-colonized homelands and older Earth-based stories. Through this collective effort, we can find important trails to help us regenerate a more Earth-centered worldview and code of conduct. This task is not easy after hundreds and hundreds of years of violence and purposeful obliteration; rather, this is intergenerational work that we take on for generations to come. We repair and heal what we can of the tattered historical tapestry as we also weave in new threads that may have been unknowingly stored away in our great-great-great-grandmother's distaff held deep in our memory waiting to be spun.

Portions of this process can be likened to the biological and psychological phenomena of atavism and genetic memory. An atavism occurs when an organism regains an ancestral trait that was thought to have been lost through generations of evolution. This might happen through a variety of organic processes, but it points to the remarkable fact that the gene sequence never really disappeared; it was simply dormant. Genetic memory is a memory that appears to have been inherited, without our having experienced it directly or even been told about it. For example, suboscine birds raised from birth in soundproof rooms will sing the songs of their ancestors. Many human savants display great artistic or mathematical prowess after an injury or disease, and scientists hypothesize that these shocks to the brain activate their genetic memories.[17]

We can lucidly but inexplicably know things if we listen deeply, because inside of us, as designed by Nature, there is a way of keeping alive what we have forgotten. The stories are in our bones, calling to our ancestral memories, waiting to be retrieved and spoken and sung to the people and the land once again. Inside our bones is life-giving marrow, which makes the components of our blood that we need to survive and is also a carrier of our DNA. Every day, billions of new blood cells are produced in our bones, and this involves hematopoiesis, a term originating from the Greek words: *haima* (blood) and *poiēsis* (to make). Our bones are a place of life, memory, and world making.

Relearning and reweaving our Earth-based story tapestries from our homelands wherever they may be is necessary, while we also humbly and respectfully sit around the fires of Indigenous Peoples and learn from their wisdom, when we are invited. Yet, it is good to remember that this Indigenous knowledge must be approached without an extractive agenda.

Clearly, most modern-day people have not been heeding the natural laws of the Earth, embedded in Original Instructions. The question now is, how can we bring forth a respect for these vital directives? While the human world has changed a great deal over our relatively short tenure on Earth, the natural laws have remained the same. Consequently, it is paramount to respect the Earth's natural laws and remember the instructions and warnings of our ancestors.

Kichwa cultural teacher and travel guide Vidal Jaquehua tells us that life originated from the Kallpa, an infinite energy within the cosmos, or Pacha. Within Hanac Pacha, the heavens, flows the celestial river called Mayu, the Milky Way—a power keenly venerated by Jaquehua's Andean ancestors. Within Mayu, the essence of life germinated and came to be. In this prehuman time, wondrous beings dwelled on the Earth, comprising animals, Uywas; the winds, Wayras; the plants and trees, Mallkis; and living deities, Taytas.

In the opposite of a hierarchical and colonial approach, as Vidal explains, the Kichwa became unified with all the beings of their environment, and through this endeavor, their relationship with the natural world and respect for the natural laws formed the very basis of their entire way of life.[18] Yes, worldviews matter immensely.

As I walked back to the hacienda late that night with the moon pouring silver light onto the path, I heard Blanca's voice echoing in my mind, sharing how Pachamama is the origin, the center of our cosmos and communities. Her words enriched my understanding that recognizing our collective Origin Story in the heart of the cosmos is vital to course-correcting our ethics and returning to Earth-respecting principles and communities. I thanked the full moon and my celestial ancestors as they shimmered down onto the Imbabura and Cotacachi volcanoes, and all of us gathered in the Valley of the Dawn.

Chapter 3:
Ancient Trees and Ancestral Warnings

At ten thousand feet above sea level, I stand in the presence of majestic ancient beings who rise from the White Mountains in California's Inyo National Forest. This mountain range is home to the bristlecone pine trees (*Pinus longaeva*), the oldest living non-clonal organisms on Earth, some well over four thousand years old.

A lifelong love of trees, and a dedication to protecting forests and partnering with Indigenous and local community forest defenders, brought me on this quest to pay homage to these old ones in my own corner of the world. What can we learn from these tree elders at this crossroads time for humanity?

California is home to some of the most spectacular and ancient trees on the planet, including the coast redwoods (*Sequoia sempervirens*) who tower high into the heavens, higher than any other tree in the world. But here, among the bristlecone pines, it isn't their height that fascinates me; it's rather their startling sculptural shape and extraordinary age. Their thick trunks twist and spiral upward as if caught in a tornado, pulled between sky and rocky earth. Streaks of sienna, beige, and gold wrap around, from base to branch, accentuating every gnarled knob and crevice.

Here, just at the top of the tree line, the stark silvery stone landscape supports very little vegetation aside from these trees. The gusty wind that whips past me sculpts the bristlecone pines, creating their distinctive swirled trunks and branches, a phenomenon known as *krummholz,* which only appears at high altitudes. Bristlecones are bedecked with green needles that cling to the branches for up to forty years, and growth rates are extremely slow, with reported vertical measurements of only fifteen centimeters for forty-year-old "seedlings"—undoubtedly, these trees have an extraordinary density and sense of deep time. As I consider how we can undo our over-driven, planet-devouring, product-demanding social constructs, the bristlecone pine's timescape compels me to slow down and drop deeper into listening and learning from the perspective of long-arc wisdoms.

Beneath my feet is permeable dolomite, a type of limestone in which the bristlecones somehow thrive. The trees have long been special to the Paiute People, as the White Mountains are the traditional homelands of the Numu, as the Paiute originally called themselves. The oldest-known bristlecone, at over five thousand years, lives in these mountains and grew to adulthood during the early Bronze Age.[1] As I continue to hike in the early morning light, I stop suddenly under one of the pines—a particular tree branch has grown swirling downward and reaches out over the trail. Instinctively, I reach back and run my hand along the incredibly beautiful silky tree limb. Unexpectedly, I am caught in a rush of emotions, realizing I am connecting with a being whose life literally goes back thousands of years. I feel shy and humbled to be able to touch this ancestor.

Why is it that the bristlecone pines have been able to survive for so long in extreme temperatures with very little rain? How is it that these old tenacious trees grow from silver rock, in a moon-like mountain environment? One key reason is that, remarkably, bristlecones can slow or completely stop their own growth when nutrients and water are scarce, allowing certain parts of the tree to go dormant, or even die, to allow other parts to live. In some instances, much of the outer bark is lifeless and the trees can then appear to be altogether dead—except for the small green needles on several branches indicating that the trees are in fact alive. Additionally, the dolomite that the bristlecones emerge from reflects sunlight, so their roots remain cool, and they need less water.

I marvel at their resilience. These ancient ones are able to not just grow but thrive for thousands of years in a parched and formidable landscape. While humanity faces extreme crises, created by rapid expansion and delusions of endless material growth, there is much to be learned from the bristlecones' survival strategies in sufficiently slowing down to match the Earth's fluctuating nutritional offerings and ecological boundaries. By reining in and moderating consumption, and reducing growth, the entire "tree of society" can live. What if our global governance, economies, and communities were based on this understanding of long-term collective survival and well-being? Bristlecone societies.

Left breathless by these trees' stoic beauty, as well as the high altitude, my mind reflects on the Akan teaching of Sankofa (the magical bird with its feet firmly pointed forward and head gracefully turned backward), which illuminates that new knowledge is crucial over the passing of time, yet must be grounded in the wisdom of what has come before. The past must not be forgotten, and needs to inform the present and future. In this sense, the bristlecone pine trees are profound keepers of time, memory, and histories as they traverse temporal realities, with their tree rings mapping concentric circles of time that reach thousands of years into the past to this very moment. In their expansive presence, I find myself recalling the ancient story of Gilgamesh.

Just as the bristlecone pine is the oldest living organism, the *Epic of Gilgamesh* is the oldest recorded story of Western civilization. Many oral traditions are, of course, much older, but to our knowledge, Gilgamesh is the oldest story from Western civilization to be written down. The bristlecones of today were slowly rising out of the White Mountain shale when Shin-eqi-unninni, the ancient Mesopotamian scholar, committed the tale of Gilgamesh to cuneiform, carefully engraving each wedge shape into a series of stone tablets. By the time the story of Gilgamesh was recorded, it had likely been shared orally for generations, and thus it is a story from the land and its people over a long period of time. Much like the bristlecones whom I walk among, stories that have been passed down for thousands of years hold within them deep knowledge about how to survive. And, like other old-time stories, Gilgamesh contains vital insights and warnings about our human relationship to nature.

I was introduced to the story of Gilgamesh in my first year of university studies, and while much of the literary analysis focused on Gilgamesh's quest for immortality, I was struck by the ecological and social implications of this ancient tale. Here was a narrative thousands of years old that encompassed the same deep-rooted structural systems and attitudes of supremacy regarding nature and women that I was intensely wrestling with at that time (and continue to do so today, with the critical addition of colonial and racial injustices). To chart a truly transformative path forward, the dominant culture would benefit from tracing, understanding, assimilating, and integrating lessons from historical recountings, and this first recorded story can provide essential markers.

Gilgamesh was a real person—a Sumerian king who lived around 2700 BCE and ruled Uruk along the Euphrates River in what is now Iraq. However, it was not until centuries later that the stories of Gilgamesh were consigned to writing. Fragments of these stories, including parts of the *Epic of Gilgamesh*, were found engraved, in poetic form, on clay tablets in Sumerian cuneiform, and they are dated to around 2000 BCE. Later, more complete versions were discovered, written in the Akkadian, Hurrian, and Hittite languages.

Significantly, at the time these tablets were engraved, Gilgamesh's ancient Mesopotamian civilization was in transition. Once a culture that honored a Mother Goddess, it was part of a widespread shift toward patriarchy, which greatly impacted humanity's codes of conduct with the land and women.

In the long span of time from the actual life of Gilgamesh until the recording of the story, many legends and myths developed about him, including that he was created part divine by the Mother Goddess Aruru. The tablets also document what is usually translated as Gilgamesh's "exploits," bringing to mind "exploitation." This is a story of a king and a culture losing their balance, responsibilities, and reciprocity with nature. In part, and astonishingly, the story narrates a people recounting their own awareness of humanity's venture into a worldview of dominion over the natural world, and it warns us of the deleterious consequences of such a stance. Let's take a closer look at the segment of the story that concerns trees because this imparts insight into one of the many attributes the dominant culture needs to dismantle and reconfigure as we seek to generate a new way forward.

As the story goes, Gilgamesh, with his hubris and superhuman strength, rules his people with an oppressive and domineering hand. Among his abuses is the practice of *jus primae noctis*, forcing brides to submit to him sexually on their wedding nights before going to their husbands. This is a patriarchal practice that continued into feudal Europe.

The people of Uruk call out to the Sky God, Anu, the most ancient deity of Sumer and the chief god of their city, to help them stop this abhorrent abuse of women. Anu responds by enlisting the help of Aruru, the Mother Goddess, who takes a pinch of clay and tosses it down into the wilderness on the outskirts of Uruk to create Enkidu, a man of strength and virility equal to Gilgamesh. Covered in thick hair, Enkidu runs wild and naked with the animals.

One afternoon, the son of a trapper becomes alarmed upon seeing the unclothed wild man with the forest animals, and he runs to tell his father. The father instructs the son to enlist a sacred temple prostitute, Shamhat, from the temple of the Goddess Ishtar to entrap Enkidu.

Enkidu spends a week making love with Shamhat and learning the ways of Uruk civilization, and at the end, he discovers he has lost his animal strength and the wild beasts now run away when he approaches them. At this juncture in the story, we see the narrators of Gilgamesh's time beginning to articulate and grapple with the loss of their culture's intimate connection to the natural world. Perhaps this is not only the oldest recorded story of Western civilization but also the earliest recorded separation-from-nature narrative. We also see that a woman is to blame for Enkidu's loss, which is part of the patriarchal narrative already well underway.

Shamhat and some shepherds teach Enkidu to eat, dress, and behave like a "civilized" person, and upon hearing of Gilgamesh's abuses of power, Enkidu vows to confront him. Enkidu enters the city during a wedding celebration, just as Gilgamesh is making his way to the bridal chamber to rape the newlywed. Enkidu blockades the door to protect the woman, and Gilgamesh and Enkidu clash in a ferocious fight. Enkidu, who has just recently traversed the passage from wild nature into the ways of civilization, comes to the defense of the bride. He maintains a vestige of the old order of Mother Goddess cultures and respect for the feminine. Gilgamesh fails to violate the bride, and although the two men are quite evenly matched, eventually Gilgamesh dominates, but not before developing a respect and admiration for Enkidu. The two quickly become close friends.

In time, the two men grow tired of living comfortably in the city, and Gilgamesh proposes to Enkidu a daring exploit: a trip to the sacred Cedar Forest, the territory of the Mesopotamian deities, to kill the forest spirit Humbaba and chop down the trees. No longer content with subjugating the women of his city, Gilgamesh wants to be master of nature. Enkidu attempts to persuade Gilgamesh to forgo the idea. Though he has spent much time in the city, Enkidu is still the more nature-minded of the two, and he knows more about Humbaba, who guards the forest per the command of the Weather God Enlil. The elders of the city also disagree with the undertaking, but Gilgamesh refuses to listen.

The two warriors set out on their six-day journey to the Cedar Forest, and each night Gilgamesh prays to Shamash, the Sun God, asking for a positive omen on their mission. However, each night Shamash sends him inauspicious dreams urging him to abandon his quest, which the king persistently disregards.

Upon reaching the Cedar Forest, Gilgamesh and Enkidu fight the forest guardian and, with the assistance of Shamash, subdue Humbaba, who pleads for his life. Enkidu, apparently forsaking his connection with the wild in his further immersion into civilization, urges Gilgamesh to finish off Humbaba, goading him with enticements of fame and glory. Gilgamesh severs the head of the faithful forest spirit and defender. But before Humbaba dies, he curses Enkidu for his part in the violation, declaring that Enkidu will soon die for his treachery.

With the protector annihilated, the two men cut down the entire Cedar Forest. With some of the felled trees, they build a raft to float themselves, the head of Humbaba, and the cedars back to Uruk on the Euphrates River. The biggest cedars are crafted into a massive gate for the city, placing Gilgamesh's extractivist prowess on full display. For an adventure, material gain, human ego, and widespread fame, Gilgamesh and Enkidu not only pillage old-growth sacred trees, they slay the divinely appointed guardian of the forest, proclaiming human dominance over nature. It is unknown whether the Cedar Forest mentioned in the tablets was located in present-day Iran or Lebanon, but it is a fact that the ancient cedars of Lebanon are nearly gone today, lost to extractivist enterprises that have continued since the time of Gilgamesh.

Back from their triumphant expedition, Gilgamesh attracts the attention of the Goddess Ishtar, who approaches him and offers to become his lover. But Gilgamesh refuses the Goddess, listing her former lovers and allegations of cruel treatment she inflicted upon them. This rejection enrages Ishtar, who demands that the Sky God Anu send the Bull of Heaven to attack Gilgamesh, or she will send a famine to the people of Earth and raise the dead. Anu acquiesces, delivering the Bull to assault the two warriors. However, Gilgamesh and Enkidu easily slaughter the divine animal. To avenge the killings of Humbaba and the Bull of Heaven, the gods afflict Enkidu with a fatal illness. As he writhes in pain near the end of his life, Enkidu expresses regret and ambivalence about joining civilization. He also curses the cedar gate for which he has committed crimes against nature and against the gods, leading to his downfall.

After bitterly mourning his friend, Gilgamesh confronts his own mortality and embarks on a quest for everlasting life. With further hubris, Gilgamesh attempts to circumvent the laws of nature. His journey takes him to the ends of the Earth, where Utnapishtim and his wife, the only mortals granted eternal life, reside. After many obstacles and trials, Gilgamesh at last arrives at the understanding that he cannot overcome the powers of nature and cannot become immortal. This portion of the tale ends with Gilgamesh reflecting on what the purpose of his journey has been as he admires the great walls of his city, Uruk—their impressiveness, but also their limitations.

There is much to learn from the wisdom embedded in this ancient story; for now, let's focus on the transmutation over a long interval from an ancient Mother Goddess society to a more patriarchal one, and the transition from a reverential and reciprocal relationship with the natural world to a dominating, extractive one. This same transition was happening in many cultures globally in a similar historical timeframe: with the downfall of the Mother Goddess and respect for the feminine, so goes the loss of respect and reciprocity practices for the Earth.

The story draws a clear lesson: when we oppose the natural order, we pay a price. In Enkidu's case, he died as a consequence of abandoning good conduct with nature and destroying the forest. Yet, even after Enkidu's death, Gilgamesh refuses to acknowledge and respect the intractable laws of nature. His search for immortality shows that he still believes he can use force and sheer will to dominate and overcome the natural world. In the end, Gilgamesh returns to his home having lost his beloved friend and failing to achieve everlasting life.

As I walk through the bristlecone forest, I reflect upon how strange and tragic it is that for as long as these trees have been growing, the ancient Gilgamesh story has attempted to communicate a warning message that generations have failed to receive. Now, Mother Earth is exhorting us in every way she can to stop exploiting the forests and the land. Trees and plants are indispensable to life on Earth. We cannot live without them.

As I continue my hike in the White Mountains, another bristlecone pine catches my attention. In its thick trunk, whirling and curling as it juts from the shale rock, I see the forms of two magnificent dancers entwined, holding sunrays with their outstretched limbs. I approach the spiraling dancers to examine the needles on a low branch, and contemplate the power these little needles hold—the power of creation, of life, of energy. Our very existence depends on plants and trees and the extraordinary feat that they can perform in their leaves or needles: photosynthesis.

Only plants and algae have the power to convert energy from the sun into chemical energy. They do this by using solar energy to transform water into carbohydrates, and through this process release oxygen into the air. This stunning phenomenon creates food for animals and air for them to breathe. Of all the planets in our solar system, and in our entire galaxy as far as we know, ours is the only one where the conversion of light into chemical energy occurs. Photosynthesis makes the life we have on Earth possible.

Not only do trees nourish all of us, they also have sophisticated and elegant mechanisms to support each other. Western naturalists and scientists have long assumed that trees in a forest compete with each other for resources such as water, sunlight, nutrients, and space for roots and branches. However, this is not the case; rather, trees share carbon, water, and glucose through an elaborate underground network facilitated by delicate threads of fungi. Tree trunks divide into roots, and those roots in turn divide into smaller and thinner segments until the tiniest of these roots weave into microscopic fungal filaments. These filaments, known as mycorrhizal networks, are vast reticulations that spread through the ground, fueled by the carbohydrates that the trees photosynthesize. They are immensely broad yet completely invisible to the naked eye—a single teaspoon of forest soil contains several miles of microscopic fungal filaments. It is through these networks that trees communicate, cooperate, and share water and nutrients. When a tree is at risk—experiencing drought, disease, or damage from humans, insects, or animals—it sends signals through this webbed network to other trees. You might have seen a young sapling in a forest, dwarfed by older, taller trees, and wondered how it can survive without immediate access to sunlight. The wondrous phenomenon is that the larger trees, including their parent trees, are pumping sugar into its roots through this mycorrhizal network.

The forest is a place of infinite kin relationships that create community and rootedness to place. How can we humans participate in this ecology as respectful relatives?

One scientist who has conducted thorough research and experiments on the mycorrhizal networks is Suzanne Simard, based in British Columbia, Canada. Although she was initially ridiculed by other scientists, her work is now not only accepted but applauded in demonstrating that trees do in fact support each other and communicate. She has inspired another generation of tree lovers by revealing that each forest has certain trees that stand as central hub trees, which are the primary nourishment-sharing trees: the Mother Trees. These are older trees with deeper roots who support the younger trees with shallower roots by passing food through the twisted miles of threaded fungi.[2]

At first, this collectivism among the trees perplexed scientists. Why should they share? They found that sometimes trees even create networks with other species of trees, a behavior that all former research would suggest is a disadvantage to their specific species. Yet, upon further observation, Simard described how forests actually work more like healthy neighborhoods, with different species acting not as competitors but as supportive neighbors. And in a forest, a strong, diverse community means more protection from ecological harms. Our human communities could learn a great deal from this forest teaching. Additionally, the elders of this neighborhood community are an integral part of its survival. When the biggest and oldest trees in a forest are cut down, the survival rate of the younger trees is greatly diminished.

But climate change is seriously threatening our global forests in a variety of ways, including through temperature rise, because photosynthesis decreases as extreme heat prevents plants from absorbing the amount of carbon dioxide they were previously able to. Couple this with increasing demand for land, lumber, and agribusiness, and our planet's ability to sequester carbon in forests is already greatly diminished. Simard has recently launched an initiative she calls the Mother Tree Project and is questioning how we as humans should be helping forests.[3] This one-hundred-year experimental project with firs, pines, and larch trees across British Columbia explores the most effective ways to protect and proliferate the forests.

Western scientists, ecologists, and forestry experts are coming around to ways of thinking that Indigenous Peoples have known for millennia.

Jaime Yazzie, a forest technician of the Navajo Nation, conducts research on tree rings in an effort to conserve the forests of her region. She approaches her research with an understanding that trees are deeply interconnected as she applies the Navajo concept of K'é, or kinship, to her work. As a university student, Yazzie was asked which trees she wished to save as part of her forestry thesis research.[4] The question shocked her because Yazzie's cultural teachings led her to understand forests as a whole integrated living community. The loss of one tree species affects the health of the whole forest. She decided to focus her research, and her life's work, on how we should change our modern way of life to save entire forest ecosystems and help them adapt to climate change.

I sit on a shelf of cool stone under one of the bristlecones, close my eyes, and imagine the magnificent underground mycorrhizal network and sense its profound vitality and knowledge. With my gaze turned inward, the elder trees continue to speak to me, and I strive to hear their antique voices that are connected to many stories from many lands.

Given the life-generating power of trees, it is no wonder they are central to spiritual worldviews in cultures globally. Trees provide shelter, food, and medicines for people across our home planet—and it is often the stories of the trees, not mere lists of scientific facts, that can illustrate the innate magic and wonder found in these sentient beings.

Our forebears did not share these tree stories as pure entertainment. They contain profound knowledge and advice about our comportment, and in our current time, they can help show us that the path out of dominionism is found in intimately reintegrating with the land, the trees, the mountains, and the stars—our relatives. For some, these stories are familiar; others may be learning them for the first time. Either way, we have the opportunity to discover, reclaim, and revitalize these living narratives and mythologies that reflect an animate worldview of our Earth-centered ancestors. I think of it as the work of remembering and mending fragile or broken story roots within the cultural mycorrhizal underground network.

Most of our antecedents come from cultures that carry traditions that honor a World Tree, which is similar to the Tree of Life, representing the whole of the cosmos, including the heavens (the branches), the Earth (the trunk), and the underworld (the roots). While each culture depicts this sacred World Tree as a species from their own region, the stories of these trees are strikingly similar in that they each tell of an ancestral community that views the World Tree or Tree of Life as the axis mundi—the world pillar, the center of the cosmos, the cosmic axis. These cosmologies express a brimming ecology in that the unseen landscape under our feet, the horizontal reality of our daily lives on Earth, and the great firmament overhead are all intertwined in wholeness. These cultures uphold a respect and reverence for nature with ethics and responsibilities to maintain a healthy relationship with the plant and animal relatives. There are many variations of these tree stories to explore, but here are some initial threads to begin the remembering and mending journey.

In Norse mythology, the World Tree is an ash growing in the center of Asgard called Yggdrasil. The branches of this enormous tree hold together the cosmos and nine worlds. Three exceedingly long roots support the tree, and they extend to three wells or springs in faraway lands, often described as parts of the underworld. One root is watered by the Norns, three powerful women analogous to the Fates or sometimes depicted as keepers of time. According to the *Prose Edda,* the deities hold court daily in this sacred tree, and many creatures live among its branches: four stags, a wise eagle, a hawk, a rooster, and a mischievous squirrel that carries messages between the eagle and a dragon or snake that chews at one of the tree's roots. While snakes nibble at the root, the stags consume Yggdrasil's leaves; hence, the tree supports life but makes a sacrifice while doing so, also signifying its mortality. The welfare of the universe is linked to that of the great tree.

In some mythologies, an oak is a World Tree, or otherwise holy. The oak is sacred to Thor and Zeus in Norse and Greek mythology. It is also sacred to thunder gods in Celtic, Baltic, and Slavic traditions, possibly because oaks, due to their deep central taproots, are among the trees most often struck by lightning in their regions. The oak is often seen engraved in jewelry and adornments, usually within a circle, as the Celtic World Tree. The Celts considered trees to be essential and magical sources of wisdom. When tribes cleared land for a new place to live, they always left a large tree, most often an oak, in the center of the community where assemblies gathered and ordained chieftains.

The Celtic Tree of Life was a doorway to the realm of goddesses, gods, and ancestors. In illustrations, the roots and branches are often woven together, depicting the cycle of life and the forces of nature working in perfect harmony to preserve balance in the universe—in fact, the Welsh phrase *dod yn ôl at fy nghoed,* which means to achieve a sense of balance in one's mind, literally translates as "to return to my trees," further illustrating the harmonious function of the tree beings.[5] The World Tree grew from the center of the Earth and sheltered all existence, and held the focal point of the community.

The Indian World Tree is a peepal (sacred fig) described in the *Rig Veda.* It is called Ashvattha and houses Agni, the Fire God. This Fire God lives in the tree in acknowledgment that trees perform photosynthesis and literally carry the Sun's fire in their very being. Stories of this tree vary widely, but some accounts describe the tree as having its roots, representing the god Brahma, upward in the heavens, and its branches downward in the ground. The world in this case is a mirrored universe in the spirit realm. Today as in the past, the leaves of the peepal tree are used in Ayurvedic medicine, and cutting down a peepal is considered sinful by observant Hindus. The peepal is also called a Bodhi tree—the same tree under which Siddhartha Gautama attained enlightenment.

Observers of Tengrism, a diverse religious tradition practiced in Turkic, Mongol, and other communities in Central Asia, describe a Tree of Worlds, often a beech, that forms the only link between the terrestrial realm, the underworld, and the heavens. Shamans commune with the tree and may, in this way, visit the celestial realm. Often, a bird sits atop the tree, ready to guide souls into the afterlife.

Similarly, the ceiba tree connects the underworld, the terrestrial world, and the celestial world in many pre-Columbian Mesoamerican communities, including the Maya. In Mayan cosmology, a colossal ceiba grows at the point where the world sprang into existence. Its roots extend to the underworld (Xibalba) and its branches lift up the sky. The ceiba is the axis mundi and a path between the realms along which the deities and the spirits of the dead travel. Ceibas vary by species and region, but some grow to 230 feet, making the tree one of the tallest in the forest. Much like the giant redwoods in California, ceibas host an ecosystem in their resplendent crowns. Brilliant magenta, crimson, and butter-colored epiphytes such as bromeliads and orchids perch on its highest branches, as do mosses, lichens, ferns, and ficuses, creating kinetic watercolor paintings across the forest canopies. The ceiba's flowers bloom rarely but extravagantly, synchronizing with bat and hummingbird pollination visits during the nectar-filled flowering days. With toucans and other birds nesting in their branches and bromeliads collecting water in which tadpoles grow into frogs, the ceibas are undeniably a demonstration of the Tree of Life. Modern Maya continue to respect the ceibas and often leave them standing when they cut other trees for lumber.

Closely related to the World Tree concept is the Tree of Life; some traditions conflate the World Tree, the Tree of Life, and the Tree of Knowledge. In many Christian traditions, two trees grew in the Garden of Eden: the Tree of Life, and the Tree of Knowledge of Good and Evil. When Adam and Eve ate from the Tree of Knowledge of Good and Evil, they were expelled from the garden and could no longer access the Tree of Life. The two trees in the garden are merged into one in the Islamic tradition.

Jewish tradition also recognizes the two trees in the Garden of Eden, but the Tree of Life, or Etz Chaim, also refers metaphorically to the Torah itself, the "fruit of the righteous," and a "healing tongue."[6] In the Jewish mystical tradition, the Kabbalah, the term refers to a diagram consisting of ten spheres showing the path to God.

The term *Fusang* has different meanings in ancient Chinese mythology, but one of those meanings is the "Tree of Life," sometimes depicted as a mulberry and sometimes as a hibiscus. The tree grew on a faraway island in the east, from where the sun rose. Ten ravens dwelt in its branches; while nine slept, the tenth guided the sun on its journey across the sky.

Asherah, a Phoenician and Hebrew variation of Athena, was often honored as a holy tree and worshipped as the Great Mother in special groves. One of her most important aspects was as the sacred Tree of Life. Her ancient symbol was the well-known sign of healers, still recognized today—the caduceus, which depicts two serpents twined around a tree or rod known as an Asherah pole. This same divine figure was also known as Anath or Atteena in North Africa, later to become the more familiar Athena of classical Greece, who was imbued over time with further meanings and symbols associated with wisdom, war, weaving, and council.

Part of the re-storying process in healing and rejuvenating a culture that lives and breathes with the trees and all our kin involves learning and assimilating what happened to our once-sacred relationship to trees, personally and culturally. Only after uncovering and acknowledging unbidden fractures in our animate and spiritual relationship with nature can we truly navigate and heal past traumas caused by our perceived estrangement from the web of life. Instead of carrying centuries of unattended grief and traumas that often induce detrimental and brutal behaviors, we need to know and name what was violated, forsaken, and forgotten in order to embrace and heal wounds. These are ruptured rememberings that need love, accountability, and repair within our interior worlds and within the broken story roots of the cultural mycorrhizal underground network.

Given the sacred and practical significance of trees worldwide, it is not surprising that the vicious cutting of trees has been used as a weapon to break a community's spirit and show superiority. The Celts of Old Europe understood that they would lose their strength if their sacred tree was cut down, and overall, the felling of trees outside the proper reciprocal protocols was a crime. Later, when the Irish had converted to Christianity (after years of violent coercion), certain sacred trees, called bile, were still respected as sites of royal inaugurations and symbols of kingship. When Brian Boru, a Christian king of Munster, defeated the Vikings of Dublin in 999 CE, one of his first actions was to cut and burn a sacred grove where the pagan Hiberno-Norse honored Thor. Brian Boru clearly intended this act to be a statement of not only political but also religious, prepotency over the Earth-centered peoples.[7]

There were many such violent fellings of holy trees during the Christianization of Europe. Donar's Oak was another tree associated with Thor and was sacred to the Germanic pagans. In the eighth century, Saint Boniface, an Anglo-Saxon missionary, brought a number of Hessians into the Catholic faith but questioned their commitment to it. He knew that the earlier Earth-centered spiritualities maintained a hold on the hearts of the new converts and that some of them continued to visit the holy trees and rivers. Intent on quashing these Earth-honoring practices, Saint Boniface cut down Donar's Oak, one of the largest sacred trees in present-day Hesse, Germany. He then used the wood to build a church dedicated to Saint Peter.[8] Also in Germany, the Christian king Charlemagne, during the Saxon wars, destroyed an Irminsul, a tree trunk or wooden pillar dedicated to the Germanic deity Irmin.

Missionaries destroyed sacred trees on other continents, including Africa. In the early 1900s, a group of Italian Catholic missionaries landed at Mombasa, Kenya, on their way to Abyssinia. In Kenya, Chief Karuri wa Gakure of the Kikuyu hosted the men for a time and invited them to hold their mass under a mugumo tree, a fig tree sacred to the local people. The Kikuyu viewed mugumo trees as sacrosanct to their identity and as places of ancestral spirits and the god Ngai. The missionaries, on the other hand, saw them as "temples of the Kikuyu paganism," and as objects to be destroyed in the name of Christianity.[9] Different groups of European missionaries in the twentieth century took it upon themselves to ravage the Kikuyu community's holy mugumo trees and replace them with churches.

This brutal felling of trees signifies a great desecration of the sacred and the web of life, and it was common for the church to take control of sacred groves and sacred trees to destroy the spiritual and cultural survival of Earth-honoring peoples in various regions of the world. The church destroyed trees, but also worked to commandeer their relationship with people. For example, consider that yews grow in many churchyards in Britain and Ireland. Earlier and congruent cultures honored the yew in sacred rites, so the church aggressively built its structures in places held to be sacred.

From the time European colonizers entered the Americas until now, they have disrespected and demolished the sacred trees and forests of Indigenous Peoples—often in violation of treaties and land agreements. Corporations and governments would not dare to destroy a church in pursuit of financial gain, but Indigenous hallowed forests and sacred sites are all too often deemed invalid and disposable.

As part of the colonial settlers' scorched-earth campaign against the Navajo (or Diné as they call themselves) in the nineteenth century, many trees were chopped down to oust the people from their land. At that time, the Diné lived in the fertile Canyon de Chelly in Arizona, and the settlers wanted them out. Led by frontiersman and fur trapper Kit Carson, they plotted to cut off the Diné community's food source and force them, under threat of starvation, to make the long journey to the reservation of Bosque Redondo, a desolate place with bad water and few trees. In Canyon de Chelly, the Diné and other nearby tribes had painstakingly cultivated thousands of peach trees, along with other crops such as maize. On multiple occasions in 1864, cavalries stormed the canyon and felled nearly all of the peach trees.[10]

We are now experiencing the full consequences of the dominant culture's commodified, despiritualized, and destructive stance toward trees and forests, which has been ongoing since the felling of the great Cedar Forests in ancient Babylonia.

Because sacred World Tree and Tree of Life stories and practices are witnessed and celebrated within a living landscape, when holy trees were felled, the peoples of Old Europe lost many of their sacred tree stories and ceremonies—striking a devastating blow to their connection with the land. No mincing words here: When the sacred trees are cut, the stories are also severed from the people. Invading armies, religious zealots, and colonial settlers have purposefully ravaged the storied World Trees, and we are very much in need of reconstituting and nourishing these sacred narratives and the teachings they convey.

Since the first narration of the revealing tale of Gilgamesh, a damaging worldview has persisted, one that conceptualizes the natural world not as a wondrous integrated symphony of relatives to be cared for and given back to, but rather as something to be dominated, exploited, and conquered. The dominant cultural notion of perpetual progress and endless material growth at the expense of nature has led to devastating forest loss in the name of timber harvesting, ranching, mining, agribusiness, and development.

Statistics on forest loss vary sharply, as those evaluating the data take into account different factors, including government reporting, satellite imagery, and whether the land is used for another purpose or left free to rewild and regrow itself, but scientists estimate that Mother Earth has lost at least one third of her forests since the Ice Age—and half of this forest loss has occurred since 1900, in just a little over one century.[11] Also within the past century, the continent of Africa has lost between eighty and ninety percent of her tropical forests along the west and the east—due to agriculture, fuel and energy needs, mining, and logging, with the latter two extractive industries perpetuated by colonial and post-colonial governments and entities who disregard the land rights of local peoples—though forests elsewhere in Africa have expanded into the savannah during that time.[12]

And the shocking destruction continues into the present century. Deforestation is rampant in tropical areas such as Southeast Asia, where precious biodiverse forests are cut to make way for oil palm and rubber plantations and to plant rice. Scientists tell us that 235,500 square miles of forest have been clear-cut in Southeast Asia just between 2001 and 2019. Worse, 73,000 square miles of this lost forest land existed in the mountains.[13] Mountain forests are filled with endemic species and sequester more carbon than forests at lower altitudes, so their loss is devastating to the planet. Data on worldwide forest loss is still coming in from the past several years, including 2019, which witnessed staggering losses from wildfires as well as destructive industries.

Yet we know that our woodlands are essential for maintaining the delicate balance of our planet's bioregions and biodiversity. Trees in tropical forests store between 90 and 140 billion tons of carbon, playing a vital role in regulating the temperature of their ecosystems.[14] By pulling moisture from the ground and releasing it into the air, trees cool the atmosphere around them and regulate precipitation patterns in much broader areas. When tropical rainforests are cut down, the local surface temperature can rise by up to three degrees. By removing the trees of the forest, we remove one of the pivotal ways in which we curb the warming of our planet. While the evidence of our reliance on forests is well documented by the scientific community, conservation efforts are often at odds with the profits of extractive industries. The Amazon rainforest, which plays a crucial role in local and global water and carbon cycles, has been under attack from agribusiness and extractive activities for decades and is on its way to becoming a savannah-like ecosystem.[15] Once the global temperature hits a tipping point, there will be little we can do, and the warming will exponentially exacerbate the problem through a chain reaction. Many trees simply will not withstand an increasingly heating climate.

In New Mexico, two scientists studied how rising temperatures break down a tree's resilience. Postdoctoral student A. Park Williams observed that warming affects a tree's ability to undergo photosynthesis—even a one-degree increase drastically impacts a tree's ability to convert CO_2 into carbohydrates and oxygen. This process taxes the tree so extremely that it dries out, becoming vulnerable to predatory insects and microbes that can smell the tree's diminished sap levels.[16]

Dr. Nate McDowell, Williams's postdoctoral advisor, simultaneously studied the effects of climate change on the local piñon pine trees of

New Mexico. By building a plexiglass cylinder around individual trees, McDowell was able to simulate rising climate temperatures and monitor the effects on the trees. What he discovered was that, if temperatures continue to climb, the Southwest's forests will be replaced by grassland and shrubland by 2100. And as these native trees with edible seeds dry out and die, not only will they cease to be able to capture and convert carbon, but they will also release carbon into the atmosphere as they decompose.[17] Even if we replant a thousand forests, if the atmosphere has already risen beyond a certain temperature, the effort will be futile, as the trees will be incapable of surviving.

But we don't have to wait for that climate tipping point and more of the serious droughts associated with forest loss. We are already seeing the destructive effects of deforestation, including many diseases that plague humans, in large part because of forest destruction. Ebola, bird flu, SARS, and MERS originated in the bodies of non-human animals, where they existed for years, harming no one. When humans cut down forests and build suburbs, roads, or livestock grazing grounds, or clear-cut forests for material products or for drilling or mining projects, the wildlife from those forests is forced into contact with humans and livestock. Species that may have never lived together are thrust into close contact, providing opportunities for pathogens to jump from wild to domestic animals or directly to humans.

Indigenous Peoples such as Levi Sucre Romero, coordinator of the Mesoamerican Alliance of Peoples and Forests, have understood the threat for years and tried to warn the world about these diseases. Sucre Romero likewise points out that forests are also incredible sources of medicines: "We know that 25 percent of the medicines [the world] uses come out of the forests and that by losing the forests we put in danger future solutions."[18]

Deforestation and loss of Indigenous rights to their land go hand in hand. Indigenous Peoples, in the main, have respectfully managed the ecology of their ancestral lands and forests since time immemorial, and they must have their land rights upheld so they can further steward the forests in reciprocal relationship and avail of them for ceremonial, cultural, and medicinal purposes. Then, possibly, Indigenous and Western scientists can work together to find cures for the increasingly prevalent diseases and pandemics. But Western scientists must approach such collaboration respectfully, as Dinamam Tuxá, coordinator and legal adviser to the Association of Indigenous Peoples of Brazil, points out: "The cure for the next pandemic might be in our lands, and what's important is that our traditional knowledge is adequately recognized." He continues, "These large pharmaceutical companies come into our communities, extract our traditional knowledge and plants without recognizing our rights."[19]

Indigenous Peoples around the world have been successfully living within, maintaining, and caring for forests for longer than modern science has even existed. At a press briefing in 2019, Victoria Tauli-Corpuz, the UN Special Rapporteur on the Rights of Indigenous Peoples at that time, asserted, "No one understands the value of forests better than Indigenous and local communities. As experts, often guided by hundreds of years of knowledge, we are uniquely suited to manage, protect, and restore the world's forests."[20] It is because of their expertise and experience that the Indigenous Peoples of the Earth not only must be consulted but must be central leaders in the task of caring for the forests of the planet, and their rights to consent or not to any forest activities must be respected. Beyond diseases, we are witnessing the severe ecological stresses that global warming has already caused, including catastrophic fires in the Amazon, Australia, Africa, Europe, and the United States.

On the other end of the spectrum, it is clear that global leaders' perspectives on forests—drawn from the neoliberal economic model that demands the commodification and financialization of nature, and the colonial system that violates Indigenous Peoples—will continue to threaten and harm Indigenous Peoples and further the climate crisis. All too often, wealthier Global North countries promote false solutions that attempt to tackle deforestation from a purely market-driven approach, such as carbon offsets, as a way to evade true climate action. Studies have shown that carbon offsetting fails to reduce emissions, causes further harm to biodiverse ecosystems, and most often violates human and Indigenous rights.[21] And these false solutions allow big polluters to continue to poison and displace communities at the sites of extraction, distribution, and processing by buying up pollution permits.[22] We need to reject any solutions that do not safeguard the dignity and flourishing of all people and the planet, and do not stop pollution at the source.

"Frantically planting trees to 'offset' carbon emissions is part of a wider movement within conservation, known by some critics as 'selling the right to harm,'" writes Tesni Clare in the *Ecologist.* "Poor and damaging behavior, such as emitting greenhouse gases or destroying biodiversity, is supposedly 'cancelled out' by paying to plant trees or improve biodiversity elsewhere."[23]

Carbon offsetting also assumes that one tree or one forest is equal to another. But ancient woodlands are the product of hundreds or thousands of years of slow interconnected development, while these new "reforested" lands amount to a series of identical non-native trees that do not foster the ever-important mycorrhizal underground network nor a broad diversity of plant and animal life with healthy habitats for these many interconnected species. An old-growth forest, with its multiplicity of mutual benefits, cannot be replaced by a barren grid of identical trees.

Market-based mechanisms are a false solution to curb catastrophic climate change and deforestation. We must instead protect and defend the Mother Trees and the forests they have nurtured for thousands and thousands of years. Likewise, we need to stop approaching forests as extraction zones and rather remember and revive a worldview that is deeply entwined with ecological well-being and the sacred Tree of Life.

In my youth, I was deeply impacted by reading the words of Heȟáka Sápa, more widely known by the name Black Elk, a leader of the Oglala Lakota people. In his elder years, Black Elk conveyed his experience of and reflections on the unconscionable 1890 massacre at Wounded Knee:

> I did not know then how much was ended. When I look back now from this high hill of my old age, I can still see the butchered women and children lying heaped and scattered all along the crooked gulch as plain as when I saw them with eyes still young. And I can see that something else died there in the bloody mud, and was buried in the blizzard. A people's dream died there. It was a beautiful dream.
>
> And I, to whom so great a vision was given in my youth,—you see me now a pitiful old man who has done nothing, for the nation's hoop is broken and scattered. There is no center any longer, and the sacred tree is dead.[24]

His words have echoed life-long in my mind, and I still cannot read them without my heart wrenching and tears brimming. While many of us continue our campaigns to stop deforestation globally, it is also time for the sacred tree to be recognized and returned to the center of our communities once again. We need to listen to these sentient tree beings who are far older than us.

Our ancestors from every land around the world had council trees in their communities where they would discuss the laws of the people and arbitrate disputes. From the Celts of Germany, Ireland, and Britain to Original Peoples of the Americas, and people in other regions globally, old trees were often the place of council meetings. The root of the words "arbiter" (one chosen to judge or counsel in a dispute) and "arbitrate" (to settle by arbitration) is the same as that of "arbor," as in the shelter of the trees, and of "arboreal," of or pertaining to trees. In Old High German, the root *ar* refers back to "rat," or counsel, and in Old Norse, *radh* also means counsel. We know that scientists have observed improvements in human well-being associated with time spent in nature, so it makes sense that if we are to have ethical codes of conduct with the Earth and each other, then *where* we hold the deliberations of how we should live and act does in fact matter.[25]

In the Valley of Carranza, located in the Basque region of Spain, council meetings to discuss local politics were held under a holm oak in Soscaño. This took place from the Middle Ages until 1740, when an assembly house was built. The holm oak was a massive tree, said to have lived for seven hundred years. The council of Lanzas Agudas was also known to meet under an ash tree, sitting on stones arranged for that purpose.[26]

Likewise, in the villages scattered throughout the Manda Wilderness in Malawi, legal matters are often debated and resolved under the shade of a mango tree. While the local chief presides over the meeting, assisted by a healer or a government secretary, the legal process is democratic, as illustrated by its public nature.[27]

Several former and current council trees recognized by Indigenous Peoples in the United States—or Turtle Island, the original name of this continent—have been faithfully documented, such as the Neenah Treaty Elm near Lake Winnebago in Wisconsin, which was so large that navigators on the lake used it as a reference point to find their way.[28]

Perhaps the most recognized council tree in North America is the Tree of Peace, the white pine chosen by Dekanawida, the Great Peacemaker, a leader of the Haudenosaunee—or as they are most commonly known, the Iroquois. There is debate regarding the time the Iroquois Confederacy was founded, but most historians (including Indigenous historians' accounts) assert it began in either 1142 or 1451 because each year featured a solar eclipse.[29] The Great Peacemaker, along with Jigonhsasee and Hiawatha, counseled peace and proposed a confederacy of the five Nations: the Mohawk, Oneida, Onondaga, Cayuga, and Seneca. They established the powerful Iroquois Confederacy, planted a white pine at the edge of Onondaga Lake as a symbol of their union, and buried their weapons of war under it.

Just as the confederacy joined five tribes, the white pine symbol possesses bundles of five needles each. The tree's four Great White Roots of Peace extend in the cardinal directions. Any other tribes desirous of joining the confederacy were welcome to follow these roots to the Tree of Peace and stand under its branches. The chiefs sat under the white pine as caretakers of the Great Peace and passed on their roles to the next generations.

As I leave the primordial bristlecone forest and descend the mountain trail, I reflect further on the bristlecones' tenacity and their survival strategies. Like them, we can give energy to the parts of the sacred tree that have survived, even in a treacherous landscape. We can give energy by getting to know and care for the trees near our homes as well as supporting the protection of forests globally from the onslaught of an extractive worldview and economy. We can elevate and respect the rights, leadership, and knowledge of the Original Peoples of forest lands. For those of us who are settlers on lands not of our ancestry, or were violently stolen from homelands through enslavement, or sought safety in a land while fleeing persecutions in other places, we can work to revive our Earth-honoring stories. They are held dormant within us, and by tending the roots and limbs that escaped previous conquests, we can bring them back to life. If we protect the forests, respect Indigenous rights and leadership, and cultivate those storied roots and limbs, we have the chance to beautifully revive and reanimate the sacred tree, the center heartbeat of the people.

Chapter 4:

A Visionary Declaration from the Amazon

"WE ARE CONNECTED TO THE SPIRIT of the jungle through dreams," Narcisa Gualinga, a Kichwa woman from the Ecuadorian Amazon, tells me as we walk through an open field at the Parque Arbolito in Quito, Ecuador, where we have met for a major event organized by her community.[1] "But when there is contamination, this connection is lost, and the spirits die. Without them, we cannot be well, our children and future generations will no longer have this life. This was taught to us by our ancestors, and we have passed this on to our children."[2]

As we collectively face the daunting reality of socio-ecological upheaval and the impacts of our climate crisis, radical inspiration and practical transformative solutions are honey-nectar for our souls. In the midst of tragic news and one disaster after another, shimmers of light and hope radiate from brilliant community-led formations. One that shines brightly emanates from the heart of the Ecuadorian Amazon, led by the Kichwa People of Sarayaku.

Dwelling in the Bobonaza River basin in a remote area of the Amazon rainforest, the Kichwa of Sarayaku are surrounded by immense biodiversity and beauty. They are reachable only by canoe, small aircraft, or one's own feet. The Sarayaku territory, according to the cosmology of the Kichwa, is sustained by three life-giving elements: the forest, the rivers, and the land. However, the territories of the Sarayaku and their ways of life have been repeatedly impaired by mining and fossil fuel companies that covet the oil and minerals buried beneath the dense forest. In this fantastically lush region, there is unfortunately no shortage of corporate entities seeking to profit from the Amazon's abundance.

Yet the Sarayaku People are resisting and responding to unrelenting colonial threats with impeccable wisdom and ferocity. Kichwa leaders of Sarayaku have worked for decades to develop their Kawsak Sacha, or Living

Forest Declaration, which aims to protect and defend their Traditional Lands from extraction and the onslaught of colonization. As the Kichwa state, "The Kawsak Sacha declaration is universal and proposes a legal recognition of the revindication for territorial rights and Mother Earth, which is necessary and essential for the balance of the planet and the preservation of life."[3] Their bold leadership demonstrates why the Kichwa people of Sarayaku are known as the Pueblo del Medio Dia, or People of the Zenith. This acknowledgement originates from an ancestral prophecy stating that, during a time of difficulty, Sarayaku would be a center of strength for territorial, cultural, and spiritual defense, a beacon of light as powerful as the noonday sun.

The Sarayaku formally launched their Living Forest Declaration in July 2018 during four days of public events, educational activities, and ceremonies in the capital city of Quito, Ecuador. The inauguration included a formal presentation to the Ecuadorian government and to international organizations and dignitaries.

I was honored to partake in the inauguration events as a representative of WECAN along with other international allies including Amazon Watch, Indigenous Environmental Network, and Pachamama Alliance. Together, we formed a requested demonstration of global support for the vision and demands set forth by Kawsak Sacha leaders.

For the launch event, a large Sarayaku community delegation and other Indigenous Peoples traveled from their jungle homes to Quito to hold activities at the Pabellón del Habitat and Parque Arbolito, an exhibition hall and park in the city center. In the open space of the Parque Arbolito, the community constructed an elegant traditional village house with a thatched roof. Here, discussions took place as well as ceremonial drumming, singing, and dancing, while chicha was passed around in a finely decorated pottery bowl. Chicha is a distinctive drink made in every Kichwa home—the women chew the yuca fruit until it binds with their saliva and creates a probiotic drink, which can be consumed immediately or fermented to make an alcoholic beverage. The drink is shared communally and has an important role in ceremonies and in hospitality: consequently, chicha was offered generously throughout the Kawsak Sacha launch. The first time a beautiful pottery vessel brimming with chicha was kindly handed to me, I felt humbled and welcomed. To drink was to taste and be touched inside by the very message and wisdom conveyed from the jungle and the communal knowledge of the Sarayaku and their ceremonies. I drank deeply.

As several of the Kichwa leaders explained to me, the goal of the Living Forest Declaration is to attain national and international recognition of a new legal category for the permanent protection of Indigenous lands. Rather than a worldview that only sees nature as a resource to be used, the Declaration recognizes the intrinsic physical and spiritual relationship between the Peoples of the Living Forest and all the beings that inhabit and compose the forest. Kawsak Sacha is a vision, worldview, and strategy, presenting all at once an ecological, political, cultural, spiritual, and economic framework. It shows how to live in a healthy and equitable world beyond conventional capitalist views of progress, development, and conservation.

Studies by the United Nations as well as many other institutions have determined that the most effective way to protect forests and biodiversity, and thus help mitigate climate catastrophe, is to support Indigenous Peoples' rights, lands, and Traditional Ecological Knowledge.[4] And yet, no international legal designation for conservation currently exists that recognizes the unique relationship between Indigenous Peoples and the forest. There are important unique recognitions and international standards for Indigenous Peoples such as the United Nations Declaration on the Rights of Indigenous Peoples, but no international standards pertaining specifically to forest protection and conservation and Indigenous Peoples.

Additionally, as the Sarayaku have made clear, the systems of natural protected areas created thus far by the Ecuadorian state and international organizations do not result in full consultation, participation, or consent of Indigenous Peoples. The current systems are not a solution, and in fact often degrade and violate basic Indigenous rights to autonomy and sovereignty.

The Living Forest Declaration proposes the protection of the Kichwa way of life, including emphasis on uncontaminated rivers, sovereign communities, and a mandate that their territory be permanently free of any type of extractive activity such as oil exploitation, mining, and logging.

Kawsak Sacha embodies countless generations of ancestral knowledge and the vision of the people of Sarayaku, including the Jatun Kausak Sisa

Ñampi, or the Living Way of Flowers project. José Gualinga, a former president of Sarayaku, eloquently described the Jatun Kausak Sisa Ñampi project in all its physical and mystical manifestations during one of the commencement's educational presentations. With this project, Sarayaku is planting native flowering trees and plant species along the boundary of their 333,000-acre territory to demarcate their lands symbolically and physically, and to convey through the language of flowers the aliveness of the land and the people who live in the heart of the forest.

After hearing José Gualinga's Living Way of Flowers presentation, I was asked to give a small video report about the ongoing inauguration activities across our social media platforms. Partway through my narration, while attempting to relay what I had learned about the Jatun Kausak Sisa Ñampi, I suddenly found myself hardly able to speak as my voice cracked and a rush of tears came tumbling forth. The idea of demarcating one's territories and lands through gorgeous blossoming flowers, and not through the barbed-wire border walls that are monuments to violence and injustices, completely and magnificently shattered me.

The formal and lively Kawsak Sacha inauguration night ceremonies began on the streets directly outside the Teatro Capitol, a historic theater in Quito. A procession of well-adorned Sarayaku leaders entered the building, accompanied by a pulsating gathering of singers, drummers, and dancers. Many of the community members wore traditional regalia, which included brightly colored feather headdresses, transporting the spirit of the birds and jungle to the urban center. In one of the traditional Kichwa dances, the men moved in a circular pattern while playing rhythmic drums and making high-pitched whistling sounds, as the women spun in a spiral dance, gracefully tilting their heads so that their long, shimmering midnight-black hair swirled outward like the arms of a galaxy. Youth climate activist Helena Gualinga from Sarayaku told me, "The women's hair is a symbol of her strength and beauty; as she dances, this shows her female power and joy and the highest reflection of all that is female."[5]

In this majestic and life-affirming ceremonial manner, all the participants followed the Sarayaku leaders into the auditorium. After everyone was settled into their seats in the great hall and introductory remarks were made, Mirian Cisneros, president of Sarayaku at that time, presented the proposal to Elizabeth Cabezas, president of the Ecuadorian National Assembly, and Julio César Trujillo, president of the Plenary of the Council for Citizen Participation. The International Union of Conservation of Nature (IUCN) Regional Office and Amazonian Indigenous federations of Ecuador then expressed enthusiastic statements of support. Following the presentations, youth and women from the Sarayaku delegation shared haunting and melodic songs, and the historic evening came to a ceremonial close with honored Elder Don Sabino Gualinga offering tobacco and a traditional Taki song from the spirits.

Here is one excerpt from the Declaration, illuminating how the Kichwa People of Sarayaku present the wisdom within their proposal—wisdom that we all can learn so much from:

> Kawsak Sacha (The Living Forest) is a proposal for living together with the natural world that grows out of the millennial knowledge of the Indigenous Peoples who inhabit the Amazonian rainforest, and it is one that is also buttressed by recent scientific studies. Whereas the western world treats nature as an undemanding source of raw materials destined exclusively for human use, Kawsak Sacha recognizes that the forest is made up entirely of living selves and the communicative relations they have with each other. These selves, from the smallest plants to the supreme beings who protect the forest, are persons (runa) who inhabit the waterfalls, lagoons, swamps, mountains, and rivers, and who, in turn, compose the Living Forest as a whole. These persons live together in community (llakta) and carry out their lives in a manner that is similar to human beings. … In essence, the forest is neither simply a landscape for aesthetic appreciation nor a resource for exploitation. It is, rather, the most exalted expression of life itself. It is for this reason that continued coexistence with the Living Forest can lead to Sumak Kawsay. This encourages us to propose that maintaining this lively space, based on a continuous relation with its beings, can provide a global

> ethical orientation as we search for better ways to face the worldwide ecological crisis in which we live today.[6]

The Living Forest Declaration also includes the following cogent point, which outlines the call for a new legal status:

> Our Concrete Proposal consists in attaining national and international recognition for *Kawsak Sacha* (the Living Forest), as a new legal category of protected area that would be considered *Sacred Territory and Biological and Cultural Patrimony of the Kichwa People in Ecuador.*[7]

The Kichwa aim to create autonomous management of their own territory, fully on their own terms, with their own proposals and their own governance structures. As was expressed at the launch event, they do not want their lands to be categorized in the colonially determined systems of conservation, so they are proposing a new legal category.

During the event, Hilda Santi, who was the first female president of Sarayaku in 2005, expressed to me the change in worldview within the Kawsak Sacha Declaration and the call for self-determination, unity, and solidarity for the protection of the Amazon:

> This is about an alternative proposal from the Indigenous world, against the "development" that brings about the destruction of our Mother Earth. Our Mother must not be sold nor destroyed. Our Mother gives us life, and within it there are many lives and beings that cannot speak with a voice, this is what we are defending. We are presenting this proposal so that the government no longer destroys our jungle, so that we are respected as first peoples with our own languages, customs and cosmo-visions. The government must accept our proposal. We don't want to suffer more abuses, more divisions. I call for unity among women of the world, Indigenous and non-Indigenous. We must unite in order to strengthen this fight to defend the lungs of the world. Global warming is making us suffer, the sun is getting hotter, our harvests are not the same

> as they were. So now we are raising our voices for the world to listen.[8]

The Declaration's inauguration was coordinated to coincide with the sixth anniversary of a historic ruling of the Inter-American Court of Human Rights. In 2012, the Court found that the Ecuadorian Government had violated the rights of the Sarayaku People in allowing an oil company into their territory without consultation or consent. This ruling has been an extremely significant victory for the community, but one that requires tireless vigilance to uphold. And this win was buttressed by a major breakthrough in February 2022 when the Constitutional Court of Ecuador recognized a new ruling that granted Free, Prior and Informed Consent to the Indigenous Peoples of the country. Industries—including those already active in the region—must now seek and receive the consent of Indigenous communities before proceeding with extractive activities on Indigenous land.[9] This ruling sets a precedent for what needs to happen worldwide—not only because it is the moral thing to do, but because it is best for the planet and all beings. Of course, there will need to be much advocacy and vigilance to ensure the ruling is respected and implemented.

Sarayaku's powerful efforts against oil extraction over recent years have already kept an estimated 100 million barrels of oil in the ground and protected primary rainforest in their territories.[10] President Cisneros spoke with me on this point, saying:

> Sarayaku has fought off oil companies for 35 years. We have expelled companies from our territory, taken our legal case to the Inter-American Court and won. So, this experience has well positioned us to present our Living Forest Proposal to the world. . . Our proposal is not only for Sarayaku, but also for all humanity and all life on the planet.[11]

To further define how the Living Forest Declaration can be a model for the world, during the commencement days, ongoing ceremonial activities, workshops, films, and information sessions elaborated in detail and multidimensional realms the principles of Kawsak Sacha, as well as perspectives on Indigenous economics, the leadership of women, and human relationships with nature and with the spirits of the forest.

One workshop focused on how the vision of Kawsak Sacha is vital for maintaining the ecological balance of the Amazon, which is crucial in the fight against climate disruption. Consequently, protecting the Amazon rainforest, the largest of the world's tropical forests, must be central to all climate change discussions and policies. During this workshop, the audience learned of the Ecuadorian Amazon's extraordinary ecology and the many non-human relatives who live in the Amazon and nowhere else—species such as Maryanne's robber frog; a large-eyed mouse opossum named *Marmosa waterhousei*; the plush-looking Western pygmy marmoset; the Spix's night monkey; the fruit-loving greater yellow-shouldered bat; the brilliant crimson and yellow orchid called *Lepanthes llanganatensis*; and *Parkia balslevii*, a fern-like plant in the legume family. In fact, ten percent of all known flora and fauna species the world over live in the Amazon. The tropical sunlight supports an explosion of plants, which then feed an immense animal world and that energy from plants and animals continues to multiply—a perfect cradle for biodiversity.

What struck me was learning about the phenomenon of Amazonian "flying rivers." Trees pull in moisture from the atmosphere and then transpire water vapor through pores in their foliage into the air, as great living humidifiers. This sylvan perspiration condenses into clouds that flow as sky rivers in courses throughout South America, over the expansive farms south of the Amazon rainforest and to the southern reaches of Chile, offering life-giving rain to the lands below. Several scientists, including Antonio Donato Nobre, explain that the flying rivers reach as far as North America; Nobre asserts that not only is the Amazon rainforest the lungs of the world, the magnificent nature system also serves as the world's heart and liver by pumping life-giving moisture through the air and purifying it.[12] Secretary of Environment of the Brazilian Federal District José Sarney Filho adds that the forest is also the world's air conditioner, lowering the global temperature and balancing the climate.[13] The Amazon literally creates the continent's rain, a Mother Forest nourishing plants—and humans and animals—growing hundreds and hundreds of miles from her arboreal embrace.

Given this information—the flying rivers, the carbon storage, and the cooling power of trees—it should be obvious that deforestation of the Amazon spells disaster for the planet, not to mention the human and non-human beings who call that region home.

Without the sky rivers, which have been dwindling over the past decades, South America would lose at least one third of its rainfall and suffer more serious droughts—already, many regions near the Amazon are experiencing longer dry seasons and less rainfall. And the cycle continues. The Amazon rainforest, which was once fire-resilient, is growing weaker due to these droughts and rising temperatures, making it more vulnerable to ever-increasing wildfires. The irreplaceable Amazon will become more like a savannah if we do not immediately take action, halt all extractive industries in the region, and fully respect and support Indigenous Peoples' leadership and directives about what is needed for their Traditional Territories. If we lose the forests of the Amazon, now at a tipping point, we will release billions of tons of carbon into the planet's atmosphere, deprive Indigenous communities of their homelands, and forever drain the flying rivers.[14] The Living Forest Declaration is critical for saving this venerable forest that is also central to the well-being of our Mother Earth's lungs, heart, liver, and air-cooling system.

During the commencement proceedings, along with Amazon Watch, WECAN was honored to participate in a panel discussion with Indigenous women of the Amazon to emphasize the connection between extractivism and exploitation of women and the land that is inherent in patriarchal and colonial systems of oppression.

These frontline leaders explained that only through women uniting will the Amazon be protected. To that end, a collective of Amazonian women, Mujeres Amazónicas Defensoras de la Selva, has organized over the past years to defend their ancient rainforests.[15] The collective includes women from seven nationalities of the Ecuadorian Amazon who are advocating for the protection of their Traditional Lands, education, health, and women's rights. While their work is thoroughly inspiring and effective, their bold stand is also perilous: throughout the launch events, women conveyed tragic and heartbreaking stories of how they are increasingly attacked—through criminalization, death threats, property harms, and sexual violence—for their resistance efforts to defend their territories.

Nevertheless, in the face of egregious violations, the women made it clear that they stand strong—determined and courageous in their ceaseless efforts to protect their territories and future generations. After the women's panel, Patricia Gualinga shared further thoughts with me:

> The decision of the women of Sarayaku was a definite "no" to extractivist activities. We do not want destruction; we want to protect life. I think Sarayaku is one of the first instances in which women have participated so much in the processes and contexts of struggle, resistance, and innovative proposals.
>
> Earlier this year, I received a death threat and denounced this to the authorities. I'm glad this process has encouraged many women leaders to denounce the threats they have also been receiving for taking a stand against the extractive industries. They are demanding respect for their rights, respect for nature, respect for women. This is a new articulation with very strong leaders of all nationalities. It is good that we are not afraid because we are united. We have united to protect each other. We are going to continue fighting according to our own vision because this strengthens the organizational processes of Indigenous Peoples, our processes of resistance and struggle, with grassroots actors coming with their own voice to say, "This is what we want."[16]

Much progress has continued with the Living Forest Declaration since the launch in Quito. Of note, in July 2022, the Kichwa People of Sarayaku convened the first meeting of the Peoples of Kawsak Sacha in a gathering they named Kawsari, which means an awakening, specifically the awakening to life.

Indigenous Peoples from the Ecuadorian, Peruvian, and Colombian Amazon gathered in Sarayaku to deepen their work to protect their territories, including how to strengthen legal and political tools to implement their self-determination and autonomy, and standards for exercising Free, Prior and Informed Consent for any activities on their lands.

Because the Amazon's ecosystems are vital to planetary well-being, justice for the people of Sarayaku is inseparably bound to the fate of us all. Kawsak Sacha is indeed a pillar of light as strong as the midday sun—a brilliant, multifaceted vision and worldview—that needs to be respected by the Ecuadorian government which has not yet accepted the Declaration, and also replicated to protect other Indigenous territories globally.

My journey to the Living Forest Declaration launch was not my first trip to Ecuador to meet with the incredible Amazonian women leaders. The Ecuadorian government had just signed a contract with a Chinese oil corporation for extraction and exploration in territories belonging to two Indigenous Amazonian communities, and these plans opened up almost one million acres of jungle to oil extraction, deforestation, and irreversible environmental damage.[17] This contract with Andes Petroleum clearly violated Indigenous rights as well as the health and integrity of the forest ecosystems, and it was a betrayal of global significance given the role of the Amazon in ensuring a healthy planet.[18] In celebration of both their leadership and their fierce dedication to protect their Traditional Territories from threats such as those from Andes Petroleum, Indigenous Amazonian women leaders organized an event on March 8, 2016, International Women's Day, and invited Amazon Watch and WECAN to stand in solidarity.

A few days prior, I had traveled to Puyo, the closest city to the Sarayaku territory in the jungle and the location of the march, with Casey Camp-Horinek, an Indigenous leader of the Ponca Nation of Oklahoma and board member of WECAN. Although we had journeyed farther than many of the other travelers, our trip was much less arduous. Our plane ride could not compare with the journeys of the women who had come from deep within their rainforest homelands—some had traveled as many as five days by foot and canoe, with babies strapped to their backs. In attendance were seven Amazonian nationalities, including Andoa, Achuar, Kichwa, Shuar, Shiwiar, Sápara, and Waorani.[19]

Before the march began, hundreds of women assembled in an outdoor community center pavilion for presentations and speeches, making their demands known. Women leaders dressed in the ornate and sacred regalia of their various nationalities sat on the stage as speakers expressed that they had come to firmly say no to the extraction and exploitation of their land, which correlates directly to genocide of their people. Many women sang songs, and Casey as an honored guest was invited to open the event with a prayer for water. As an Indigenous Elder from Turtle Island, representing

Indigenous Peoples of the North, she shared her people's traditional songs and told the story of how her people have been impacted by the very same issues of fossil fuel extraction and fracking on Ponca Nation land in the United States. She also spoke of hope, and how women in the Global South and North are uniting and organizing, standing up, and saying "No More!"

Afterward, Leila Salazar-López, executive director of Amazon Watch, and I shared news of a petition we had organized on behalf of the women of the Ecuadorian Amazon. People from over sixty countries had already signed it in solidarity. The petition addressed to the Ecuadorian government called for "No Extraction in the Amazon! Women of Ecuadorian Amazon and International Allies Reject Oil Concessions, Stand for Rights of the Earth and Communities."

After the assembly, women chanting and carrying signs streamed out of the pavilion and made our way through the streets of Puyo in the intense humid heat. As we prepared to march, I was approached by a woman from the Sápara territory. She was admiring the banner I had brought with me, all the way from California, which read "Keep fossil fuels in the ground!" She asked if she could help me carry the banner as we marched. I happily agreed, but as she grasped the end of the banner opposite me, the center sagged under the weight of the sturdy fabric. Instantly, Rosa knew just what to do and took her handmade spear, seven feet or so in length and used for hunting and fishing and slid it through the sewn sheath at the top of the banner, careful to send the blunt side through to me. The sign stretched out perfectly across her spear. As we joined the other women filtering their way into the processional, Rosa told me, "With this spear, we march as warriors!"

The demonstration spanned several blocks, and our chants of "Defend the forest, don't sell it!" could be heard echoing through downtown Puyo. It was a stunning sight to behold: over five hundred women, of different Amazonian nationalities, marching through the streets, babies fastened with woven shawls to their backs and voluminous jungle leaves lifted above their heads both to represent the beauty and mystical power of the jungle and to act as makeshift sunshields. They held gut-wrenching photographs of forest destruction and oil spills in riverways, as well as handmade signs that displayed demands in Spanish such as "No more persecution against women defenders of Mother Earth!"[20]

National press arrived and interviewed women leaders. Tourists gazed down at us from their hotel balconies, some with looks of perplexity stretched across their faces, but others cheered us on, lifting their fists in solidarity.

Earlier that day, word arrived that a clandestine meeting was taking place between Andes Petroleum and representatives from the Ecuadorian government: they planned to illegally invade Sápara lands while the tribal leaders were occupied in Puyo with International Women's Day events. The Sápara quickly drafted a missive and delivered it to the meeting, denouncing the plot. This move derailed the unlawful incursion, and the Sápara delegation returned to celebrate their triumph. Sápara leader Rosalia Ruiz announced, "Right now the oil company is trying to enter our territory. That is our homeland, this is where we have our chakras (gardens), where we feed our families. We are warriors, and we are not afraid. We will never negotiate."[21]

At the other end of town, after hours of winding our way through the city in searing Amazonian heat, the march turned back into a cluster as we gathered for a final assembly. One by one, women stepped up to the microphone, sharing strategies and pledges to never stop their defense of their forest homelands—some speaking from their hearts about the importance of women in the fight for Indigenous rights and climate justice, and others singing songs from their territories.

As chicha was passed around as part of a ceremonial offering, the women also paid tribute to a devasting loss. Although the mood of the assembly and march had been spirited and festive, a heaviness hung over us that day. Less than a week before the International Women's Day convening, Berta Cáceres, a Honduran environmentalist, had been assassinated in her own home. Word of her death had spread quickly through the international community, sending grief, and considerable shock and outrage, along with it.

Part of the Lenca Indigenous community, Berta Cáceres had led campaigns resisting illegal logging, unjust landowners, and the US military since her time as a student. The daughter of Austra Bertha Flores Lopez, also an activist, Berta fought for women's rights and LGBTQ+ rights as well. She also co-founded COPINH, the Council of Popular and Indigenous Organizations of Honduras, with the purpose of supporting Indigenous Peoples' rights.[22]

In 2006, Cáceres was asked to become involved in the movement to resist an illegal attempt to construct four hydroelectric dams on the Gualcarque River by the Chinese company Sinohydro, the World Bank's International Finance Corporation, and the Honduran company Desarrollos Energéticos (DESA). The Gualcarque is both sacred to the Lenca people and essential to their livelihood, and developers had disregarded international agreements by failing to consult Indigenous and local people on the project. With her community firmly beside her, Berta sprang into action, petitioning the government to cancel the twenty-two-megawatt Agua Zarca Dam, coordinating nonviolent civil disobedience actions to call for the project to stop, appealing to the Inter-American Human Rights Commission, and setting up a peaceful but firm road blockade so that DESA could not reach the dam.

All this time, Berta and her organization endured harassment, violence, death threats, and public defamation, as both her own government and DESA painted the land and water defenders as "violent anarchists bent on terrorizing the population through their protests at the project site,"[23] as court files later proved. Berta personally reported being followed and threatened with sexual assault. Still, she persevered. Her friend, Mexican environmental activist Gustavo Castro Soto, recalled that Berta used to often say, "They are afraid of us because we are not afraid of them."[24]

But the violence escalated tragically in 2013, when Berta's colleague Tomás García was shot dead while protesting the hydroelectric project. As voices around the world condemned Tomás's murder, Sinohydro pulled out of the project and the World Bank withdrew its sponsorship.[25] Meanwhile, as their campaigns slowed the dam project to a halt and the companies involved lost a great deal of money, Berta and COPINH continued to gain the attention and support of the international community, and in 2015, Berta received the Goldman Environmental Prize for her tireless and courageous environmental activism.[26]

Tragically, Berta's campaign success also made her a target. On March 2, 2016, her home was broken into in the middle of the night, and she was shot and killed at the age of forty-four. One of Berta's daughters, Berta Isabel Zúniga Cáceres, accused DESA of orchestrating her mother's death, saying that it is "very easy to pay people to commit murders in Honduras, but those who are behind this are other powerful people with money and an apparatus that allows them to commit these crimes."[27]

It would be many months, after much pressure from international human rights organizations, including Amnesty International, before the Honduran authorities mounted a full investigation leading to the arrest of eight people, including two men from DESA. Seven of them were sentenced to prison in 2019, but the alleged corporate ringleader was still to be tried.[28] Some closure for Berta's family and colleagues finally came in July 2021 when Roberto David Castillo, the former president of DESA, was found guilty of engineering her murder.[29] He had paid members of the military to keep tabs on Berta for years before finally having her executed. After the verdict, Berta's youngest daughter, Laura Zúñiga Cáceres, encouraged the global community of land defenders, saying, "This is a collective victory, and my message to other communities in similar situations is this: the fight is hard, but in the end, as my mom said ... we are going to triumph, and we are going to dismantle the violence against our people."[30]

The heartbreaking fact is that the story of Berta Cáceres is not unique. Each year, new reports are released proclaiming the previous year as the deadliest on record for environmental land defenders. According to Global Witness, a record 227 land defenders were murdered in 2020 (of those recorded), with the logging industry the worst offender, and a third of those slain were Indigenous Peoples protecting their ancestral homelands. Each week since the Paris Agreement, an average of four land or water defenders have been assassinated.[31] One of the worst offenders is Honduras, Berta Cáceres's homeland, where over forty environmental leaders were murdered in the five years following her death.[32] Because not all of the killings are reported, the number is surely higher, and many more defenders have been arrested, threatened, or attacked. Former United Nations Special Rapporteur on Indigenous Rights Victoria Tauli-Corpuz referred to the consistent, brutal murdering of Earth defenders as an "epidemic."[33]

My friends and colleagues from Sarayaku also spoke about how they are regularly threatened. At the Living Forest Declaration launch in 2018, Noemí Gualinga, nicknamed "Mother of the Jungle," conveyed her reflections on the persecution of Indigenous women land defenders: "Lately, women are the ones who have been fighting on the frontlines but without leaving our partners, the men, behind. I think the persecution of woman leaders has become an objective of the companies. The oil companies want to find a way to enter our territories, but since they haven't been able to so far, they're starting to target the women, trying to scare and terrorize them. And this is the case for many different nationalities, Waorani, Kichwa, Sápara, and Achuar. It is a psychological persecution."[34]

In an interview I had with María San Martín, Human Rights Defender Visibility Coordinator for Front Line Defenders, she offered some advice on protecting women land defenders: "Many organizations and women defenders have also highlighted the importance of supporting and enhancing the capacities of networks of support between and for women defenders themselves, which can save lives, can empower with a huge impact, and can protect women so that they can continue their work." She added, "We all must voice the need and demand for application of a gendered perspective at all levels—and we need to understand [Women Human Rights Defenders] in their local context, with their families and in their communities, and to try and divert support in that way through protection mechanisms grounded at the local level. We also need to echo their own voices and their own stories. We can help to build the alternative narratives that are really showing what women are doing and why, and how they are trying to construct, both socially and economically, other ways of living."[35]

I dedicated myself again to the cause of women land defenders as I stood, with tears running down my cheeks, at the tribute to Berta Cáceres in Puyo. As part of this fervent homage to women in all of their diversity on International Women's Day, we collectively commemorated and bid farewell to a courageous, formidable woman whom we will forever love and revere. Hundreds of women stood in solidarity to declare that Berta had not died in vain—we would continue her fight. A phrase I heard more than once that day was "They tried to bury us, but they didn't know we were seeds," and I will always think of Berta when I hear it.[36]

While the details are specific to each region, people, and conflict around the world, the overarching story remains similar. Women are standing at the forefront of movements to defend and protect the health of our water, land, air, and communities. They put their bodies on the line, take on governments and corporations, and face extreme harassment and vilification, all so that present and future generations may still hope to inherit a sustainable and livable planet.

The violation of the rights of women and land defenders speaks in a profound way to the derangement of our times. It echoes a dangerous worldview of domination and exploitation that also sits at the root of the degradation of our Earth's natural systems. It is imperative that we voice the need and demand for the application of a gendered perspective to environmental protections at all levels. We need to deepen our understanding and strengthen our resolve to stand in solidarity with the courageous and dedicated women, very often Indigenous women, who continue their frontline work, despite violent threats and assaults. We must defend the defenders. It is because of this imperative that events like the International Women's Day event in Puyo are so crucial to lift awareness and demand change.

In the last six years, WECAN has been campaigning and engaging in advocacy for a policy that potentially could make a significant difference for women forest defenders in the Amazon: the Escazú Agreement. This new multilateral accord guarantees public access rights on environmental matters and justice mechanisms as well as the explicit protection of human rights and environmental defenders in the Latin American and Caribbean (LAC) region. As a further security, it cannot be changed by governing entities.[37]

The full name of the agreement is the Regional Agreement for Access to Information, to Public Participation and Access to Justice on Environmental Issues in Latin America and the Caribbean, but it is most often known as the Escazú Agreement, after its Costa Rican city of origin. It was adopted by several LAC countries in 2018 and ratified in 2020; the participating countries have since begun the implementation phase of the legally binding multilateral accord after the first Conference of the Parties on the agreement in 2022.

This Agreement is important because, as we have seen, there has been a lack of political will at the national and international levels to implement policies that ensure the protection and rights of land defenders. Although the Escazú Agreement has not received much international attention yet, it could potentially present a promising path forward in protecting both Mother Nature and her frontline defenders, and it can set a precedent for other countries as well.

Worldwide, women are bravely on the frontlines protecting the Amazon, the Boreal, the Congo Basin, and Indonesian forests globally from the marauding Gilgameshes of today.[38] It is our collective responsibility to stand with these land defenders and their communities; support them in any way that we are able with our voices, funds, time, care, and bodies; and follow their leadership. We can support such efforts as the Living Forest Declaration; Free, Prior and Informed Consent; and the Escazú Agreement to stand together in resistance to the exploitation of women, Indigenous Peoples, and Mother Earth while uplifting these legal instruments, which can protect people and planet. There is a saying in the Global South that articulates the change that must now come: "Neither the land nor women are territories of conquest."[39]

Freeing the world from the colonial and dominator context is central to the worldview shift we must call forth. I think back on the words of Patricia Gualinga during the Kawsak Sacha ceremonies in Quito and her insight into the worldview that the Living Forest Proposal invokes:

> The Living Forest Proposal goes far beyond the demand of rights, it's about trying to reach the deepest consciousness of human beings, trying to see life in a different way and to see the integrity of conservation. Kawsak Sacha is about the essence of our existence as human beings and this planet Earth, and the beings that regenerate ecosystems and give life to the planet.[40]

Part II

Dismantling Patriarchy, Racism, and the Myth of Whiteness: Ancient Mother and Women Rising

Chapter 5:

She Rises

FROM THE MARRIAGE OF TWO MOUNTAIN RANGES, the Sierra Madre de Oaxaca and the Sierra Madre del Sur, emerges Oaxaca's dramatically rugged landscape near the southern tip of Mexico. And it was here that my father's work brought our family for several weeks when I was eight years old, and where I received a special and unexpected gift.

As a young child, Oaxaca generously offered me stunning experiences with nature and culture that filled my imagination with wonder and magic. Only later would I learn of the region's powerful history and immense biodiversity. Two thirds of Mexico's bird species fill Oaxaca's skies and trees with stupendous color and song, and the area is visited by legions of migrating birds. Among the oaks, pines, and plumerias roam white-tailed deer, coatis, margays, pumas, and a myriad of reptiles and amphibians. Near Oaxaca City grows the Árbol del Tule, a sacred tree of the Indigenous Peoples of this area—at least 1,140 years old, this Montezuma cypress has the widest girth of any tree in the world. To describe this land of abundant flowering and nourishing flora, the name Oaxaca was chosen from a Nahuatl word meaning "The Place of the Seed."[1]

The region is home to sixteen Indigenous communities, each with distinct languages and traditions they managed to carry through ongoing years of colonization, so that today, one third of Oaxaca's residents speak an Indigenous language.[2] The Zapotecs and Mixtecs are the largest Indigenous groups, with ancient roots in Oaxaca's history, and the Zapotecs speak of ancestors who emerged from caves or from Mother Earth herself, often taking the forms of jaguars or trees before transforming into human beings. Archaeological evidence shows that Zapotecs were in the area at least by 500 BCE, having built Monte Albán, a staggering civic-ceremonial center with elaborate stone pyramids. The Mixtecs arrived centuries later, and the Aztecs also expanded their empire to Oaxaca before the arrival of Spanish colonizers.

When news of the violent and growing Spanish invasion reached Oaxaca in 1521, the Zapotec leaders made the strategic decision to propose an alliance with their Spanish oppressors, saving many Indigenous lives in exchange for their sovereignty. In this forced entente, Indigenous Peoples in Mexico suffered constant discrimination under Spanish rule, as the invaders barred them from leadership roles and seized any land they saw as valuable.

Fortunately, the valleys of Oaxaca cradled and protected some of the region's Indigenous communities, affording those who survived Spanish imperialism the opportunity to speak their native tongues and practice their traditions despite the enforced indoctrination of Spanish schools and churches. The cultural seeds and descendants of these communities survive and thrive strongly to this day, speaking to the powerful resistance of Indigenous Peoples.

One blazing hot day during our stay, my family went shopping for food at an open market brimming with brightly colored textiles, vegetables, and fragrant fruits laid out on blankets. Dazzled by the women's colorful clothing and the bright red hanging chili pepper bundles, I left my mother and sister buying freshly made corn tortillas, and drifted to a display of ribbons neatly dangling like a rainbow curtain from a wooden rack. I closed my eyes as some three feet of every color of ribbon gently twirled in the breeze, caressing my face, and I was enveloped in the spicy aromas of the market, the sounds of cooking, and people talking. When I opened my eyes, a beautiful Oaxacan grandmother stood before me wearing an expression of deepest kindness. She pressed something into my hand—a silver-black clay figurine of a woman holding a crescent moon in one hand and a stalk of corn in the other.

The little figurine was made in the ancient Oaxacan tradition of *barro negro* (black clay) pottery. The naturally dark clay is gathered in a valley south of Oaxaca City and is cleansed of impurities but not otherwise altered. Since the Monte Albán period over two thousand years ago, Indigenous Peoples have molded this clay by hand, forming pots, jars, and ceremonial and decorative objects. If an artisan polishes the surface of the item with a quartz stone just before it has finished drying, the item will emerge from the firing process with a glossy black finish. If the polishing step is skipped, the finished product will emerge matte and silver-black, as was my figurine.

I cherished the little statuette and its smoothness in my hand. But the moon-corn woman was not a child's toy—she was powerful, and even at a young age, I knew this. The gift was a seed planted in my mind, waiting for a time to sprout. Later, as a young woman, the figurine's power and majesty, and connection to both the Earth and sky through the corn and the moon, made me wonder about her story, the dignity she exuded, and the people who had created her. When I looked at her, I instinctively knew something fundamental had been forsaken and forgotten in the world around me, in our relationship to the land in daily life and culture.

The statuette in many ways foreshadows my work today. It is a visible manifestation of the sacred Female Principle—and our human relationship and responsibilities to the cycles of the seasons and the harvests, and the heart of life itself. Today, in my work at WECAN, I often ask, how can we as women in all of our diversity connect more deeply to defend the Earth in the face of the climate crisis and oppressive colonial, racist, and misogynistic structures? How can we accelerate systemic change in these paradigms, which are the root causes of so many interlocking crises? Responding to these questions reveals our strength as women, feminists, and gender-diverse leaders, through our relationship with the lands where we live and our communities. The inquiry also leads to learning about the women who have been engaged in this work for justice, health, and liberation for generations in the face of violence and all odds.[3]

Due to ubiquitous gender inequality, women are disproportionately impacted by ecological and social crises, and yet, simultaneously, women are often the backbone of social and environmental movements and are the caretakers, defenders, and protectors of land and water. Through my work, I have had the great honor and pleasure of meeting many women leaders, among them Phyllis Young, former councilwoman for the Standing Rock Sioux Tribe. She was a lead organizer of the Oceti Sakowin camp during the Dakota Access Pipeline resistance, which was the largest gathering of Native American Nations in the United States in more than a century.

The Oceti Sakowin camp assembled in 2016 to protect the water and to resist the construction of the Dakota Access Pipeline (DAPL), which now carries 750,000 barrels of oil per day from the Bakken oil fields in North

Dakota to southern Illinois. A clear example of environmental racism, the pipeline route purposefully bypassed the mostly white cities, and runs underneath the Mississippi and Missouri Rivers and Lake Oahe, threatening the water supply of the Standing Rock Indian Reservation, as well as destroying cultural and burial sites during construction. The oil surges through the pipeline even though the legally required Environmental Impact Statement process has not yet been completed, and Tribal consent was never obtained for the project, disregarding Indigenous rights and treaties.[4]

Along with many, I was honored to stand in nonviolent direct action and peaceful prayer in North Dakota along the Cannonball River during the Indigenous-led resistance to DAPL, and was subsequently arrested in Washington, DC, in a nonviolent action to stop the pipeline. As part of the resistance effort, Michelle Cook, a powerful Diné lawyer and founder of Divest Invest Protect, invited WECAN to co-direct Indigenous Women's Divestment Delegations, in which we organized for Indigenous women leaders to have in-person exchanges with representatives of European and US banks, insurance companies, and other financial institutions who had backed DAPL and other fossil fuel pipelines. The delegations aimed (and still do) to expose injustices and divulge to financial institutions exactly how their fossil fuel financing and underwriting violates human and Indigenous rights, while also driving climate disruption. These delegations have successfully put pressure on institutions to divest funds from fossil fuel extraction and infrastructure, as well as to systematically examine their policies regarding Indigenous and human rights and the climate crisis.[5]

At the Oceti Sakowin camp in the summer of 2016, Phyllis spoke to me about the strength of the Indigenous women who formed the core foundation of the Standing Rock movement and generations of Indigenous resistance before. "I am 'Woman Who Stands by the Water,' and my other name is 'Woman Who Loves the Water,'" she said. "I was given those names by my people because it's been my life struggle to protect the water. I grew up on this river; I was removed and displaced when I was ten years old . . . I came back here, and I live on the river, and I am telling the Army Corps, 'You'll never displace me again. You'll never put me somewhere where I don't belong' . . . We have been on a campaign for life, and it is our life struggle to maintain this river where we have lived all of our lives, under international principles of treaty that govern our relationship with the United States."[6]

Phyllis Young is one of many remarkable and courageous women I have met who are speaking out and taking action, aware of the inextricable connection between social and environmental injustices.[7]

The climate crisis threatens everyone, but especially women who experience climate change with disproportionate severity because their basic rights continue to be denied in varying forms and intensities across the world. Enforced gender inequality reduces women's physical and economic mobility, voice, and opportunity, making them more vulnerable to mounting environmental stresses. According to international studies, approximately eighty percent of climate-displaced people are women.[8]

Indigenous, Black, and Brown women, and women from the Global South overall bear an even heavier burden from the impacts of the climate crisis because of the historic and continuing impacts of colonialism, racism, and inequality, and in many cases, because they are more reliant upon natural resources for their survival and/or live in areas that have poor infrastructure.

Sexual violence against women in many Indigenous communities is a grave threat perpetuated by fossil fuel and mining industries. In the US and Canada, man-camps, which are expansive trailer units sometimes housing thousands of industry workers in oil and gas drilling regions, have resulted in extremely high rates of rape and abuse of local Indigenous women and girls.[9]

These patterns of abuse reinforce the connection between violence against women and violence against the Earth. The current dominant society is based upon power over and exploitation of women, Indigenous Peoples, People of Color, and the land. Our society's systems and ideologies encourage these forms of domination. Here is precisely where we need to dismantle detrimental cultural patterns and imagine and build a new worldview. To move toward justice and healing, we need to face these compounded systems of oppression through a comprehensive approach. Working to change policies and laws is fully necessary, but we need to do so with new social constructs of gender and racial equity, respect for human rights, and a deep understanding and respect of Mother Earth's natural laws and ecological systems.

Women in all their diversity must stand at the forefront of policy making and climate action because they are simultaneously the most adversely impacted by harms, and yet are indispensable actors and leaders of equitable and effective solutions.[10]

What inspires me so profoundly is that women, against all odds and great challenges, are demonstrating every day that they have unique and essential solutions and skills to offer at this critical turning point. When women have the opportunity and space to cultivate and grow their power and agency, communities experience systems-level benefits, from improved community health to expansion of local economies to thriving local ecosystems.[11]

As one of copious examples, women patrol the forests near Ghunduribadi, a village in the Indian state of Odisha, to stop illegal logging. They also harvest food from the forest. When the men used to guard the forest, they would fight over the logging rules and allow poachers to escape. As a result, the forest suffered—but now in the hands of the women, the forest is flourishing.[12] Renowned Indian Earth advocate and seed sovereignty activist, Dr. Vandana Shiva has stated, "We are either going to have a future where women lead the way to make peace with the Earth or we are not going to have a human future at all."[13]

In the Global South, women produce forty to eighty percent of the food for their households, meaning their position allows them to be leaders in sustainable agriculture.[14] To address systemic problems, women around the world are also advocating and implementing models of collective ownership of their seeds, plants, and forests, and they are working toward localizing their economies. Women are modeling small-scale solutions with powerful, far-reaching impacts, which are essential to achieving balance and disrupting centralized and monopolized practices.

Marginalized women in the Global North have also been rising up for decades. After Hurricane Katrina devastated a large area around New Orleans in 2005, many Black women, who were disproportionally affected and exposed to danger after being left without homes, brought visibility and activism to interconnected issues of environmental and climate justice, racism, health, and economic disparity. INCITE! Women of Color Against Violence, a national feminist organization, strove to protect the affected women while also centering marginalized voices in the recovery effort in New Orleans.[15] LaToya Cantrell's leadership in the Broadmoor Improvement Association turned her neighborhood into a model for disaster recovery; then, in 2018, she was elected the first woman mayor of New Orleans. Jacqueline Patterson, the founder and executive director of the Chisholm Legacy Project: A Resource Hub for Black Frontline Climate Justice Leadership and Collette Pichon Battle of Taproot Earth are two leaders who also are brilliantly forging paths forward and breaking new ground every day.

Recent history demonstrates that when women are elected, they often pass legislation on environmental and social issues. Research covering 130 countries showed that nations with a relatively high percentage of women leaders are more likely to ratify international environmental treaties.[16]

In particular, women from the most climate-impacted communities have a long history of resistance as well as knowledge of the land; thus, their voices must be centered. These women carry vital responses and solutions, and in short, there will be insufficient large-scale action on interlocking crises until women's struggles, solutions, and leadership emerge at the forefront.

There are thousands and thousands of brave women working ceaselessly to do everything possible to protect all that we can and to change our current tragic trajectory as crises escalate. As one example, at the UN climate talks, outstanding feminist leaders of the Women and Gender Constituency continue to lead the climate justice movement with some of the most critical and transformative policy frameworks. Here, to offer some well-deserved praise, I wish to share the names of just a few leaders globally so their stories can inspire us in our own effort and causes—and because it is crucial that we make visible more and more women's work and narratives. Women's collective efforts are a living prayer—a fierce and devoutly loving call to action for our survival and the survival of those yet to be born.

Sônia Bone Guajajara was elected to the Brazilian Congress in 2022, and immediately after this victory, she made history when she was appointed by President Luiz Inácio "Lula" da Silva as the minister of the newly created Ministry of Indigenous Affairs. For many years prior, she was the national coordinator of the Association of Indigenous Peoples in Brazil and organized the first March of Indigenous Women in Brasília in 2019 that spawned a series of ongoing and growing marches. Guajajara, who is a member of the Guajajara People in the state of Maranhão, believes society has much to learn from Indigenous Peoples, particularly a sense of collective identity that would align them toward balance with nature and with each other. At the UN Permanent Forum on Indigenous Issues in 2019, she said, "It's very urgent that the whole world listen to the Indigenous Peoples' voice from Brazil. . . . Our sacred territories are the ones giving the clean water, the fresh air, not only for us but for the whole planet. We truly need to stop this model of developing the economy and destroying Mother Earth."[17]

Sharon Lavigne was a special needs teacher in St. James Parish, Louisiana, until both of her neighbors died of cancer. St. James Parish lies within "Cancer Alley," the eighty-mile stretch of land along the Mississippi River infested with more than 150 petrochemical plants that are poisoning the air and water of these majority-Black communities. She explains, "They promised us jobs. Instead, they pollute us with these plants, like we're not human beings, like we're not even people. They're killing us. And that is why I am fighting."[18] Lavigne, though she had no background in activism, responded immediately when she heard that a highly toxic plastics plant was to be built near her community, in what was zoned as a residential area. A natural leader, she spoke out at parish council meetings and gathered community members in her home, which eventually led to her forming RISE St. James and to publicize their campaign through media work and letters to members of the parish council. Lavigne and RISE St. James won the fight, and the plant was not constructed. For her efforts, Lavigne was awarded the Goldman Environmental Prize in 2021. Her struggle continues, as she is currently working to stop yet another petrochemical plant from being built in the area.

Ruth Nyambura is a Kenyan political ecologist working at the nexus of gender, economy, and ecological justice. Along with her work as campaign coordinator of the Hands Off Mother Earth campaign, which opposes false solutions to the climate crisis, Nyambura focuses her research and activism on Africa's food systems through a Black feminist, decolonial, and anti-capitalist lens. Part of African Ecofeminist Collective, she explains that "food sovereignty is a deeply political project that those of us in the food justice movement take seriously. We are not just interested in a technologically different way of farming or providing food and jobs for people using alternatives like agroecology, but we realize that a technologically different way of farming from industrial agriculture will be useless if questions of power, by the class-gendered-racial, rural-urban, young-old are not constructively addressed."[19]

Nemonte Nenquimo is the leader of the Waorani community of Pastaza, a remote and magnificently biodiverse region of Ecuador at the upper headwaters of the Amazon. To unite the diverse local Indigenous communities against the common attacks of neocolonialism and extraction, she co-founded Ceibo Alliance, a nonprofit organization that encourages Indigenous communities to confer and aid one another when there are threats to a community's ancestral Amazonian lands or the rainforest. Following the guidance of her Elders, Nenquimo and Ceibo Alliance created a map of their territories, showing the irreplaceable treasures of the forest: medicinal gardens, sacred sites, creeks, and hunting grounds. As Nenquimo explains, "More than anything, we learned that the [oil] company doesn't see the forest. They don't see us. They see what they want to see. They see oil wells where we see gardens. They see money where we see life."[20] When the Ecuadorian government announced their plans in 2018 to auction off seven million acres of Indigenous forest territory to oil companies, Nenquimo sprang into action, launching a powerful digital campaign and taking the Ecuadorian government to court for violating the tribe's right to Free, Prior and Informed Consent. The courthouse judge tried to rush the trial, without a translator, so Nenquimo and the other women strategically broke into thunderous song, shutting down the court because no one could hear above the powerful chanting.[21] The trial continued the following month according to proper procedure, and Nenquimo won her case. The Ecuadorian government indefinitely suspended the auctioning of Waorani land to fossil fuel companies. The case set a legal precedent for Indigenous lands in Ecuador.

Greta Thunberg is a renowned Swedish activist, best known for orchestrating a school strike outside the Swedish parliament at age fifteen to protest climate inaction. She started organizing regular strikes called Fridays for Future, and her fiery speeches at UN conventions laid bare the hypocrisy of world leaders who promise change but never deliver. As a young adult, she continues to speak out. At the September 2021 Youth4Climate summit in Italy, she repeated the empty words of various international leaders, punctuated by "blah, blah, blah," further explaining: "This is all we hear from our so-called leaders: words. Words that sound great, but so far has led to no action. … As they continue opening up brand new coal mines, oil fields, and pipelines, pretending to have ambitious climate policies while granting new oil licenses, exploring enormous future oil fields, and shamelessly congratulating themselves while still failing to come up with even the bare minimum and long-overdue funding to help the most vulnerable countries deal with the impacts of the climate crisis. If this is what they consider to be climate action, then we don't want it."[22] Her voice sliced the air as she pointed out how international leaders are using "clever accounting" to make it look like they had reduced emissions when they had not. The climate crisis, she stated, was just a symptom of the larger crisis of inequality that dates back to colonialism. Thunberg's bold words and her Fridays for Future campaigns have helped lift up thousands of other youth climate activists around the world.

Fijian feminist Noelene Nabulivou is best known as the co-founder and political advisor of Diverse Voices and Action (DIVA) for Equality, a collective of lesbians, bisexual women, trans and nonbinary individuals, and those who are marginalized in some other way, such as by disability or poverty. Hailing from a culture where LGBTQ+ individuals often face discrimination, Nabulivou and DIVA work to transform the Pacific region's approaches to gender and sexuality, as well as economic and climate justice—an issue especially urgent for her vulnerable island nation. Nabulivou affirms that all people and all social issues are inextricably connected, and she calls on governments to phase out fossil fuels, embrace renewable energy, respect the Earth's limits, and support the world's poor, who have been hit hardest by the climate crisis.

Dr. Vandana Shiva has been a global leader in the environmental and climate justice movement for decades. She has written more than twenty books and appeared in several documentaries, speaking out against the agrochemical biotechnology company Monsanto and agricultural injustices. As a pillar of the ecofeminist movement, she argues that while women are better at managing biodiversity and productive and sustainable agricultural practices, they are pushed out by the "patriarchal logic of exclusion."[23] Dr. Shiva is the founder of the Research Foundation for Science, Technology and Ecology (RFSTE), which spawned Navdanya, an Indian NGO that promotes fair trade, organic farming, biodiversity conservation, and seed saving. Navdanya means "Nine Seeds," referring to the crops that make up India's food security. In 1999, Dr. Shiva and RFSTE sued Monsanto—a company she continues to fight on many fronts to this day—for its predation upon the farmers and lands of India. The case was taken up by the Indian government, and the court eventually invalidated Monsanto's patent on cotton seeds.

Geographer and environmental activist Hindou Oumarou Ibrahim comes from the Wodaabe, a traditionally nomadic farming and herding community in Chad. Her people have been severely affected by climate change, particularly by the diminishing of Lake Chad, a critical water source for four African countries. For Ibrahim, climate activism is tied to Indigenous rights, as projects connected with fossil fuels and mining displace the peoples whose livelihoods are tied to a particular area. As president of the Association for Indigenous Women and Peoples of Chad (AFPAT), Ibrahim emphasizes that Indigenous groups must have the right to own and manage their own lands in order to ensure their ways of life continue. She also stresses the importance of combining Indigenous and Western sciences to discover climate solutions.

As a member of the Lubicon Cree First Nation in northern Alberta, Canada, Melina Laboucan-Massimo has directly felt the effects of the tar sands development, much of which is being perpetrated on Lubicon lands without permission, violating treaties. Through advocacy efforts, articles, and documentaries, Laboucan-Massimo has been sharing her community's stories and telling the world about how the fossil fuel industry has harmed the Lubicon Cree First Nation and spoiled their water and land. She is also a champion of renewable energy, establishing the Pîtâpan Solar Project, a solar installation in the town of Little Buffalo, where she grew up. Laboucan-Massimo is the founder of Sacred Earth Solar, which builds solar projects in frontline Indigenous communities, and she is co-founder of the women-led organization Indigenous Climate Action. She hosted an acclaimed and informative TV program, *Power to the People*, which narrates the story of eco-housing, renewable energy, and food security issues in North American Indigenous communities.

These women and gender-diverse leaders, along with countless others, shine as North Stars through a time of great disruption and help encourage us on our own paths. Around the world, women are coming together to fight for climate justice knowing that our love and dedication to life is more powerful than the corruption and greed of predatory capitalism, institutionalized patriarchy, and structural racism at the core of our perilous global predicament.

There is also a special element that women bring to our advocacy, which is sorely needed in these trying times and should in no way be devalued or dismissed—and that is emotional and spiritual intelligence. Women have been told too often that personal, emotional, or spiritual expressions in our work efforts are inappropriate for analysis of pressing social and ecological issues. Yet women sharing and leading from their hearts and brilliant minds is precisely what is missing from the sterile and stifling processes of most policy forums and decision-making institutions. Women speaking their full truths in public spaces and places of established "power" is central to the success of our collective struggles for a healthy future. It is vital that decision-makers at financial institutions, for instance, directly hear the gut-wrenching stories of personal and community abuses experienced by frontline women because of resource extraction projects financed by their institutions—and equally important for them to learn of the stellar solutions these women deliver with passion and ingenuity. I recall the well-known systems thinker Joanna Macy relaying many years ago the story about how she had been criticized at a meeting with industry leaders for "getting too emotional." Her response was perfect: "If we women can't get emotional about the planet being destroyed, what is it exactly we should be getting emotional about?"[24]

Women are rising collectively to say colonization must end, sacrifice zones and environmental racism must end, violence against women by extractive industries must end, violations of Indigenous and human rights must end, and the destruction of Earth and water must end.

That day at the Oceti Sakowin camp back in 2016, Phyllis Young and I sat on a small hill looking out over the thousands of people who had come to protect the water and stop a pipeline. She let me know this marked a new time and told me: "We really believe that the age of man is over and that women have to step up and acknowledge their circular power. There is more spirituality exuded from that energy. We can afford to be more spiritual in the struggle that we continue to embark on, and that's for the water because we represent water, and all living things are born in water. Babies, human beings, know how to breathe and swim before they come into this world. Those are miracles of life."[25]

Given the clear evidence of women's exemplary and successful leadership in many arenas, from being central to peacekeeping to ensuring healthy communities and ecosystems, why is it that mainstream society is still not supporting, centering, and amplifying their voices?[26] Since women are leading adroitly, why are they undervalued, underpaid, and underrecognized? Why is there still such significant violence, femicide, misogyny, discrimination, and enforcement of unequal gender norms? How do we address the historical misogyny of the dominant society? These hard questions arise because of an outdated, inimical worldview.

While some important progress has been made to address gender inequity in different regions of the world, deeper cultural transformation will require us to explore the roots of patriarchal formations and worldviews. In this urgent era of interlinked crises when we need more women in all their diversity at the helm, it is quite pressing that we acquire a basic understanding of the historical foundations of misogyny so that we can challenge and rectify it. As we enter this knotted maze, the discourse is about challenging the patriarchal syndrome that has been devastating across the entire gender spectrum, as well as for the sacred ecological systems.

The primary historical narrative about ancient cultures and religions that most of us have learned in the dominant culture originates in social constructs that are explicitly devoid of living-Earth cosmologies and egalitarian societies. As we know, common foundational stories tell of an abstract male God who is at the center of the belief system—a singular supreme deity who is no longer in the mountains, rivers, forests, fields, and glens, but is transcendent and all-powerful. In this worldview, man's superiority—mirroring God's singular supremacy—abounds in the human social establishment and strata of patriarchal civilizations, towering with hubris above women and nature.

Fortunately, in the past several decades, to counter the dominant culture's aberrant historical narrative, scholars from a variety of fields have extensively researched and revealed the centrality of Goddess, Earth-centered, egalitarian peoples of Old Europe and the Eurasian steppes, particularly in the Paleolithic and Neolithic periods, predating the rise of patriarchy. There are parallel arcs of the rise of patriarchy globally in regions other than Old Europe and the Eurasian steppes, but to keep within the scope of my research and personal ancestry and experience, I am primarily focusing on these regions.

Beginning in the 1970s, archaeologist Marija Gimbutas documented thousands of pottery, bone, and stone Goddess figurines and relief images carved in rock dating back at least thirty thousand years.[27] Through her research and that of other scholars, we are offered a stunning array of Goddess statuettes with markings and symbols that demonstrate these female figurines are deities of water, plants, bears, deer, flowers, birds, sun, moon, and stars. The figurines sit with dignity on thrones straddled by wild animals. Adorned with elaborate nature-themed headdresses and decorated with intricate, mystical patterns, they hold snakes, cats, birds, and other animal kin. These deity figurines reflect a worldview of peoples who were intimately connected to a living universe and upheld extensive knowledge of Mother Earth. This wisdom continues to speak to us today through small female statuettes, as well as cave drawings, chiseled rock images, and earthen works in the landscape. As Gimbutas tells us, "Archaeological materials are not mute. They speak their own language. And they need to be used for the great source they are to help unravel the spirituality of those of our ancestors who predate the Indo-Europeans by many thousands of years."[28]

I have pored over thousands of these artifacts, in person and in photographs and archaeological drawings, inspired by the exquisite expressions of people's profound love and reverence for Earth's natural cycles and their relationship to the greater forces in the web of life. The images in clay, stone, and bone bespeak a worldview that celebrates the Feminine Principle as life-giver and regeneratrix, and also reveres the holiness of the cosmos in ecologically sophisticated detail.

With Gimbutas leading the way, guided by her archeological, linguistic, and comparative mythology expertise, along with other scholars from a variety of fields, it has been clearly demonstrated that these older European cultures were non-patriarchal and egalitarian, and very much attuned to an animate world.[29] Yet, these flourishing Goddess-centered peoples not only have been silenced in the telling of mainstream history until this new research in the 1970s was circulated in some educational spheres but were vanquished as thriving civilizations and communities thousands of years ago.

How and why Earth-centered peoples who venerated the Feminine Principle were overtaken by patriarchal societies—gradually in some cases, and more swiftly and violently in others—is still a topic of academic disagreement. Some scholars claim that nomadic communities underwent numerous changes when they settled, as many eventually did, and in the process, their nutrition, approach to public health, independence, and economic systems changed, as did their cultural norms.[30] In this theory, it is believed that gender roles became differentiated in sedentary life, with women eventually taking on roles seen as less valuable or powerful.

However, there is strengthening evidence for the analysis offered by Gimbutas who attributes the violent disruption of matrilineal peoples, beginning in the fifth millennium BCE, to waves of Proto-Indo-Europeans who worshiped a sky God and overtook the more peaceful Goddess-oriented communities of Old Europe.[31] Similarly, the patriarchal Mycenaeans from mainland Greece overthrew the Goddess-honoring Minoans of Crete in the second millennium BCE.[32] Since this time, the transition to patriarchal societies has been occurring over thousands of years to various degrees and in differing timeframes globally.

Most Mother Goddess-honoring societies have been identified as egalitarian; some have been referred to as matriarchal. As we reintroduce the wisdom and cultural knowledge of our ancestors from Old Europe, it is important to recognize matriarchies are not the direct opposite of patriarchies and are not hierarchical, with women dominating men. When Goddess-honoring and matriarchal societies were subjugated by monotheistic, patriarchal conquerors or colonists, their communities were radically disrupted. These once-egalitarian cultures became markedly hierarchical as eventually their Goddess temples, sacred sites, and groves were destroyed—and with these assaults, violations and diminishments also befell the women.

In the monotheistic societies that govern most of the world today, God is viewed as being above creation. He has dominion over all, including nature and the seasons, long associated with and embodied in a plethora of sacred Goddesses. With the rise of patriarchy, we can witness the purposeful eradication of an animate worldview, wherein nature is alive and has agency and humans live within a sentient relationship to the natural world. Bound within the patriarchal context, the dominant culture's story of separation from nature begins, and that soul-crushing narrative of abandonment from the sacred Earth has not stopped.

With the advance of patriarchy in Old Europe and across the Eurasian steppes, dominion over the Goddess of the land and dominion over women were firmly fixed together and formed the very underpinnings of the new religions that overwhelmed the ancient Goddess peoples. The Proto-Indo-European assault was a significant phase in the patriarchal besiegement of Old Europe before the invention of the Abrahamic religions and the Roman colonization of the tribal peoples of Europe whose spiritual and cultural practices were still deeply entwined with an animate Earth. Then ultimately, over time, Christianity and the Roman Empire furthered the patriarchal agenda, with women and the land violated every step of the way.

Of course, this is a broad historical recounting, but before I leave it, it is important to say that no matter the onslaught of the Proto-Indo-European incursions and those that followed, the Goddess and her stories and symbols were not entirely unseated. The Old European spiritual traditions were fully embedded in the marrow and soul of the people, and thus were not easily removed; rather, slowly and over time, the cultural symbols of diverse peoples and their meanings were often fused together. While, to be sure, much was destroyed, still the old ways were cleverly blended and concealed in the dressings of the new religious orders. And in the midst of this collision of cultures, many Goddess-honoring, Earth-centered peoples resisted these waves of assaults, both directly in battles for sovereignty and for land, and quietly, hidden in the forests and valley folds, and under the cover of night where the old ways were secretly practiced—and where today, they still remain even if only in cherished fragments.

Chapter 6:

Tracing and Healing the Assault on Women

IN 1702, secretary of the Paris Academy of Sciences Jean Baptiste du Hamel wrote, "We discover the mysteries of nature much more easily when she is tortured by fire or some other aids of art than when she proceeds along her own road."[1] This was the common attitude toward the feminine and nature by the time of the Enlightenment in Europe as we trace the escalation of patriarchy from ancient to more modern times. In this mindset, that continues today, the feminine and nature were and are forces to be violently subdued, harnessed, extracted from, and exploited in the patriarchal paradigm.

Feminist Marxist scholar Silvia Federici powerfully highlights the link between the onset of capitalism and violence against women and marginalized peoples in her book *Caliban and the Witch: Women, the Body and Primitive Accumulation.* She describes how bodies ceased to be sacred and simply became commodities, instruments of labor in the patriarchal context. Female reproduction was assigned to the control of men, policed as necessary for the production of workers, and was no longer recognized as an act of self-determination and sentience. This mindset has hardly changed as we see around the world the fierce ongoing struggle for women's reproductive rights.

In the late Middle Ages in Europe, the commodification of women's bodies intensified as the rape of poor women was largely decriminalized and they were attacked more frequently, without consequences. This violence was one part of desensitizing the population to the horrors that would proliferate next: witch burnings.[2]

The major witch hunts of Europe took place from the fourteenth century to the eighteenth century, when patriarchy was well rooted, and women were firmly regarded as second-class citizens with very few rights. In fact, the witch craze could only happen because patriarchy was the social norm; it created the perfect and tragic groundwork—again illustrating the necessity of examining how worldviews dictate our personal, social, and political lives. The Goddesses had been long deposed by the Christian Church, which now portrayed women as evil temptresses responsible for the fall of man and stated, "The husband is the head of the wife, even as Christ is the head of the church: and he is the saviour of the body."[3] Scholars disagree on how many women were killed as witches—from 60,000 to 100,000 to much higher figures of 500,000—but we can conclude for certain from trial records found in village archives that witch killing was normative in Europe for at least four centuries.

Who were these so-called witches? The Catholic and Protestant Churches defined witches as people who communed with evil spirits and would do their malevolent bidding. Witchcraft was also defined as a form of heresy, as witches were assumed to have renounced Christianity and entered into a pact with the devil in order to gain inhuman powers such as shapeshifting, compelling men to fall in love with them, flying, and sickening or killing people or livestock.

In reality, however, those persecuted as witches were regular people, mostly peasants and mostly women: healers using natural medicines, widows, outspoken people, wise village elders, midwives, and women who held spiritual knowledge and divination practices. A woman could be suspected of witchcraft for missing church, having a birthmark, spending too much time in the forest, using herbs in traditional remedies, participating in seasonal rites, dwelling with other women or by herself, and having pets and enjoying their company. Witches were targeted and murdered for many reasons, but all of these cases fundamentally formed part of a war against women by the ruling classes and the churches—a campaign that cemented structural patriarchy by terrorizing women into silence and submission.

Women deemed as witches became the ultimate scapegoat for challenges that people faced in their everyday lives, from bad harvests and natural disasters to common illnesses and family misfortunes.

According to the *Malleus Maleficarum*, or *The Hammer of Witches,* a treatise on demonology written in 1486 by the inquisitor Heinrich Kramer and Reverend Jacob Sprenger, women were deficient in faith, evil and corrupt by nature, more susceptible to demonic whispers, and more lustful than men. In the *Malleus Maleficarum*, Kramer advocated that witches be burned at the stake, a punishment at that time reserved for heretics, and tortured until they confessed their alleged crimes—and it is striking to note the gendered ways in which these women were tortured: rape, shaving the head, burning and tearing the genitals, cutting off the breasts. These are horrifying displays of the dominant culture's overt misogyny and sense of ownership over female bodies.

Peasant women were also scrutinized for signs they might be organizing against the ruling classes and the church, and those with popular influence were often arrested under pretenses such as blasphemy or lewdness, seen as indications of witchcraft. Here it is important to note that the most aggressive witch hunts occurred within "periods of great social upheaval shaking feudalism at its roots—mass peasant uprisings and conspiracies, the beginnings of capitalism, and the rise of Protestantism."[4] The privileged classes sought to consolidate power and enforce their hierarchical, church-centric worldview on the populace to protect their own position—and their wealth. It is evident that this same struggle exists in the United States today, as the news is inundated with right-wing groups constantly seeking to suppress any progress made or attempted by feminists, BIPOC groups, and LGBTQ+ communities. Fearmongering, bigotry, and hate continue to be a lucrative industry, and a means of consolidating power.

Silvia Federici states that "like the slave trade and the extermination of the indigenous populations in the 'New World,' *the witch hunt stands at a crossroad of a cluster of social processes that paved the way for the rise of the modern capitalist world.* Thus, there is much that can be learned from it concerning the preconditions for the capitalist takeoff."[5] During that time, increasing privatization and land enclosures had dispossessed farmers and others who had lived from the land, creating populations of beggars that threatened the emerging capitalist order. Begging began to be labeled as ungodly, so the poor, depicted in this way as forsaken by God, became easy targets for accusations of witchcraft.

As Federici aptly points out, the female body, like the land, was also being enclosed through the church's policing of female sexuality and reproduction.[6] Similarly, knowledge was enclosed as the populace was instructed to take advice only from the male priests, preachers, and kings (and, by extension, local lords) who were said to rule under "divine right."

Previously, everyday people would have sought knowledge from village elders, often wise women or healers who were intimately familiar with the natural world and its workings, having learned from generations of passed-down teachings, mother to daughter, and through empirical observation and experimentation. The church shunned scientific discovery—especially from "superstitious" village women—and Earth-based knowledge was deemed "pagan" and those who shared it were arrested as witches.

The church also reaped immediate economic benefits from executing some women of wealth. In many cases, widows were the only women in patriarchal European societies who legally owned land and property. When the church, particularly the institution of the Catholic Church, murdered a woman as a witch, that woman's land became the property of the church. As a further incentive, older independently wealthy widows were also the very women who were most likely to raise their voices against the injustices the populace faced under the privileged classes.

In *The Enemy Within: A Short History of Witch-Hunting*, John Demos points out the element of betrayal that lay at the center of the witch hunts. While witch hunts can be tied to the larger catalog of human atrocities, they differ because the witches were members of one's own community—a witch hunt was a conspiracy to find, frame, and execute a perceived hidden enemy within.[7]

The element of betrayal furthered patriarchy by keeping women scared, quiet, isolated, and confined to the home, dependent upon and in submission to their menfolk within patriarchal nuclear family structures and performing highly essential but unpaid labor. This fear and betrayal severely disrupted the ancient circle of women, as sisterhood was now perilous: if your friend was taken as a witch, she may utter your name involuntarily while being tortured. Additionally, the elites may have suspected you of organizing against them if you gathered or shared information, so women could not meet or share knowledge without fear. As a further layer of terror and deception, some community members labelled individuals as witches to avenge a social or financial slight.

One legacy of the witch hunts is that much women-held knowledge was also exterminated at the gallows or stakes, a victim of the patriarchy's growing control over female power and autonomy. This includes knowledge of herbs, healing, contraception, and abortion. Working as midwives, herbalists, and healers brought income and agency to many women; indeed, they functioned as extremely effective doctors for their communities, and they learned their trade from other women and from Nature herself, cultivating a deep intimacy with the Earth. The witch hunts stripped these women of their livelihood and influence, their birthright and calling.

But the elites and the church subscribed to a double standard: when the poor suffered, the church preached that life on Earth was short and that temporary misery was inconsequential, but the upper class called in physicians—male, of course, and approved by the church, often even employed by them. The burgeoning universities did not admit women, and men took over the medical profession in Europe, even participating in witch trials by giving "professional" assessments as to whether a person had been touched by witchcraft or whether a particular woman was a witch.[8]

The witch trials furthered the relationship between doctors, the church, and the ruling class. Barbara Ehrenreich and Deirdre English pointedly summed this up in their book *Witches, Midwives & Nurses,* stating: "The suppression of women health workers and the rise to dominance of male professionals was not a 'natural' process, resulting automatically from changes in medical science, nor was it the result of women's failure to take on healing work. It was an active *takeover* by male professionals. And it was not science that enabled men to win out: the critical battles took place long before the development of modern scientific technology."[9] The exclusion of women was a political decision by privileged men, a move that would establish male control over medical schools and the development of theory and textbooks.

Over time, women were relegated to the role of nurse, subservient to a male doctor who supposedly knew best. This trend has continued into modern times, though we are today seeing more male nurses and more female doctors, especially in younger generations. A 2017 survey found that sixty percent of physicians in the US under thirty-five were female.[10] Also that year, more women than men entered medical school. Given the immense amount of power doctors have, these developments are hugely significant for women's health, since women are sorely lacking in medical research, which still presents white male bodies as the default.[11] The legacy of misogyny in the medical profession, unfortunately, intersects with racism. Research has found that in the US, Black and Indigenous women are two to three times more likely than white women to die from pregnancy-related causes.[12]

During the height of the witch hunt, women who sought refuge in the church were not immune to patriarchal scrutiny. Female mystics, so revered and common in the Catholic Church of the early thirteenth century, had by the fifteenth and sixteenth centuries become rare. They had been subjected to increasing investigation with the rise of the Inquisition (a body within the Catholic Church charged with rooting out heresy) and its influence in differentiating between genuine spiritual phenomena and heresy. While this affected both women and men, the Inquisition and universities, interested in proving and disproving sanctity for academic reasons, were strictly male institutions. They claimed women were more susceptible than men to demonic possession; thus, female spirituality was increasingly pathologized and suspected. Female mystics were often expected to manifest painful physical symptoms of spiritual achievements, which devastated their bodies. In addition, the Inquisition employed torturous methods to elicit confessions of heresy.[13] Many holy women who did not pass the Inquisition torture tests were executed.

Some scholars have blamed the witch hunts on mass hysteria, saying either the "witches" themselves were women who had gone mad or that a panicked insanity had spread among the peasants.[14] But in reality, this was a long, meticulously planned war on women carried out by the church and the state to enshrine the patriarchy and destroy female agency and leadership—and this campaign has been papered over by centuries of invented "histories" that blame the peasants, malicious village gossips, and antiquated superstitions.

From the downfall of the Goddess cultures of older times to the witch hunts only several hundred years ago, those of us with ancestry in Europe and broader regions are the direct descendants of the women who survived the witch burnings and the assault against our Earth-centered roots.[15] What does this mean in our bodies and psyches? There is a remembrance aching to be brought into light, and a need to acknowledge that we carry these generational traumas that need healing. Stories of this deep-seated memory must be revealed for it is an ancestral responsibility to tell this history. Most often the witch burnings of Europe, which were also brought to the New World as the Salem witch hunts, have been dismissed as relics of a less-enlightened time that have no bearing on us personally or society at large today. Yet our cultural norms and personal lives are still impacted and shaped by these violent events, from deadly domestic violence incidents to the twisted co-option—and monetization—of pagan traditions into patriarchal religious holidays, from the isolation and subjugation of women in nuclear families to contempt for time-honored herbal healers. And because the patriarchal paradigm was foisted upon cultures worldwide, women and people from marginalized communities continue to be killed as witches in parts of Asia, Australia, Latin America, and Africa today.[16]

Women everywhere continue to be oppressed in the patriarchal dominant culture that descends from the same extractivist, misogynist worldview that allowed the European and North American witch hunts to materialize. What can we expect from a culture that believed it was acceptable to torture and burn women, and has never fully acknowledged this horrific past? What does it mean to come from a people who turned against their own womenfolk: mothers, daughters, grandmothers, neighbors, village healers, and wise women? Through my own personal experiences and countless conversations with women and some men, I have learned that the trauma and grief stemming from the loss of the Earth-centered, Goddess cultures and the terror of the European witch hunts have not been integrated nor deeply understood, and yet this intergenerational memory firmly resides in our psyche, ceaselessly calling to be known, grieved, liberated, and remedied.

British psychotherapist Cali White explains that women carry generational trauma from the European witch hunts and our ancestors' strategies for survival, and this trauma affects our relationships and behaviors today. With her sisterhood, the Silver Spoons Collective, White curates an exhibit called *I Am Witch ~ Tales from the Roundhouse,* which aims to educate people about the "Burning Times" though interactive exhibitions.[17] She describes the patterns women today have inherited, saying, "The scars we still carry show up in many ways—fears of being seen or heard, experiences of betrayal, mistrust of other women, feelings of disconnection to nature, irrational fears, and struggles to feel at home in ourselves."[18]

We cannot afford to pretend the witch hunts did not happen. It is vital to surface the way they were used to further patriarchy and capitalism.

In so doing, we can more thoroughly unwind the twisted narrative and worldview from our minds and bodies. Some of this process is keenly internal and personal, and some of this work toward reconciliation and healing must happen collectively and culturally, through processes such as ensuring we support and uplift the voices and work of women, especially women who are marginalized in our societies; retrieving our Earth traditions, including our wise women and herbal medicines, which are our ancestral birthright; elevating care work while ensuring equality of access; surfacing conversations and education about the burning times while preventing historical diminishment or erasure; making visible ongoing witch hunts today and stopping them; and ensuring women around the world in every culture are respected as they collectively celebrate the living Earth—spiritually, physically, and reciprocally.

Though much deeper efforts are in order, there have been some societal steps to course-correct from the witch hunts: three hundred years after the repeal of the Witchcraft Act, the 2,500 people slain as witches in Scotland's "satanic panic" have received a pardon for crimes they did not commit.[19] The Anna Göldi Museum, which opened in Switzerland in 2017 to commemorate the last alleged witch executed in Europe (she was beheaded on June 13, 1782), educates visitors through exhibitions dedicated to witchcraft, remembrance, and rehabilitation.[20] Healing from the witch hunts is ultimately a collective journey because it has impacted people globally and thus, cannot be bypassed as we seek to generate well-being in ourselves and communities.

The dominant culture adulates and empowers white cisgender men, and structurally discriminates against women and individuals all along the gender spectrum. Consequently, another component of this mindset is that societies that are patriarchal often insist on gender binaries, as currently seen in the US. In the first three months of 2022, more than two hundred bills targeting LGBTQ+ rights were filed in state legislatures; including Florida, where teachers of grades K–3 are prohibited by law from speaking to their students about gender identity or sexual orientation, meaning that teachers in non-heterosexual relationships cannot even acknowledge their partners and spouses, and children with LGBTQ+ parents must not speak of them at school.[21] Meanwhile, according to the Trevor Project, an organization that supports the mental health of LGBTQ+ youth, forty-two percent of LGBTQ+ young people considered taking their own lives in 2021; for youth growing up in gender-affirming families and communities, those numbers are significantly lower.[22]

In contrast, many pre-colonial and pre-patriarchal cultures respected (and some continue to respect) diverse genders, not only gender binaries. European colonizers were perplexed that Native Americans acknowledged more than two genders, and that gender identity did not necessarily align with the sex assumed at birth.[23] Most Native American Peoples recognized three, four, or five genders and had special dignified names for them in their languages. Today, many Indigenous individuals who are neither exclusively male nor exclusively female use the English term "Two Spirit," identifying themselves as both Indigenous and nonbinary, but also to highlight the importance of their spiritual relationship to their communities.[24] The term initially entered English from Ojibwe and has, in turn, been translated into many Indigenous languages, while others continue to use their own terms.[25] European missionaries, boarding schools, and government officials have incessantly attempted to force a binary system on Indigenous Peoples.

While some regions of the world have made varying degrees of progress on gender issues, discrimination toward gender-diverse individuals and groups aggressively continues. The World Health Organization did not drop "gender identity disorder" from its list of diseases until 2019, and trans, nonbinary, and gender-nonconforming peoples continually face legal inequalities and societal violence.

As for women in all their diversity, the fundamental modern worldview still paints them as "the lesser sex"—and this intensely embedded ideology results in sexist behaviors and policies. Geopolitics reporter Annalisa Merelli discusses a 2020 UN report showing that almost 90 percent of people around the world hold biases against women—and these views are shared by both men and women. The study used seven indicators to determine whether a person held sexist views, and asked "whether or not men make better political leaders; women and men have the same rights; university is more important for men than women; men should have more rights to a job than women; men make better business executives; physical violence by a partner is ever justified; and, finally, whether or not women should be granted full reproductive rights."[26]

From the responses, the study concluded that a striking 90 percent of men and 86 percent of women are prejudiced against women. Predictably, this translates into gender inequality in the countries with more misogyny and higher rates of violence against women. A study conducted by Inter-Parliamentary Union found that 82 percent of women parliamentarians in thirty-nine countries have been subjected to psychological violence, ranging from sexist remarks to death threats, rape threats, or threats of violence to their family members, in person or via other channels such as social media.[27]

In the midst of this assault, women have been fighting back against patriarchal constructs in various ways. In recent years, the #MeToo movement exposed the predatory behavior of many prominent men and opened a dialog around issues related to consent, the way women's bodies and minds react to unwelcome sexual advances, and the social and sexual conditioning of the dominant culture.

Tarana Burke, a Black activist from New York, founded the #MeToo movement in 2006 after working with survivors of sexual violence as a way to encourage women to stand up and speak up for themselves. The movement went viral in 2017 in North America, and quickly spread around the world, when several women came forward with rape and sexual assault charges against powerful men—men who had believed their privilege would always shield them from accountability and punishment.[28] The dominant society had taught such men and their victims that "boys will be boys," that women owed sex to men, and that women who accused such men would be disbelieved, publicly shamed, threatened, and possibly penalized financially.

This time, however, many rapists and sexual assaulters were exposed, censured, fired, and arrested. Women used the #MeToo hashtag to tell their own stories of sexual assault and rape, which has provided much-needed public discourse and awareness-building.

In 2018 in Chile, feminist students rose up to protest the epidemic of sexual violence and harassment in universities and institutions across the country. In Nigeria, journalists extensively publicized ongoing instances of university professors demanding sex acts from female students in return for grades, leading to the introduction of an anti-harassment bill in the

Senate, which passed in July 2020. American feminist activist and V-Day founder "V" (formerly Eve Ensler) debuted a one-woman play in the US in 2018 based on her memoir, *In the Body of the World*, exploring women's experiences of disconnect from their bodies due to traumas such as illness, war, and rape.

Overall, the major success of this ongoing movement is forcing society to evaluate the social conditioning that leads to the maltreatment of women. The work, of course, must continue because the backlash is already in full swing. On June 24, 2022, the US Supreme Court overturned Roe v. Wade, removing the federally protected right for people with uteruses to obtain an abortion. This is yet another effort to control women and gender-diverse peoples who can give birth, and to deprive them of their bodily autonomy and agency.

This effort to dismantle patriarchy also requires attention to toxic masculinity that has harmed both women and men alike, and we see incredibly important work emerging in the past years as a new era of men also work to heal themselves from patriarchal constructs. Just as women have been gathering in circles for years to heal, men are gathering to rescue the sacred masculine. The dominant culture has idealized and forcibly defined manhood as physically powerful, emotionally stoic, materially successful, combative, assertive, dominating, and possessing supremacy to take and do what they want. This toxic ideation of manhood and masculinity is wholly destructive not only to women who experience violence that stems from this taught cultural norm but also to men.

Recovering and uplifting healthy and beneficial masculinity is part of transforming the patriarchy because misogyny is built upon an entitled and dominating masculinity that seeks to control and consume women and the Earth. Instead, we can support and embrace a masculinity that welcomes balanced emotional and spiritual expressions, and holds ancient knowledge systems of the Earth and cosmos. Just as there was a time of revered Goddesses, there was a time of Gods that existed in balanced relationships. Let's welcome back these sacred deities in their majestic egalitarianism and ecological wisdom. The transformation of our time is not about putting men down, but rather about lifting women up, and everyone across the gender spectrum healing from the tragic disruption and imbalance of patriarchy.

Chapter 7:

Listening to Black and Indigenous Women and Debunking the Myth of Whiteness

A HUMAN STORY that recognizes all genders and tells their histories is crucial for a sustainable and just society. When our cultural narrative is severely edited and stories of our identities are largely excised or distorted, our viable options and dreams as a people and as a civilization are greatly diminished.

In this same vein, to unpack gender issues foundationally and productively, we must also surface racial issues—without which, it is impossible to dismantle patriarchy. We cannot talk about gender inequality without talking about the inextricable and intertwined relationship to racism and the additional disproportionate impacts of oppression on Indigenous, Black, Brown, and Asian women, and gender-diverse people.

While feminists have for decades worked toward gender equality and equity, many women of color have repeatedly pointed out that feminism is only valuable if it is anti-racist. Focusing on the United States, most history schoolbooks tell us that American women gained the right to vote in 1920—but it is important to highlight that, though the Nineteenth Amendment claimed to grant *all* American women the right to vote, in practice it only benefited white women. Though Black Americans were technically allowed to vote, arbitrary laws, discriminatory tactics, threats, and state violence largely prevented them from doing so, especially in the South. Black women, whose significant presence in the women's suffrage movement is erased from most history books, could not vote until decades later.[1] Likewise, Native and Asian Americans gained voting rights in 1924 and 1952, respectively, but faced obstacles at the polls, as did Latin Americans.

Throughout the 1960s, Black activists, including women such as Amelia Boynton Robinson, the "Matriarch of the Voting Rights Movement," gradually broke down the barriers to the voting booth, culminating in the passage of the Voting Rights Act of 1965.[2] Today, voter suppression through tactics such as ID laws, removal of polling locations in Black neighborhoods, limitations on polling hours and mail-in ballots, and gerrymandering continues to quell the voices of Black citizens. In the US, schoolbooks that simply declare women gained the right to vote in 1920 are clearly whitewashing history.

The history books also gloss over, or entirely omit, most of the horrors and injustices Black women in North America faced in the past and continue to face today, from being violently raped by white men to having their children ripped away from them and enslaved, from enduring beatings for not working fast enough to being abused and insulted by white women. White doctors, such as the "father of modern gynecology," James Marion Sims, performed medical experiments on Black women in the 1840s, often without anesthesia or pain medication because they were enslaved, forcing them into the most unthinkable, unconscionable tortures.[3] Today, Black women face immense racism every day from macro- to microaggressions, yet despite this, they have persisted and risen brilliantly in every sphere of society.

For us to attempt to hold the full gendered and racial story, an intersectional approach is needed. Intersectionality refers to the way multiple kinds of discrimination overlap to create complex webs of compounded marginalization. Black feminist scholar Kimberlé Williams Crenshaw coined the term "intersectionality" in her 1989 work, "Demarginalizing the Intersection of Race and Sex: A Black Feminist Critique of Antidiscrimination Doctrine, Feminist Theory and Antiracist Politics," in which she argues that Black women are often erased in discussions of race, which focus on Black men, and discussions of gender, which focus on white women.[4]

Non-white women experience both racism and sexism, and the oppressions multiply when combined. The gendered experiences of disabled women, poor women, trans and gender-nonconforming women, and queer women are different from those of abled, middle-class women who are straight and cisgender, and they are often left behind in feminist discourses that lack an intersectional approach. As so many women of color have asserted, feminism is only valuable if it serves, empowers, liberates, and seeks solutions for all.

Due to the lack of an intersectional lens, and, unfortunately, due to racism and prejudices among white feminist leaders and organizers, the first and second waves of feminism left multitudes of women behind, tarnished the word "feminism" in many circles, and created initiatives that mainly benefited white, straight, cis, abled, middle-class women. This occurred despite the fact that marginalized women were involved at all stages. But their contributions were, and often can continue to be, ignored.

The second wave of feminism began in the United States and encompassed the 1960s and 1970s, focusing on the legal aspects of gender equality, including factors such as family law, domestic violence, reproductive rights, workplace discrimination, and more. Again, the most recognizable name from that era is a white woman, Gloria Steinem. Few people know of Dorothy Pitman Hughes, the Black activist who co-founded *Ms.* magazine with Steinem, or Florynce Kennedy, a Black feminist who often shared the stage with Steinem and advocated for rights for Indigenous Peoples, disabled people, the poor, gays and lesbians, the elderly, sex workers, former inmates, and People of Color.

Because the first and second waves of feminism did not serve Black, Brown, and Indigenous women, many rightly saw no point in joining or staying in a movement that dismissed their life experiences as invisible, unimportant, or inconvenient—a movement that therefore furthered white supremacy. Many Black women did, however, start their own movements and organizations that supported their needs by focusing on issues of race, gender, and class together, such as the womanism movement and the Combahee River Collective organization.[5] Based in Boston, the Combahee River Collective argued that the white feminist movement and the Civil Rights Movement were not addressing issues vital to Black women, and in 1977 they developed a pivotal document, the Combahee River Collective Statement, which outlines concepts of identity politics.

The fact is that many white feminists were (and some continue to be) completely ignorant of—or chose out of privilege to ignore—the unique lived experiences their Indigenous, Black, and Brown sisters endure every day. Most had no Black, Brown, or Indigenous friends and spent their lives in all-white neighborhoods, and their white middle-class privilege influenced their responses to the reality of racism; consequently, white, primarily middle-class feminism was the result.

Even worse, there are white feminists who did—and do—see the discrimination faced by their sisters of color, but knew that incorporating a racial lens to their feminist activism would threaten their white privilege. White women would have to share space, attention, and resources with Black, Brown, and Indigenous women. As Black American author and activist Rachel Elizabeth Cargle remarks, "If there is not the intentional and action-based inclusion of women of color, then feminism is simply white supremacy in heels."[6]

Certainly, feminist movements have become more inclusive and transformational over the years, but like all growth, it is a work in progress, and white women still have critical labor to do to deeply reflect inwardly on racism, and to truly listen to, center, and amplify the voices and work of Black, Brown, and Indigenous women.

The third wave of feminism was led by a more diverse group of women, including bell hooks, Rebecca Walker (who coined the phrase "third wave feminism"), Kimberlé Crenshaw, and Judith Butler, and it took a more intersectional approach, bringing LGBTQ+ rights and Black rights under its umbrella. While some in these movements say we are still in the third wave, others point to a fourth wave, beginning in 2012 or 2014 and using the internet as its stage.[7] It is intersectional and is strengthening a growing global feminist movement. Some of the most potent, just, and brilliant analyses and solutions to current social and ecological crises are being generated in this vibrant global space of feminist movements. As poet, womanist, and activist Audre Lorde famously said, "There is no such thing as a single-issue struggle because we do not live single-issue lives."[8]

Many Black, Brown, and Indigenous women and Two Spirit and gender-diverse leaders have helped me in my ongoing journey to broaden my understanding of racial and gendered struggles and realms of multiple feminisms. This quest is a continuous and healthy process of learning, self-reflection, discomfort, dismantling, decolonization, and reorientation for me. I have been deeply moved by Audre Lorde for many reasons, including a focus on intersectionality in her critique of racism within the second-wave feminist movement. Lorde maintained that individuals are a mix of different identities, all of them fundamental to one's experience, and that feminists must affirm the differences between the women in their networks in order to overcome oppression and create lasting changes.

Audre Lorde was a path-maker in intersectional feminism, embodying the concept even before the term was coined in 1989. I am reminded of an excerpt from her 1981 keynote speech at the National Women's Studies Association Conference in Storrs, Connecticut, called "The Uses of Anger: Women Responding to Racism":

> When women of Color speak out of the anger that laces so many of our contacts with white women, we are often told that we are "creating a mood of hopelessness," "preventing white women from getting past guilt," or "standing in the way of trusting communication and action." . . .
>
> To turn aside from the anger of Black women with excuses or the pretext of intimidation is to award no one power—it is merely another way of preserving racial blindness, the power of unaddressed privilege, unbreeched, intact. Guilt is only another form of objectification. Oppressed peoples are always being asked to stretch a little more, to bridge the gap between blindness and humanity. Black women are expected to use our anger only in the service of other people's salvation or learning. But that time is over. My anger has meant pain to me but it has also meant survival, and before I give it up I'm going to be sure that there is something at least as powerful to replace it on the road to clarity. . . .
>
> I am a lesbian woman of Color whose children eat regularly because I work in a university. If their full bellies make me fail to recognize my commonality with a woman of Color whose children do not eat because she cannot find work, or who has no children because her insides are rotted from home abortions and sterilization; if I fail to recognize the lesbian who chooses not to have children, the woman who remains closeted because her homophobic community is her only life support, the woman who chooses silence instead of another death, the woman who is terrified lest my anger trigger the explosion of hers; if I fail to recognize them as other faces of myself, then I am contributing not only to each of their oppressions but also to my own, and the anger which stands

> between us then must be used for clarity and mutual empowerment, not for evasion by guilt or for further separation. I am not free while any woman is unfree, even when her shackles are very different from my own. And I am not free as long as one person of Color remains chained. Nor is anyone of you.[9]

In addition to these powerful words that stir deep within my being, I would like to highlight Lorde's quest into West African spirituality. Lorde reminds us that Black lives were Indigenous lives in their homelands before they were violently stolen away and sold into slavery.

Audre Lorde invoked mythological symbols from West African Indigenous spiritual traditions, often including them in her poetry. She visited West Africa and intentionally reconnected with the Goddesses of her ancestors, particularly those from Yoruba traditions, working, as she described, to recreate a spiritual tradition that would honor Black people.

Lorde's seeking of Goddesses was in keeping with the radical feminists of her time, many of whom sought—and continue to seek today—a female-honoring spirituality outside patriarchal religions. Feminists of Color resisted white colonizing spiritualities that did not make room for collective liberation, creativity, and sociopolitical justice, and that cared little for the trauma of oppressed peoples.[10] As Black women friends and colleagues have shared with me, connection for them to African Goddess traditions is a path to decolonization, liberation, celebration of Black womanhood, and reclamation of spiritualities white people have often deemed primitive or superstitious.

As a Black American, a woman, and a lesbian, Audre Lorde was barred from the opportunities enjoyed by whites, males, and straight people. She had to fight for acceptance in a racist, sexist, and homophobic society. In the realm of the West African Goddesses, Lorde identified a home where she could be embraced fully as herself; as she explains, she discovered roots and power.[11]

Lorde's spirituality was unapologetically political. Through reclaiming ancestral memories that connected her to the land from which her ancestors were uprooted, she was reclaiming Black history and values that could inform and guide Black Americans. She invoked matriarchal deities in a patriarchal culture, and the power of Black creativity in a racist society. In her poem, "The House of Yemanja," she called out to the Yoruba Mother Goddess, Yemanjá: "Mother I need / mother I need / mother I need your blackness now / as the august earth needs rain."[12]

Lorde offered her visionary song out to Black women, reminding them that their ancestors were African warrioresses, priestesses, and heroines. She spoke to all women that had been erased or placed in a subservient position by patriarchal religions, reminding them of the African reverence for the great Female Principle. Her tremendous courage and brilliance are feeding still the august Earth with buckets of rain.

I now turn to Paula Gunn Allen, a powerful Indigenous woman leader who described herself as an Indigenous feminist. However, before saying more here, I also want to acknowledge that several of my Indigenous colleagues and mentors do not use the term "feminist." They have explained to me that this word does not portray their knowledge systems; rather, they have a concept they call "Mother Law." In this understanding, women are the law in traditional Indigenous society (more on this topic later). The point here is that, in our discourse to rebuild healthy cultural systems, we also need to make space for differing and complementary views.

Paula Gunn Allen had Lebanese, European, and Laguna Pueblo ancestry and was a luminary in the field of Native American literature and poetry, as well as women's spirituality. She also penned essays, biographies, a novel, and a study of Indigenous women's traditional roles. She compiled anthologies of Native Oral Traditions and short stories by Indigenous authors. Like her colleague, Laguna writer Leslie Marmon Silko, Allen was heavily influenced by the stories of Spider Grandmother and the Corn Maiden.

Allen's book of essays *The Sacred Hoop: Recovering the Feminine in American Indian Traditions*, published in 1986, explores Indigenous communities before and after colonization, focusing on the roles of women. According to Allen, most pre-colonial Indigenous communities, including the Laguna Pueblo, were gynocratic,* matrifocal, and/or matrilineal, and all egalitarian. She invokes the deity Thought Woman, who thought the world into existence, to uplift feminine centrality in Indigenous traditions.[13] Allen

* Gynocratic meaning that women made tribal decisions; matrifocal meaning they acted as leaders; matrilineal meaning that tribal members traced their kinship through the maternal line.

provides examples of female Indigenous leaders and expresses that Native American literature is feminist in its very foundations.

In 1992, Allen published *Grandmothers of the Light: A Medicine Woman's Sourcebook,* in which she recounts Goddess stories from Indigenous communities across North America, as well as stories Indigenous female healers have shared for generations. At the same time, Allen informs us that due to the violent influence of colonization and Western patriarchy, masculine domination has misconstrued the telling and understanding of Indigenous worldviews:

> I became engaged in studying feminist thought and theory when I was first studying and teaching American Indian literature in the early 1970s. Over the ensuing fifteen years, my own stances toward both feminist and American Indian life and thought have intertwined as they have unfolded. I have always included feminist content and perspective in my teaching of American Indian subjects, though at first the mating was uneasy at best. My determination that both areas were interdependent and mutually significant to a balanced pedagogy of American Indian studies let me to grow into an approach to both that is best described as tribal-feminism or feminist-tribalism. Both terms are applicable: if I am dealing with feminism, I approach it from a strongly tribal posture, and when I'm dealing with American Indian literature, history, culture, or philosophy I approach it from a strongly feminist one.
>
> A feminist approach to the study and teaching of American Indian life and thought is essential because the area has been dominated by paternalistic, male dominant modes of consciousness since the first writings about American Indians in the fifteenth century. This male bias has seriously skewed our understanding of tribal life and philosophy, distorting it in ways that are sometimes obvious but are most often invisible.[14]

Perhaps as significant as Paula Gunn Allen's works is the legacy she left, in that she inspired further feminist studies of Indigenous cultures and influenced the development of Indigenous feminism, an intersectional theory centering on the fact that the US and many other Western countries "are settler colonial nation-states, and settler colonialism has been and continues to be a gendered process. Because the United States is balanced upon notions of white supremacy and heteropatriarchy, everyone living in the country is not only racialized and gendered, but also has a relationship to settler colonialism."[15]

As Allen emphasized, the world of Indigenous women was desecrated by colonization, when Indigenous children were torn from their families, taken to boarding schools, forbidden to speak their languages, converted to Christianity, indoctrinated with patriarchal attitudes and gender roles, inculcated with the idea of private property, and very often abused physically, sexually, and psychologically.

The horrors of these boarding schools became more public in May 2021 when the remains of 215 Indigenous children were discovered at Kamloops Indigenous Residential School in British Columbia, Canada. Later that year, the Cowesses First Nation discovered 751 unmarked graves near the Marieval Indian Residential School in Saskatchewan.[16] And there are many, many more bodies of Indigenous children callously brutalized, neglected, starved, and discarded across the US and Canada now being revealed. An ongoing investigation in the US has turned up fifty-three burial sites and five hundred reported deaths at Native boarding schools, and these tragic numbers are expected to rise, most likely into the thousands.[17] In Canada the number is already over four thousand and still counting.

Indigenous feminist theories seek to uplift female agency within their Indigenous cultural frameworks and philosophies, and given that Indigenous Peoples are land-based, Indigenous women seek their rights to tend, protect, and defend their ancestral territories. Another important component of Indigenous feminist work is focused on healing because Indigenous women and Two Spirit people have experienced egregious intergenerational trauma because of settler colonialism. Powerful work in this sphere is being led by organizations such as Indigenous Climate Action in Canada, and it involves healing from colonialism and extractivism through their Healing Justice initiative.

Returning to Mother Law, I have also been influenced by Mohawk Bear Clan Mother Louise Wakerakats:se Herne, who is living on her Traditional Lands in Akwesasne, New York, and by her views of women's leadership and the role of matriarchs. During a small gathering in June 2020, Louise Wakerakats:se Herne stated that "as Indigenous women, we are not feminists—we are the law." With her permission, I offer her insightful and instructive remarks:

> I want to do my best to share with you where it is I get my authority from. It comes from a culture, a language, a land, and the ladies that came before me. To be a clan mother, you get your authority from your clan family, and clans are kinship networks based upon women leaders. At the head of every clan is a woman, and she has her finger on the pulse of the people, and she has grown into leadership because she belongs to a family of many women, to a mother that had many daughters. And I am a third-generation clan mother. My mother was a clan mother; my grandmother was a clan mother. So, one of my four daughters will be a clan mother someday, or one of my granddaughters. If I didn't have daughters, it would go to my sister's daughters, my nieces. So, part of that leadership comes from being the center post of a society, and we know that the place that we are in is a point in history in which a lot has happened. Our culture was almost obliterated through colonization, so we are in a rebuilding stage, a reconstruction stage.
>
> We are trying to rebuild the clan mother as an authorized authority, and also, I often like to say, "the law," and I like to say that as Indigenous women we are not feminists—we are the law, based upon the primordial principles of women being so connected to the cycles of the earth and to the cycles of the moon, we have a natural primordial pattern within us that is endlessly through the generations being connected. So, as we take a look at this day and time and consider the amount of colonial attacks that have been on our people, our children, our men, our language, our culture, our land, which is still happening, I decided that as part of my leadership, and I have been a clan mother going on twenty years now, and being a mother of six children and soon-to-be twelve grandchildren,

I take seriously my position. And on any given day I could be cooking, I could be cleaning, I could be writing a proposal, I could be handling finances, I could be correcting men.

When men step out of places of honor, we have to be gentle, but we do have to correct them. And I think as women, nowadays, we tend to diminish our authority by hanging on to the old narratives that a woman is a possession of a man. In my culture, the children of a mother do not take their father's last name; they take their mother's clan, so we are a matrilineal, matrifocal society, and all the details of the society are taken care of by small concentric circles of women who are taking care of the details, and the Elders say that the women are the backbone of our Nation and the men are the jawbone of the Nation. They are the ones that speak for us. However, the man can't speak unless he is told what to say through the women's voice that allows him to be that leader. As part of my position as a clan mother, I have the authority to raise a man to a chieftainship title. I also have the right of recall. I can depose him. I can replace him with another leader if he has any shenanigans going on or if he has a self-agenda. For me, I pay serious attention to what the men are doing. I have sons, too, so I am really gentle about making sure that they have pride and that I don't diminish them and their value, but it is about balance, and it is about understanding each other's responsibilities and roles in society.[18]

It is clear through historical and lived experiences in our dominant culture that colonial, patriarchal, white supremacy has permeated every realm of our collective existence. Consequently, it is essential to continue to unveil the power structures of the white supremacist worldview which, galvanized by the burning fuse of unbridled capitalism, has furthered white supremacy constructs of colonialism, slavery, and extractivism. In short, white supremacy is an ideological disease that is viciously harming people and planet, and consequently it must be recognized, named, and deconstructed—and this is the responsibility of white people.

Tracing the origins of white supremacy is a vital component of transforming current detrimental gendered and racial ideologies—an enterprise that requires ongoing humility, learning, listening, courage, and advocacy as well as political, personal, spiritual, and emotional education.

This is a thick and burdened story to tell, to be sure—but we cannot rebuild a liberated world if we bypass how we have arrived in this moment. Given the enormity of this task, I am only touching upon a few key nodes throughout these chapters—recognizing and respecting that these realms of discourse and thought patterns require much further discussion and inner work. In all instances, the stories and voices that have been suppressed or marginalized can no longer be excluded if we are to generate the healthy and just world we seek.

For many white Americans, working toward decolonization and dismantling racism means delving into histories and mindsets that are not easily navigated or reconciled. Most of us in the dominant culture are descended from colonized peoples ourselves, yet we simultaneously benefit from the discrimination and colonization of "non-whites." This conflicting composition can make it challenging to enter truly transformative anti-racism and decolonization processes without some semblance of solid ground in understanding who we are.

For instance, the modern construct and idea of whiteness shrouds the unique mix of peoples' ancestral cultures, oppressing and distorting authentic lineages; thus, as I have mentioned, I have found it to be crucially important to retrace our family histories and learn about our cultural backgrounds, from long ago until today. Why did our ancestors leave their homelands? Why did they become colonizers or flee their own persecution? How did they settle in their new lives? Tracking our origins is essential in the enterprise to be responsible to and for our ancestors and provide real acts of reconciliation, reparations, and healing.

There are many tributaries along this journey, and for now, in addition to stressing the need for taking immediate actionable responsibility for anti-racism and reparations work, I want to focus on two worldview streams for people who identify as white. My experience tells me these will remain an impediment to necessary transformation unless we can understand them. The first is understanding the invention and myth of whiteness, which has been deeply explored by leaders such as James Baldwin, Maurice Berger,

Roxanne Dunbar-Ortiz, Kim F. Hall, and Nell Irvin Painter. The second is working to heal from the loss of Earth-centered cultural identities and pre-colonial lineages—a disconnect that causes a deep-simmering depression that can lead, and has led, to violence, and a fearful, and at times fanatic, attachment to the idea of whiteness.

In essence, whiteness is a social construct based on power and exclusion, and a false narrative that persists as a core principle of the colonial exploit. To illustrate this fabrication, we need only observe how the classification of whiteness in the Unites States has changed over time. Before arriving on the shores of North America, settler colonists were not called "white"; they were English, Norwegian, German, Scottish, and so on, with ancestral origins in Earth-based European tribal peoples. By embracing the myth of whiteness, settler colonists excised their cultural identities and rebranded themselves as a superior white force in opposition to the Indigenous Peoples of Turtle Island, whom they were displacing and killing, and also the Black peoples they had enslaved.[19]

Not everyone with light skin was automatically welcomed into the fold of whiteness. Irish immigrants fleeing a devastating famine, brought about by English colonialism, were subject to racist derision and originally coupled with Black Americans, with whom they worked, and were devalued in their labor and their personhood. However, Irish Americans soon realized their skin color would allow them to blend in with other settlers of European descent and join these whites in deriding and persecuting Black Americans. Although most of the Irish had been abolitionists back home due to their own history of oppression at the hands of the English, they firmly, and at times violently, blocked the abolition movement in the US.[20]

Roxanne Dunbar-Ortiz, author of *Not "A Nation of Immigrants": Settler Colonialism, White Supremacy, and a History of Erasure and Exclusion,* points out that Italian Americans achieved their status as white people in part by linking themselves to Christopher Columbus, an Italian Catholic like them, who was credited with being the first European in the Americas—therefore, Italians belonged on this land, even more than the whitest of their colleagues.[21] The largest wave of Italian immigrants arrived during the US Industrial Revolution. They came from the then-impoverished southern regions of Italy and Sicily, and most of them primarily had farming and manual labor expertise. Other European settlers mocked them for their particular skill sets and lack of academic education, as well as their Mediterranean complexions and features.[22] Italian Americans were consistently denied employment or advancement and were made to sit with Black Americans in church. In 1891 in New Orleans, eleven were arrested—with no evidence—in connection with the murder of a police officer, and then lynched by a massive white mob.[23] This atrocity prompted President Benjamin Harrison to acknowledge the contributions of Italian Americans, including Columbus, and for the Sons of Columbus Legion, a group of Italian Americans in New York, to campaign for Columbus Day to become a permanent national holiday.[24] Italians were now officially on their way to being white.

The violence and persistence of white supremacy depends on an inherited sense of privilege. Many white Americans are blinded by purposefully ingrained privilege built upon systemic racism; and many are all too aware of the existing supremacist paradigm and fervently support policies and politicians that maintain it, fearful of losing privileges and societal white domination. Many cling to white supremacy even as the Black Lives Matter movement—as Black organizations and leaders have for decades—lays bare thousands of acts of deadly anti-Black racism.

The Black Lives Matter movement, originating from a long history of abolition work and work against police brutality, began in 2013 when the man who murdered Black teenager Trayvon Martin was acquitted for the crime. People began using the hashtag #BlackLivesMatter, created by Alicia Garza, Patrisse Khan-Cullors, and Opal Tometi, on social media platforms to highlight instances of police brutality and other violence against Black people, including Eric Garner's 2014 death by a police officer's chokehold; the shooting death of Michael Brown by police in Ferguson, Missouri; the choking to death of George Floyd by police in Minneapolis; and the shooting of twenty-six-year-old emergency medical technician (EMT) Breonna Taylor by police who broke into her home on a no-knock warrant.

Under the banner of Black Lives Matter, activists have organized to raise awareness about the egregious injustices Black Americans face. The Black Lives Matter movement seeks freedom and justice for all Black people, and they also affirm women's rights, LGBTQ+ rights, and rights for disabled people, poor people, and immigrants.[25]

Founded on the genocidal colonization of Indigenous Peoples and enslavement of African peoples, US democracy has never been an actuality for all, and we see how structural colonization, racism, and patriarchy continue to undermine the dream of a fully inclusive and operating democracy. This issue grotesquely materialized in the insurrection on the US Capitol building that took place on January 6, 2021—an event motivated by racism, misogyny, xenophobia, and right-wing conspiracy theories, as evidenced by Confederate flags and Nazi symbolism; racial slurs aimed at Black police officers; and the insistence that Donald Trump, a personification of male white-supremacist entitlement, had won the election.[26]

Additionally, we are collectively in a primal struggle because extractive corporations depend on white supremacy in order to not lose their sacrifice zones, which they rely upon for their polluting industries and for exploiting the Earth and marginalized communities. Their dirty work has to take place somewhere—away from white people, particularly wealthy white people. Thus, coming to terms with white supremacy directly impacts the health of frontline communities and all natural systems.

This is why we need our movements to hold an intersectional justice lens, in which everyone, with their unique struggles and solutions, is included. It is only by uniting together, that we can build a different and equitable future.

In addition to the obvious advantages white people experience due to institutionalized racism, there is another less talked about aspect of why white Americans cling to the myth of whiteness, knowingly or otherwise. Beneath the surface, and more difficult to detect in the superficiality of mainstream society, there is a cavernous severance from white peoples' pre-patriarchal and pre-colonial ancestry. The lack of accessible land-based knowledge, ancestral rootedness to place, and origin stories that form the core umbilicus of who we are as human beings in a living cosmology has been devastating at every level of life experience. Sacred old stories and knowledge have been cast down under patriarchal and colonial conditions.

Consequently, and tragically, white supremacy acts as a misbegotten shield against the mammoth loss of ancestral, Earth-based culture. An unquestioned and unexamined propped-up white identity, along with the promise of the American dream, veils often-disregarded and deep-seated fears of not belonging, of not having an orientation to meaning in one's life, of not having community, and a sense of orphanage from the Earth and cultural practices that bring communion with the land. This malady is not only true for white Americans, but to many people globally caught in the colonial, white supremacist paradigm. Humans yearn to be rooted to home and place, to be entwined with the land and their human and more-than-human kin, to be in Earth-centered community.[27] This fierce and primordial longing is one I have explored extensively in my prior writings through the topic of reconnecting culture with nature because it is such a deep part of the work the dominant culture must engage in.

Some of the most insightful and knowledgeable mentors I have had regarding white supremacy, regarding the Americas specifically, are Indigenous leaders. Several have told me they became experts on European white trauma, behavior, racism, and lack of respect for nature not out of any great sympathy for white pain or to center white people, but because their survival in the face of white supremacy depended on them intimately learning about white European colonial history and power structures. Knowing who these people were and what generated their violent colonial ideology was and is a part of their cultural survival strategy. One Indigenous colleague shared that her Elders had told her that the only sense they could make of the behavior of the colonial settlers who came from Europe—their perverse conduct, selfish behavior, and greed—was that they had a mental illness.

Unraveling, dismantling, and working to heal that mental illness is imperative. White people learning of our own ancestral origins is one part of the endeavor to reach into the deepest tendrils of the illness and injustice as we work to diminish and reverse the effects of conquest, racism, and patriarchy. This kind of work is no easy task, regarding both the substantial research effort and the emotional labor. But if, as we each discover more of our history, we share it with others, we can also share the burden of discovery. This is true also of the emotional pain that we will inevitably encounter in facing the shameful and egregious acts and behaviors in our histories. We don't have to do this alone, nor should we. Together, we also can begin to create both a personal and collective cultural tonic and gain much-needed strength by learning of the older, hidden, and longing-to-be-freed beauty of our pre-patriarchal and pre-colonial lineages when our antecedents were indigenous to place and held cosmologies of a living universe. This is a critical component of the discovery work, as we will need the strength and wisdom of our Earth-based ancestors to see us through.

Ancestral work is an internal and external undertaking, through honest, in-depth education. The stories and knowledge of our Earth-honoring ancestors, before their cultures were overtaken by patriarchal and separation-from-nature narratives, offer us essential wisdom and belonging. The point of the inquiry is not to look nostalgically to the past and try to relive what once was, nor to lament that contemporary life is hopelessly lacking, nor to hang our heads forever in shame or grief from a violent or forgotten past. Rather, we need to revive and embody knowledge of when we were place-based peoples, and begin to remake Earth-centered communities today in new ecologies as we take responsibility for our reconciliation and reparations work.

It is important to note that if we have not lived in our ancestral lands and our traditional cultures for generations, we are not going to magically become indigenous to those places just from learning about our ancestry. We can instead embrace the complex reality of learning about and embodying our ancestral roots as a way of personal and cultural healing, which also can help us respect where we currently dwell. Even the attempt to retrieve some of our lost lineages will put us in a better internal standing to respect the Indigenous Peoples on whose stolen lands we are settler guests. We will not be hungry ghosts needing to devour because we are empty inside. Nor will we hang on to the veneer of white identity or supremacy. No matter how far back we need to search to find decolonized lineages who honored the Earth and the feminine, this work is vital because these offerings from our ancestors are entirely needed today. Learning suppressed or forgotten histories as best we can is part of decolonization—we must reclaim our stories and identities and rescue them from patriarchal and colonial narratives. And, as our ancestors taught, the stories are meant to always be in motion and ever evolving, that is what keeps them alive and how they speak to our core of being in the present. While the bones of a story are shared precisely generation after generation, we are forever to dress the bones anew to meet our particular times and ecologies in an ever-evolving, relational world.

Chapter 8:

Worldviews of Our Ancestral Lineages

THE REINTEGRATION OF OUR ANCESTRY and cultural identities may involve not only learning social and political histories, languages, geographies, and ecologies but also entering mythical dimensions that should not be devalued. Specifically, they can be understood as other ways of knowing. They arise from collecting the threads of our ancestral origins, even as they may come to us in our dreams or intuitions—for the stories are in our bodies still, held fiercely and tenderly deep in the marrow of our bones. The stories also reside in the landscapes where our ancestors walked and in the food traditions, dances, customs, and songs lingering just at the edges of colonial spaces. These various assemblages of knowledge carry the magic of the old traditions because that is where our ancestors hid them when patriarchal, monotheistic elites forced them to relinquish their Earth-centric rituals and ceremonies. The Earth-based wisdoms are alive in the Goddess figurines lovingly hand-crafted long ago, in folk tales, in cave paintings, in antique songs and texts, in old family stories, in the dream world, and in the rivers, forests, mountains, and skies wherever we may live. As we rekindle these knowledges and bring them to light, they can help re-harmonize us and our various communities. Like a sprouting ancestral seed, if approached with respect, alert thinking, intuition, and love, these sources of information can guide us to healthy relationships with the land and each other.

Ancestral work is powerful in that it can address the roots of both patriarchy and racism. It's as if our long-ago ancestors hold out their arms to us, fastening us back to our Mother Earth–umbilicus, to each other and the entire web of life. This is the remembering and reintegration of a land-based identity that existed before the adoption of deleterious social constructs, and it wholly needs to be liberated from the vicissitudes of a patriarchal and colonized history.

Were the dominant culture's pre-colonized ancestors perfect people? Did they live in a utopia, free of human foibles and community struggles, and live socially and ecologically ethical in every way? Of course not. But we can surely learn from their land-based knowledge and expand our horizons by leaning into our long-ago indigenous roots, and in so doing, learn more deeply how to re-embrace an animate and relational worldview.

Finding the story filaments of my own rooted heritage is an ongoing odyssey still far from complete, with threadbare parts, given suppression and conquests. I am an American with three grandparents from Ukraine, one grandfather from Poland. All my grandparents from Ukraine departed as toddlers, their parents fleeing the Russian-instigated anti-Jewish pogroms in the early 1900s. They carried many sorrows and traumas with them as they sailed away from their homelands, their other family members, friends, languages, and communities, forced by antisemitism to seek refuge in a new land.

This is how my family found themselves among yet another wave of settlers and immigrants, in what we call the United States of America and Canada, and what my Indigenous friends have taught me to call Turtle Island. My great-grandparents immigrated to Canada and the US in fear for their lives, and to seek safety for themselves and their young children.

When they arrived on the shores of Turtle Island, the gatekeepers of these new lands were not the original Indigenous Peoples. There was no consciousness or acknowledgment of whose original land it was by my family, or of the impact of their arrival as part of a five-hundred-year history of persecution of Native and First Nations Peoples who were experiencing genocidal violence like what my family had escaped. As far as I can tell from conversations with my grandparents, they exhibited no curiosity about Indigenous Peoples. Rather, they endeavored to partake in settler-colonial entitlement, which was fully embedded in the culture of their new home. Their attention was fully focused on antisemitism, which was a continued threat to their well-being, further heightened by the Holocaust in Europe, as they strove to assimilate into mainstream society.

Embarking upon the journey to learn about both my more recent and my long-ago ancestors from Ukraine and the Eurasian steppes, as well as lineage trails to ancient Hellenic Greece and the Near and Middle East, has been both a challenging and healing experience. There are no neat and tidy through lines here. From my own research and from stories my father told me, I learned my ancestors were of various heritages in Ukraine and other regions many generations back.

Every part of our ancestral story from ancient times to modernity is of import—a braided cord of histories, lands, and peoples we carry with us. Of keen interest for me is the healing that comes from restoring and renewing Earth-centered roots through pre-patriarchal and pre-colonial explorations. To offer some encouragement, I share a taste of some initial findings.

From ancient times, I have begun to track my lineages of both agrarian and nomadic peoples when they were immersed in Goddess traditions celebrating the wild, fecund Earth. Through the research of Marija Gimbutas, mentioned earlier, as well as other scholars, I have learned about Earth-honoring matriarchs who were healers and wise seers, whose ceremonies bound them to the cycles of nature, the moon, the waters, and the land. In this search, I learned that in the Cucuteni–Trypillia culture, which flourished from 5500 to 2750 BCE in the countries known today as Moldova, Romania, and Ukraine, thousands of female icons were discovered. One Cucuteni–Trypillian artifact is of a circle of ceramic female figurines, each uniquely ornamented, and depicted as assembled in council seated on thrones. According to archeologists, this artifact, found in a ceremonial container, indicates that this ancient community revered a Mother Goddess, and looked to women as leaders. Women rode horses, led their communities, conjured ceremonies, and lived in one of the most sophisticated, egalitarian urban centers of Old Europe.[1] Learning this, I wasn't only intrigued, I felt pieces of myself being returned to me, somehow making me more whole.

Along another ancestral trail, I found the Seated Woman of Çatalhöyük, shaped from reddish baked clay, which was uncovered in a large Neolithic village in Anatolia, near present-day Konya. Like most of the Goddesses discovered in Çatalhöyük, formed of marble, limestone, basalt, alabaster, schist, calcite, and baked clay, the Seated Woman is an aspect of the Mother Goddess. She sits, nude, on a throne with feline armrests; her body is round, with luxurious thighs and full breasts. Along with figurines, Çatalhöyük contains vibrant murals depicting the Goddess in three ways: a maiden, a mother giving birth, and a mature woman. The murals also show males with prominent phalluses, local animals, and hunting parties. The sacred feminine and the sacred masculine are both clearly revered, and from burial sites and buildings found at the area, researchers have deduced that Çatalhöyük was an egalitarian society in terms of both gender and class.[2]

One research pathway that particularly beguiled me is that of the Deer Goddess. Some of my ancestors venerated the deer who provided them with abundant food, clothing, shelter, and life as noted by art historian and scholar Esther Jacobson, who pursued the sacred deer tradition across southern Siberia, Central Asia, and southern Ukraine.[3]

The sacred Deer Mother presided over the cycles of nature, such as the winter solstice, which marks the return of the sun when the days gradually become longer, bringing spring and new life. The night of the winter solstice was called "Mother Night," since that was when the Deer Mother cultivated the seeds, which were sleeping deep under the chilly soil, so that they could grow and flourish when spring came, providing life-giving food for humans and animals.[4] In the Baltic regions of Lithuania and Latvia, this gift-giver is Saule, the Goddess of light and the sun. On the winter solstice, Saule flies through the night on a sleigh drawn by a team of antlered reindeer, catching her tears in a golden cup. The tears turn into amber, and she casts these sun-like nuggets of amber down to the people on Earth, along with apples.[5] Of significance is the fact that male reindeer lose their antlers in winter while female reindeer keep theirs. Thus, it is the Deer Mother who carries the spring sun through the sky upon her magnificent crown of antlers, flying over the Earth and blessing her children. In calling forth this story, in my dreams I open my hands to catch some of these Deer Mother amber tears and wear them now as a necklace to remember my ancestors.

In further seeking how to approach this ongoing heritage quest, I learned from Great-Grandmother Mary Lyons, Ojibwe Elder from Minnesota, "Know why your ancestors left their homelands if you are not from the lands you stand on. Honor them, sing their history." Maybe those of us who are not living in our places of origin can find something we are longing for by remembering our long-ago ancestors and singing their stories that have been silenced for too long. This ancestral singing out is to regenerate a vibrant memory of who we are as people in a story initiated ages ago. A story that weaves us into multiple realities, worldviews, and ways of knowing that can provide a foundation for a reimagined present and future.

Hearing the Goddesses' names started my awakening. The vocalization of the ancient celebrated names stirred reverberations, long-lost memories flooding my body. How can sounds, or even words that are read and whispered in the mind, that perhaps you have never knowingly heard before, be suddenly recollected, or give you the impression of remembering something you had known long ago? This is another kind of memory, a different way of knowing. This happened to me when I first heard the names of several primordial Goddesses spoken aloud. Their epithets were incantations encoded with knowledge. Inanna. Nana Buluku. Spider Woman. Demeter. Tonantzin. Kuan Yin. Hestia. Lakshmi. Ostara. Hutash. They pulled at me from within, fully embracing every cell in my body, beseeching me to remember.

The Goddesses' names were conveyed to me, one by one, with a long pause after each, by a poet who visited my university in Oregon. The experience occurred at the same time as a host of women scholars, academics, and authors were working to reclaim suppressed knowledge of when the sacred feminine was once universally venerated. Paula Gunn Allen, Audre Lorde, Carol Christ, Mary Daly, Susan Griffin, Barbara Walker, Riane Eisler, Merlin Stone, Alice Walker, Mary Helen Washington, Carolyn Merchant, and so many others were bridging the feminist movement with the living Earth, and the rise of patriarchy with the downfall of matriarchies and the Mother Goddesses.

I heard the poet, whose name unfortunately I do not recall, speak at a symposium featuring feminist scholars, held in a hall not far from the university campus. The venue was filled over capacity, standing room only, with mainly young women like myself desperate for sense-making in modern society, and eager to hear what these leaders breaking new historical ground had to offer. Their discourse was completely radical in comparison to our required humanities course in which we were studying male histories, male historians, male philosophers, and male gods. The humanities class was meant to prepare our young minds to understand how people in different historical contexts dealt with foundational inquiries about human existence. As was to be expected, we only studied the history of the dominant culture and its civilizations—there was no gendered or racial inclusion in the pedagogy, nor any discussion about human relationships to nature.

The event was organized by several of the female professors at the college, who decided to venture past the traditional curriculum to organize an unaccredited symposium on women's spirituality and feminist history. The scholars and movements emerging at the time revealed a previously suppressed and rich landscape of an extensive history when women and the sacred Feminine Principle were culturally esteemed. This is when I began to learn about not only the forsaken names of many Goddesses but also the purposely eradicated histories of cultures who honored women and the Earth.

The symposium audience was fully silent and spellbound as the poet with her bell-toned voice called out: Mudungkali. Mawu. Ceridwen. Danu. Saraswati. Nut. White Buffalo Woman. Coatlicue. Aja. Eingana. Hekate. As I heard their names recited, the sacred roll call exploded like a storm in the core of my being, and just as we momentarily see features in the landscape when lightning strikes at night, I experienced flashes of memory as the Goddesses' names lit up the darkness. Each lightning strike seemed to give me a larger sense of what had been lost, what was now imploring to be remembered, and I longed to draw those memories into the full light of day.

Unsurprisingly, since that first occasion, as I have researched the stories of these Goddesses, I have seen how their narratives and meanings were captured and colonized by the dominant culture that summarily diminished, twisted, and partially erased them. As colleagues have done along the way, I have tried to free my understanding of these deities as much as possible from later religious, patriarchal, and racist interpretations—which is an ongoing process. These Goddesses originate far back beyond recorded history, and are the cosmological forces of life itself. When we listen with our full selves, most especially when in the wilds of nature, to their antique names and the surviving strands from once-intact cultural tapestries, we have the possibility to learn what our forebears understood about our human relationship to the world around us. The old knowledge, coupled with historical texts and artifacts, exists in the realm of all time, and is thus as alive as ever, in forests, mountains, starlit skies, creeks, and caves only waiting for us to embrace, learn, and renew. Here are just a few threads to begin reweaving these cultural tapestries.

The ancient Sumerian Goddess Inanna was known for gifting the city of Uruk with knowledge and culture. The high priestess Enheduanna, who lived in the twenty-third century BCE, was the first poet in history whose name was recorded, and her devotion to Inanna manifested in volumes of poetry, praising the Goddess in the Sumerian pantheon as the Queen of Heaven. Enheduanna thus described Inanna in her poem "The Exaltation of Inanna":

> Queen of all given powers
> Unveiled clear light
> Unfailing woman wearing brilliance
> Cherished in heaven and earth
> Chosen, sanctified in heaven, you
> Grand in your adornments
> Crowned with your beloved goodness
> Rightfully you are High Priestess
> Your hands seize the seven fixed powers
> My queen of fundamental forces
> Guardian of essential cosmic sources
> You lift up the elements
> Bind them to your hands
> Gather in powers
> Press them to your breast
> Vicious dragon you spew
> Venom poisons the land
> Like the storm god you howl
> Grain wilts on the ground
> Swollen flood rushing down the mountain
> You are Inanna
> Supreme in heaven and earth.[6]

Perhaps, in reading those words, the raw power of Inanna grips you as it does me. Inanna was honored for her fierceness as well as her benevolence because these were not seen as opposing forces. Enheduanna describes the

Goddess as a vicious dragon who also howls like a raging storm—imagery that would have been seen as decidedly negative for female deities in later times. She is life-giver, healer, warrioress, and divine.

Other stories told of Inanna include her journey to the underworld, her sacred marriage with Dumuzi (a vegetation God), which was re-enacted by the king and high priestess during festival times to enhance the fertility of the land, and a creation account involving a Huluppu tree. Inanna was revered in numerous shrines and temples throughout Mesopotamia, though her main temple was at Uruk. Around the time of Hammurabi's reign (circa 1810–1750 BCE), Goddesses began to be replaced by Gods (for example, the Goddess Nisaba became the God Nabu), women's rights and position in society deteriorated (ancient Sumerian men and women had enjoyed relative equality), and Inanna's stories and image underwent many transformations.[7] She continued to be revered by the Assyrians as the Goddess Ishtar. Inanna was also identified with the Hittite Goddess Sauska, the Phoenician Goddess Astarte, the Greek Goddess Aphrodite, and the Roman Goddess Venus.[8]

In his article "Inana and Šukaletuda: A Sumerian Astral Myth," Jeffrey L. Cooley recounts a story in which Inanna appears as the planet Venus.[9] Šukaletuda seeks Inanna in the sky. She stretches herself across the heavens at one point, and she rises from the underworld and "sets" in the mountains. Inanna's movements in much of the story correspond with the specific course of the planet Venus. Here it is important to understand that the Goddesses not only were (and are) associated with symbols of Nature, but they *were* actual aspects of Nature; Goddesses were and are Earth, Sun, River, Desert, Forest, Flower, Deer, and in this instance, Inanna was and is Venus. She is the Morning and Evening Star, and the stories of Inanna reflect her traverses across the heavens. This understanding is crucial to a cultural shift as we regenerate an animate relationship to our living planet and Earth-based worldviews. Venus is not some distant dead-matter rock in the sky, but a living relative, a revered part of our cosmological constellation, disappearing into the underworld for most of the night, yet wondrously appearing in our dawn and twilight skies, glimmering majestically above both the eastern and western horizons.

One of the most influential Goddesses in West Africa and among Afro-Caribbean peoples is Nana Buluku, celebrated by the Fon and Ewe Peoples in Benin and Togo, and by surrounding communities.[10] She is also known as Nana Buku, Nanan-bouclou, Nana Bùrúkù, and Olisabuluwa. She is the mother of all deities and is said to have created the universe and birthed the Sun and the Moon.[11] In the state of Ketu in Benin, an ileeshin staff, made of raffia, cowrie shells, and leather is used to honor Nana Buluku. The staff is pure Àjẹ́, which is a powerful feminine energy of creation, destruction, healing, and many other forces. Nana Buluku continues to be a beloved deity among Afro-Caribbean practitioners of Candomblé as the head of the Vodoun pantheon.

Spider Woman, or Spider Grandmother, is an Earth Goddess honored by many Indigenous Peoples on Turtle Island, including the Hopi, Diné, and Zuni. She embodies wisdom and industry and is credited with teaching her people how to weave cloth on a loom as a spider weaves a web. She can appear as a mature woman or as a spider, and she has a spider's home underground; in the Diné tradition, however, she lives on a red sandstone formation called Spider Rock in Canyon de Chelly. Her involvement in the creation of the world and of humans differs between communities. In Hopi tradition, Spider Grandmother thought the world into existence as she was weaving a web. Then she sang and sang, along with the Sun God Tawa, until the humans they had molded from clay came to life. In other narratives, Spider Grandmother is a guide, leading humans through four worlds, spinning a web for them to stand on when, at one point, a flood threatens to sweep them away. Spider Woman creates medicines for her children and is an endless source of sage advice and practical teaching.

There is also the Teotihuacan Spider Woman, or the Great Goddess of Teotihuacan, a pre-Columbian Goddess of the Earth. She wears a nose piece from which extend curving arachnid fangs, and spiders appear behind her and on her clothing in many murals. She often holds a shield adorned with spider webs, showing her as a weaver of the web of life. She also is depicted in a series of murals in present-day Mexico, wearing a headdress featuring a jaguar's face and an owl. The jaguar reinforces her influence and power, as a large, majestic animal relative, who was and is one of the most important animals in Mesoamerican mythology. Jaguars can also swim, demonstrating the Goddess's relationship to water, and in some depictions water flows from her lower body. The owl symbolizes her connection to the night world.

Demeter is best known as the Greek Goddess of agriculture and the Great Grain Mother. She is the mother of Persephone, whose abduction by Hades caused Demeter, in her grief, to leave the Earth barren. This is a later version of an earlier layered story of the seasonal cycles and plants, and the movement from spring/summer to autumn/winter—and it is this annual loss of her daughter to the underworld that causes the crops to die at the autumn/winter cycle.[12] Demeter's name is likely derived from *gê mêtêr,* or "Mother Earth," and she gives fertility to the soil and is often pictured holding corn or wearing corn in her headdress.[13]

The Eleusinian Mysteries were secret rites, including several in which initiates honored Demeter and Persephone, evolving from earlier agrarian spiritual practices. Little is known about them because most initiates respected their secrecy, but the story of Persephone's journey to the underworld, and Demeter's quest to recover her, were a part of seasonal teachings and ceremonies. In 379 CE, the Christian Roman emperor Theodosius I extinguished a number of pagan sites, including the Telesterion, where Demeter was honored and where the Eleusinian Mysteries were celebrated.[14]

Demeter was revered in sacred structures, but also in sacred groves. In one illuminating story, King Erysichthon orders his men to cut down Demeter's grove, particularly the central oak tree, which is adorned with wreaths that serve as offerings for the supplications the Goddess has granted. His men refuse to touch the holy oak, so Erysichthon grabs an axe to commit the crime himself, killing a dryad (a tree spirit) who curses him with her dying breath. At the first axe blow, Demeter appears in all her divine wrath and majesty. She graciously dismisses the terrified servants but curses Erysichthon with perpetual hunger. After that, no matter how much he ate, he was always hungry; in the end, Erysichthon consumed his own body.[15] Not only does this story warn us about the danger of destroying sacred sites, but it also contains a grisly portrayal of human hubris and taking from nature without respect in the all-consuming hunger of the king.

Tonantzin, or Tonantsi, was a Mother Goddess revered by the Huastecan Nahua Indians in east central Mexico, descendants of the Aztec and Toltec civilizations. In Nahuatl, her name means "our Sacred Mother," and she has many divine facets as the land, animals, seeds, and soil. She is the provider of food, of her people's livelihood. The Earth is not simply dirt for humans to manipulate, but an active provider of sustenance and shelter.[16]

Today, there is syncretism of the pre-Columbian spiritual understandings of Tonantzin with the more recent Catholic Virgin of Guadalupe. As the story goes, in 1531 CE, the golden dark-skinned Virgin appeared to a newly Christian villager on the hill of Tepeyac near Mexico City and asked him, in Nahuatl, to convince the friar to build a church on the site where she stood. The Virgin of Guadalupe told the villager to fill his garment with roses as proof to the friar that he had seen her. When the villager opened his garment to the friar, the image of the Virgin was imprinted on it. The Virgin of Guadalupe became the patron saint of Mexico. Another interpretation of the story exists: The apparition was actually Earth Mother Tonantzin, who had been celebrated in that very spot for centuries, showing her people how they could continue to honor her without being persecuted by the Spaniards.[17] Her seasonal rite, Tlakatelilis, takes place around the winter solstice and involves complex ritual dances, ceremonial items made from tropical marigolds, and offerings of chickens. Aside from the presence of crosses and images of Mary and Joseph, the ceremony bears little resemblance to European Christian festivals, demonstrating that some of the ancient cultural threads and elements survive and thrive today from pre-Columbian times, and Tonantzin still lives among the people.

Kuan Yin, also called Guanyin, Kannon, Avalokitesvara, and several other names depending on the region, is venerated by many peoples all throughout East and Southeast Asia. Her name means "She Who Hears the Cries of the World" because she knows when humans are in distress, and in her dedicated love and compassion for her people, seeks to relieve them—she is the embodiment of mercy and benevolence. In her portrayal as Avalokitesvara, she was originally seen as male by Indian Buddhists, who believed only males could attain enlightenment (part of the arc of patriarchy in this corner of the world). Chinese Buddhists later began presenting Kuan Yin as male, female, or androgynous, since the *Lotus Sutra* explained that bodhisattvas are not limited to a single form; the shift also implied that women could attain enlightenment.[18] Kuan Yin is one of the most beloved figures in Buddhist art and is often portrayed as an Asian woman wearing a white dress or robe. Several images show her with a willow tree branch in one hand and a delicate vase in the other—it is said that to be truly compassionate, one must be flexible like the willow, which can bend far in the greatest of storms and then bounce back without breaking. The vase holds the nectar of compassion that the Goddess sprinkles upon her worldly children. In other artwork, she appears as an armed warrioress because responding to the cries of suffering humans may mean fighting injustice.[19]

Sharing these stories is a way of passing along the inspiration I received from the poet at that symposium: we need to conjure the forsaken names and presences of the Goddesses of our ancestors. We need to bring them into the present moment where they can help us renew a respectful relationship with an animate Earth and cosmology.

As the poet continued her oration, I sat upright in my chair, exhilarated. As mentioned, until that point, I had primarily heard stories of male Gods and male spiritual leaders, fully prominent and ubiquitous in historical textbooks and religious writings. While I had learned about the Goddess Athena in school, it was only stories from the classical era, nothing of her Earth-centered origins; other Greek female divinities were often presented only as consorts of the male Gods. They were not figureheads, nor the main actors of the stories. Their leadership or spiritual teachings were not quoted, and their rites were not studied. And we rarely if ever heard about any deities from non-white, non-Western traditions.

In the patriarchal order, not only are the Goddesses and women catastrophically forsaken but so is Nature. The Tree of Life was uprooted and redefined as the tree of forbidden fruit; gleaning knowledge from nature became a grave sin. The snake was once a Minoan Goddess who protected the household and brought healing; in Egyptian cosmology, she purified water. With Western religions, the snake became the form the devil took to trick Eve into eating from the Tree of Knowledge of Good and Evil. These and other aspects of nature, sacred to Earth-honoring peoples, were denigrated. The story of a dangerous and evil natural world also has paved the way for exploitation and extractivism of the land since the sacredness of life has now been relegated to malevolence.

In contrast to this worldview of domination, misogyny, and exploitation, it is important to highlight the egalitarian nature of earlier societies that expressed a balance of female and male power, as reflected in the balance between the Great Mother and Great Father who created the world. Under patriarchal influence, the Jewish God Jehovah, a name which was originally pronounced "Ho-Hi" and contained the Hebrew masculine and feminine pronouns, became a strictly male God, as revealed by antiquities scholar Michaelangelo Lanci.[20] In cosmologies around the world, many female and male deities were always portrayed together equally: Freya and Freyr in Norse mythology, as well as Hjuki and Bil, whom children would later sing about as Jack and Jill; Hellen and Helios in Greek mythology; Belle and Balder in Western Europe; Izanami and Izanagi in Japan; Isis and Osiris in ancient Egypt; Nu Kua and Fu Hsi in ancient China; and Al-lat and Al-lah in Arabia. Indigenous Peoples the world over today continue to honor both Mother Earth and Father Sky.

As we work to renew our world, it is helpful to realize that while the Abrahamic religions came into being around the ninth century BCE, the Goddess traditions, and the recognition of Mother Earth as sacred, date back to at least 25,000 BCE according to archaeological findings. And Indigenous Peoples around the world have much older dates for their Earth-centered, egalitarian traditions, going back tens of thousands of years. Thus, patriarchal religions are relatively new in the larger arc of the human story, as various scholars have noted. They need to be put into proper perspective as a profoundly unfortunate and errant turn by the dominant culture that has driven a dangerous course—yet a course that can be uprooted and realigned.

This is why so many thought leaders from diverse fields, traditions, and cultures the world over are collectively saying that we must restore the balance in humanity by uplifting the Feminine Principle; if the Masculine Principle were oppressed, we would uplift the masculine to restore balance. Some people may see the imbalance of feminine and masculine as a trivial issue and claim that it does not matter that the Feminine Principle no longer appears in dominant creation stories or that it does not matter that women and Two Spirit people and all the many genders have been written out of our histories. Yet these cultural, historical, and religious stories keenly shape our realities in the political, familial, personal, and spiritual realms.

Here, I also want to acknowledge the deeply necessary and healing work to renew and reclaim our sacred Gods. While we have been violated and oppressed by a domineering male pantheon for far too long, these once majestic, Earth-centered, and beneficent Gods have also been uprooted from their original cosmologies and overtaken by the ills and violence of patriarchy. It is also time to bring back balance and beauty to the sacred masculine as we uplift the Feminine Principle. Imagine the healing our world will encounter when our full relationship to our living Earth and all the histories and knowledge of who we are as humanity are retrieved and respected in mainstream society—a transformation I see happening slowly but steadily, moving from the outer rim toward the center of daily life. Imagine the healing our world will experience with a reclaimed and renewed intimacy with the Earth as the one who gives us life and continues to sustain us. The late cultural historian and religious scholar Thomas Berry calls on humanity to restore our connection with Mother Earth, emphasizing the impact this would have on society:

> We might now recover our sense of the maternal aspect of the universe in the symbol of the Great Mother, especially in the Earth as that maternal principle out of which we are born and by which we are sustained. Once this symbol is recovered the dominion of the patriarchal principle that has brought such aggressive attitudes into our activities will be mitigated. If this is achieved then our relationship with the natural world would undergo one of its most radical readjustments since the origins of our civilization in classical antiquity.[21]

Just like the moon-and-corn-carrying figurine I was so kindly gifted in Oaxaca, the feminine power of life speaks through the millennia because this is the life force of Nature herself. Within this calling, I hear my long-ago ancestors, from a time before they were colonized and the women violated, from a time when the land was held as sacred. I invite these beautiful and erudite voices and stories to return so that my ancestors know I have not forgotten them—they and their mystical wisdom have waited unacknowledged in the spirit world for a very long time.

I will continue to trace strands of my ancestral story back to a time when Mother Earth was exalted by my various lineages, to when my nomadic mothers on the Eurasian steppes cut the grass sod tapestry on which they birthed their children, then dried the sod and rolled it into a tight bundle to be carried by those children so they would always have a relationship to the very land where they were born. Even while traveling on their long nomadic migrations, each carried their bundled-up birthplace, the exact spot where they first touched Mother Earth.

As I hold the clay moon-and-corn icon from Oaxaca, given to me so many years ago, I also think of the Goddess figurine Laussel from France uplifting in her hand a crescent moon horn, marked with the thirteen lunar cycles of the year, and I recall the Scythian Amazon women from ancient times wearing their moon headdresses and carrying crescent-shaped shields adorned with the symbol of the moon huntress. The moon reminds me of forces that are so much greater than human toiling and exploits, and how no one can take the moon away from women as we rise like the silver sphere in our many phases. I am reminded of generations of women before me who have gazed at the same moon as I do today, contemplating the seasonal cycles and our own monthly menses that so intimately connect us to the continuum of life. I think of the next generations, our descendants who will come after us, and how we must leave a better world for them.

The presence of the moon brings me into my body and to a resacralization of my flesh and bones, going back woman after woman, mother after mother, grandmother after grandmother, connecting our bodies to the moon cycles and calling us out from our amnesia into remembrance. We carry ancestral knowledge that connects us to the web of life, relationships that extend far back through time; beyond patriarchal stories, beyond deleterious social constructs, beyond separation from nature narratives. Embraced in silver moonlit skies, we women know we are connected to the cosmos as life-givers—a knowing that we belong to the ongoing sacred systems of life. This relationship is much older and more powerful than any oppression that has been constructed—we women hold the moon in our wombs, a birthright that cannot be colonized. This is a knowing and presencing vital for all of humanity's well-being. This is what I learned when the ancient Goddesses and matriarchs whispered in my ears.

And the poet speaks again: Satana. Djanggawul. Tabiti. Nokami. Asintma. Pachamama. Hina. Baba Yaga. Laima. Mari. Anahit. Mandarava. Anu. Beira. Xi Shi. Amaterasu. Isa. Freyja. Atete. Oshun. Mamlambo. Yemaya. Al-Lat. These Goddesses and the living cosmos they embody are calling us to return home to our ancestral umbilicus that reaches deep into our bone marrow and Mother Earth, and far into the skies forever and ever again.

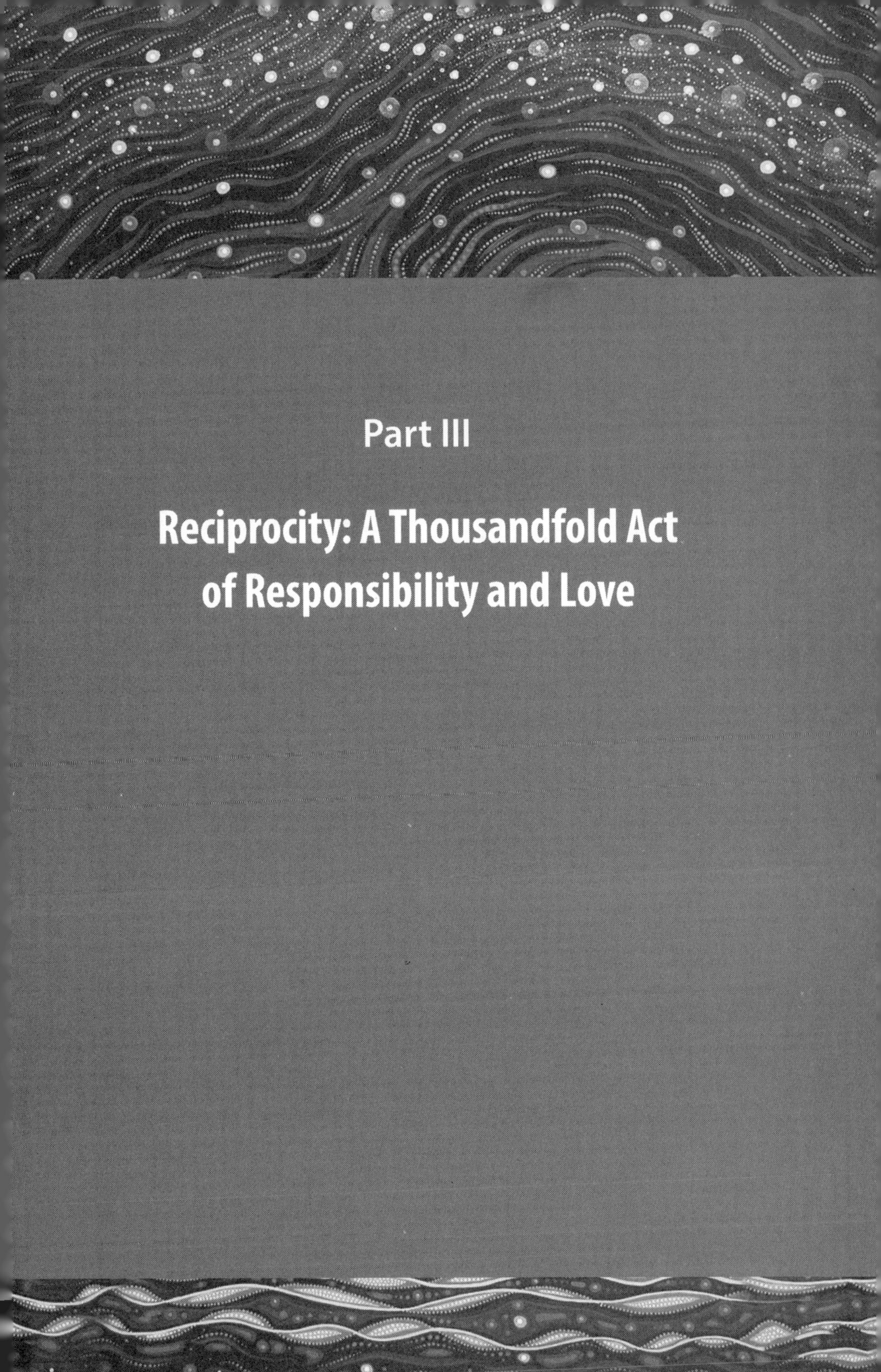

Part III

Reciprocity: A Thousandfold Act of Responsibility and Love

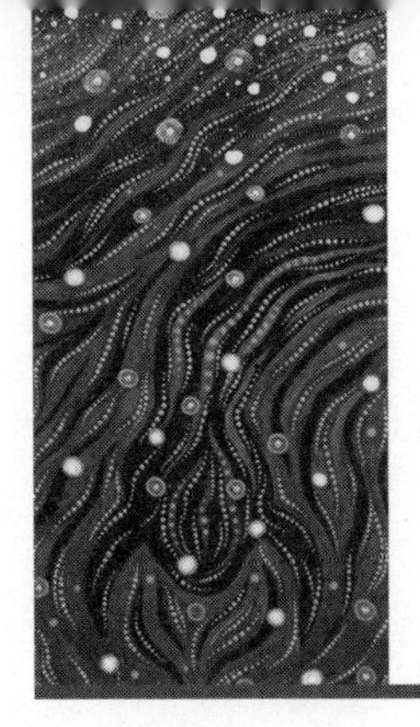

Chapter 9:

Offering and Tending to the Land

In the amplified silence that emerged from the blue-sky day, in the quiet place inside us that experiences events unfolding as we sensed they would, I understood that this was the last time I would ever see Sonia, my ninety-one-year-old grandmother. She seemed to be almost floating in her bed in the in-between-world light that emanates from people who are beginning to make their journey to the spirit world. She lay quietly, her hair full-moon white, her breath falling slowly like winter's first snow.

One last time, I had come to see and love her. Grandmother Sonia and I did not speak for a long while, and I gently stroked her fine skin and hair, marveling at the deep lines and crevices upon her face and hands that reminded me of ancient stone glyphs, each earned mark telling a story.

We often do not know when we are experiencing last times—the last moments with a sweetheart, unaware that we will never again intimately touch their familiar skin that has somehow become a part of us; the last time we sit with a friend who suddenly has only weeks to live because of an aggressive illness; the last time we walk in a forest beloved from our childhood before it is destroyed by fire. Afterward, we wonder if we were fully present, aware, and taking in that last moment. Now, with my grandmother, I tried with all my senses to be fully attentive.

When Sonia's eyes were clear and she was decidedly in her body, I asked her to share with me her favorite childhood story. She was silent for a moment, as if casting her net wide, when she finally said, "I will tell you a childhood story about my own grandparents who lived near Kyiv."

She began without hurry, proudly stating that her grandfather had been a ranking officer in the cavalry and loved horses more than anything else. He was also the mayor of a small village in the countryside outside Kyiv in the mid-1800s.

The images and words of her tale brought to life the country where our ancestors were buried and had become a part of the soil and grasses, where our lineage mothers for generations had birthed babies, where our families had worked the fields whose gifts of Ukrainian wheat, barley, and buckwheat became their bodies, and where our nomadic ancestors of old rode horses across the grasslands. Sonia was no longer simply my grandmother, but a spirit voice echoing the ancient sound of the Eurasian steppes, petitioning to not be forsaken.

Every springtime, she explained, the people of her grandfather's village celebrated the winter thaw and the arrival of the sun's warmth. At this seasonal turn, the villagers welcomed once again the time of planting the fields. A young maiden from the village was chosen each year to symbolize the emergence of the sun's return in lengthening days. Dressed in beautifully stitched garments with floral patterns, she wore a decorated wildflower garland upon her head. She then rode the most handsome horse from the cavalry around the outskirts of the village, and most especially around the newly planted fields. Traditionally, the mayor escorted her, so this esteemed duty was bestowed upon Sonia's grandfather. While musicians sang and people danced and feasted, the lovely maiden circled the fields and the community. She then "planted" an intricately painted egg in the newly sown fields, an offering to the land calling forth a good harvest, an offering built upon generations of elaborate knowledge practices about how to live in a mutually enhancing manner with the land.

When Sonia died a few weeks later, I dreamt of her as a garlanded girl mounted upon a magnificent white horse, riding through layers of clouds shaped like wheat fields high above the Kyiv skies.

The springtime rite of the gift of the painted egg to the grain fields is a simple but significant example of reciprocity with the Earth, part of a long tradition of such customs by place-based peoples. Traditional cultures around the world value reciprocity practices as a quintessential cultural foundation—deeply rooted demonstrations of respect for the land that can take the form of physical offerings, songs and prayers, and extensive tending to the Earth that feeds and nourishes us. These sophisticated customs and practices reflect a very different worldview from simply taking what we want from the Earth or placing a seed in the ground and expecting the Earth to produce food for us. Old-time reciprocity practices reinforce a different approach and mindset, one that demonstrates our interdependent relationship with the land, and the understanding that Nature generously grants us gifts every day. Reciprocity reminds us of the need to respond with proper conduct, healthy ecological practices, and respectful etiquette that shows our appreciation, love, and care for Nature's well-being and magnanimity.

In Earth-respecting cultures, caring for the land and tending to the ecological systems are a way of life, part of an intimate exchange that recognizes our kinship with Mother Earth. Kinship denotes responsibilities and obligations. Just as we care for humans who are related to us, as my grandmother Sonia cared for my mother, and in turn my mother for me, we have a responsibility and duty to care for all animals, the trees and plants, the soil, the waters, and the air. In turn, these relatives literally sustain our lives, teach us, and inspire us. But true reciprocity is not a chore—it is an act of thousandfold love and an expression of respectful relations. My ancestors offered ceremonial eggs to the Earth with an attitude of appreciation and knowledge that an exchange was in order, and I strive to learn from this custom. There is nothing that we can offer to Mother Earth that doesn't already belong to her. So, in part, what makes the egg a sacred offering is what goes into the painting: the devotional and intentional act of our human ingenuity, effort, and artistry, along with our prayer, praise, and song. These reciprocity customs and practices are born out of one's understanding of place and honoring of the profound generosity of the natural world, and how we can participate in this sacred exchange.

Yet with a conscious connection to the land broken for many of us in the dominant culture, we have become orphaned forgetters and have lost our responsibilities and duties to the web of life. Instead of reverently burying gifts of ritual eggs, we bury enormous mounds of toxic garbage and waste in the now desacralized Earth.

Mistreating our life-giving Mother Earth is a relatively recent human phenomenon, and exists primarily in the dominant society. Outside of mainstream norms, there is much to be learned about restoring land reciprocity. Some of the most important environmental and cultural expressions of reciprocal relationships with the Earth are Traditional

Ecological Knowledge systems and stewardship of the land practiced by the world's Indigenous Peoples and land-based local communities. These practices are elegant geospatial protocols, which are mutually beneficial with nature and create lifeways that nourish the entire socio-ecological community. Looking to models of reciprocity with the Earth, both ancient and modern, is an opportunity to shift how we understand and view the land.

Scientists and historians have for centuries described the pre-colonial state of diverse landscapes as "untouched wildernesses," but that is far from the case. There are longstanding practices of Indigenous and place-based peoples shaping and managing the lands where they live—including the ancient forests of Europe, the plains of North America, the great rainforests such as the Amazon and the Congo Basin, and many other regions. These reciprocity practices were, and are, mutually enhancing to humans and nature. As has been shared with me by Indigenous leaders, these practices are an Indigenous science that engages all of our human faculties in a living dynamic landscape. They include listening to the Earth's language of seasons and migrations; studying how to harvest plants, farm, and hunt while also enhancing the ecosystems; and learning to give back to the land ecologically and spiritually. For thousands and thousands of years, Indigenous and land-based peoples attuned to the collective needs of the animals and plants around them as well as their human communities; in turn, the people have been sculpted and transformed by the natural world, creating a seamless, integrated community.

For many years after the colonization of the Hawaiian Islands, the Hāʻena community of Kauaʻi saw the coastal areas of their homeland dominated by commercial fishing operations. These enterprises not only threatened their livelihood but also prevented the Hāʻena from performing their sacred task of caring for the sea, its inhabitants, and their community members. Their care-taking practice is called *kuleana,* a term signifying both "rights" and "responsibilities," and simultaneously refers to the parcel of land a family tends.[1] Hāʻena fisher families also care for a particular part of the shore, developing an elaborate reciprocal relationship with the flora and fauna by offering stone shelters to the fish and respecting their spawning times. They also rotate harvests and strengthen relationships with their neighbors by sharing what they catch, as fishers generally do not enter their neighbors' parcels except by invitation. All governance of the shore is done locally in this protocol—with respect to the ocean waters and the living beings within reflected at every level and with every decision. When a marine species is spawning or experiencing low population, overseers stop its harvest. The overseers, called *konohiki,* also care for the human community, ensuring everyone has enough to eat, checking that boats and nets are in good condition, and coordinating collective fishing techniques that involve large circular nets.[2]

Hā'ena fishers understand that if they care for the ocean, by taking only what they need for themselves and their neighbors and letting the other fish go, the ocean can continue to provide for them—even the sharks chase fish into the nets of conscientious fishers, and the fishers reciprocate by sharing the fish with the sharks. Some Hā'ena families carry the knowledge and recognition that these sharks are the spirits of ancestors, aiding them as nonhuman kin.[3]

As the government began managing the fisheries and Hawai'i became a US state, this sacred reciprocal relationship was disrupted. Commercial fishers saw the open ocean as a vault of silver-scaled commodities rather than a home inhabited by sentient beings. They thought nothing of taking as much as they could carry, depleting fish populations and leaving insufficient food for the Hā'ena and others living on the islands (and of course, over-fishing continues to be a major global crisis). Deprived of their subsistence, and seeing their traditional ways of life disregarded, many Native Hawaiian fishers, at first, joined the commercial fishers.

A few decades ago, however, a large group of Hā'ena and other Hawaiians came together under the banner of *lawai'a pono,* or responsible fishing, to petition for their rights not only to fish the coastal areas under local governance but also to tend to the marine life and habitats as they had for generations past. The Hā'ena had not forgotten that the fish, lobsters, and corals were relatives who needed their care. In 2015, over the protests of commercial fishers, the Hawaiian governor designated the Hā'ena coastline as a Community-Based Subsistence Fishing Area (CBSFA), allowing the Hā'ena to resume many of their reciprocal protocols.[4] Since the time of reclamation, under the traditional stewardship of the Hā'ena, most marine species have once again thrived (with a brief dip due to heavy freshwater flooding and landslides in 2018).[5] The Hā'ena are also able to pass their ancestral knowledge of fish species, reciprocity, and habitat stewardship on to the next generations. During the COVID-19 pandemic, the Hā'ena and others increasingly relied on subsistence fishing for food, and many other Hawaiian communities have taken steps to establish CBSFAs on their coastlines.[6]

Returning governance of stolen land to Indigenous Peoples such as the Hā'ena is an act of justice and reparations; however, it is also a gift to the rest of the world. The web of life thrives when land, waterways, and forests are managed under the stewardship of Indigenous Peoples and local communities who have long-standing land and water knowledge and a commitment to reciprocal lifeways.

The role of humans on Earth is to act as a keystone species, says Diné activist and scholar Lyla June, explaining that "a keystone species is a species that if you take it out, the whole thing unravels."[7] At the moment, most of us are neglecting this role. In many discourses regarding the environment, there is a focus on sustainability, but June asserts this is not enough: "One of the women I'm learning from for my doctoral pre-research is saying—I don't like the word *sustainability*.[8] We're not just going to sustain ourselves; that's a low standard. I'm going for *enhanceability*. The ability to enhance wherever I walk. The ability to make it better than when I found it."[9]

Illustrating our role as a keystone species, scientists can now prove with soil cores that Indigenous stewards of the land from thousands of years ago altered the plant ecology of their environments. A soil core from Kentucky shows a relatively rapid shift from mainly nonfood trees to nut trees and edible plants around three thousand years ago, an example of a collaboration between Mother Earth and humans to produce more food for people and animals, while enriching the land.[10]

The American bison (or buffalo) is another keystone species whose presence has been uniquely and tragically missed in the ecosystem of North America.*

Their population was largely obliterated by colonial settlers who killed them both for sport and to cut Indigenous Peoples off from food sources, effectively starving them off their Traditional Lands. Bison had survived

---

* The terms bison and buffalo are used interchangeably; in North America the scientific name is bison.

ice ages and roamed the grassy prairies of North America for hundreds of thousands of years at least, grazing, turning over the soil, curtailing certain grasses while stimulating the growth of others, and etching out habitats for communities of insects, rodents, burrowing owls, and many others. The animals' thick coats warm them against the winter snow, and their muscular shoulder humps protect their spines as they plow the snow with their massive heads, exposing patches of nourishing grass for themselves and other species.

The bison were essential to maintaining the astonishing biodiversity of the vast plain, and the land has suffered greatly from their absence.[11] Eradicating the bison had the intended effect of harming Indigenous Nations, both materially and spiritually, who had cultivated a mutually beneficial relationship with the bison for millennia. Bison are at the center of ancient cosmologies of several Indigenous Peoples, including the Blackfoot, appearing in symbolism, teaching stories, songs, and ceremonies, and when they disappeared a part of Blackfoot culture died with them. "Imagine what would happen to Christians if all Christian crosses and churches were gone," explains Dr. Leroy Little Bear, a Blackfoot and Kainai Elder, scholar, and professor. "The disappearance of the buffalo had a similar devastating effect on our people. Our youth now hear our buffalo songs, stories, and watch our ceremonies, but they do not see the buffalo roaming around."[12] Dr. Little Bear led a group of Kainai and other Indigenous Peoples living in the US and Canada in a series of "buffalo dialogs," starting in 2009, in which they renewed their relationship with the land through a commitment to bring back the bison. In 2014, the transboundary Buffalo Treaty was signed, a promise that human and bison would "nurture each other culturally and spiritually."[13] Today, in the areas where the bison have been brought back to once again graze and wallow, conservationists have noted significant improvements in the land and its biodiversity.[14]

The Rosebud Sioux Tribe is also working to restore their beloved bison herds. "They looked after us for thousands of years, and right now, it's our turn to look after them," says Wizipan Little Elk, who heads the Tribe's economic arm.[15] On October 22, 2021, sixty bison were released to their new home, where members of the Rosebud Sioux Tribe welcomed them with ceremony and song. As explained by the Lakota Sioux community, they were actually welcoming home their long-lost family members. According to their Origin Story, humans and buffalo emerged together from the Wind Cave in the Black Hills, which is where this group of new bison was raised.[16] As the bison thundered out of their corral and onto the plains, they carried the past with them, and also the future—a renewed path that is healing previously stolen and severed relationships, and one that offers cultural and ecological well-being. As of the spring of 2022, the Rosebud Sioux Tribe oversee more than seven hundred and fifty bison, who are expected to multiply for years to come.[17]

Heading south some six thousand kilometers from the Great Plains in the US to South America, we can learn from Indigenous Peoples who for thousands of years have farmed the breathtakingly biodiverse Amazon rainforest. Their traditional reciprocal practices include encouraging the growth of specific trees, cultivating plants to produce more abundant fruits, and domesticating new crops, such as the Brazil nut, the Amazon tree grape, and the cocoa bean, which thrive in the forest today.[18]

In the Amazonian savannah region called Llanos de Moxos, prehistoric Indigenous communities constructed thousands of raised "forest islands," where they farmed squash, maize, wild chiles, and yuca.[19] Crops such as the peach palm were cultivated to produce much larger fruits than those untended by human hands.[20] Ancient Amazonians also composted their food waste into the soil, adding more nutrients and reusing it rather than cutting down more trees for new farmland.

The present-day diverse communities who inhabit Xingu Indigenous Park in Brazil, including the Wauja, have for generations practiced controlled burning to restore the soil. Today, with fires raging across a less-humid, more-vulnerable Amazon, they have adapted the way they care for the forest by regularly clearing dried vegetation and planting trees in ways that discourage the proliferation of flammable plants.[21] And, it must not be overlooked that, over the past years, many of the fires in the Brazilian Amazon have been set intentionally, and with impunity, by cattle farmers and illegal loggers, under previous President Jair Bolsonaro, whose political rhetoric painted the forest as a lucrative resource for all Brazilians, and disrespected and violated the rights of Indigenous Peoples. To provide a green light for these burnings, he withdrew funding to environmental and firefighting organizations. One major consequence was that land defenders had to risk their lives if they wished to confront loggers and groups setting fires, as military police no longer accompanied them.[22] This meant that Indigenous Peoples' care for the forest involved not only ecological and spiritual reciprocity practices, but a literal fight for their lives and the life of the Amazon. With the election of President Luiz Inácio Lula da Silva in 2022, there is significant hope for these conditions to change in favor of respect for Indigenous Peoples and protection of the Amazon.

In the 1990s, the Ashaninka community of Apiwtxa, who dwell along the Amônia River near the Brazil-Peru border, reclaimed a portion of their homeland that had been deforested and devastated by industry. As a People deeply immersed in an ancestral reciprocal relationship with the Amazon and who also wish to maintain their independence from the capitalist and extractivist world, the Ashaninka rewilded their land. They planted native vegetation and food crops, tended the areas where fish spawn, and fostered endangered species—an effort that in 2017 won them the Equator Prize for sustainability and conservation of biodiversity.[23] The Ashaninka are also engaged in a struggle against logging companies in the region; as just one example of their courage, they once blocked a group of tractors with their bodies.[24] As the sacred Amazon hurtles toward a tipping point due to capitalism and colonialism, the Ashaninka, alongside other Indigenous Peoples and frontline groups, show their reciprocity with the land through life-and-death resistance.

An ocean away and over five millennia ago, Mesolithic-era hunter-gatherers in Europe likewise tended their environment in a mutually enhancing manner; wherever they wandered, they planted hazelnut trees, an enormously nutritious food producer. Peoples of Old Europe roasted the nuts and formed cradles, tools, and homes from the wood.[25] More than any other crop or tree, many Mesolithic European populations depended on the fast-growing hazel, which was also honored as their sacred Tree of Life. Planted generation after generation by local communities, the revered hazels that flourished across the continent nourished not only humans—deer, squirrels, and many other animals ate the nuts, as did the birds who nested in the branches, creating a more abundant ecosystem.

European hunter-gatherers of this time took only a few hours each day to hunt game for the community and gather edible plants.[26] They worked far fewer hours than most people today, allowing more time for relationships, play, ceremony, and learning—and in this sense, healthy time to reflect on the larger frame of existence. They also farmed the extensive forests, sculpting them with specialized tools to cultivate forest gardens. And just as many Indigenous Peoples do today, they used fire to encourage certain trees and plants to grow, to delineate the boundaries between their homesteads and the forest, and to create an enticing area for wild animals to browse for food near their homes. Mesolithic Europeans may have even partially domesticated wild deer in these areas, just as the Sami in Scandinavia do with reindeer. As in other parts of the world, these regions were knowledgably stewarded and not what most call wilderness areas.[27]

Our forebears in regions globally also had to navigate new ways of interacting with their environments due to natural disasters and climatic shifts. Around 12,900 years ago, the global temperature dropped abruptly. This was a period called the Younger Dryas, which lasted over one thousand years. Near Eastern and European hunter-gatherers lost many of their fruit and nut trees as well as the game animals they hunted. It was at this juncture Europeans turned strongly to grain cultivation, some of which likely came from Southwest Asia and some of which were domesticated native grasses and legumes—fast-growing seasonal and predictable foods that they could rely on as the temperatures warmed and the Earth entered the Neolithic era.[28] But our Mesolithic and Neolithic ancestors did not just plant a field with a single crop like many farmers do today—these were lands where various grains and legumes were encouraged to mingle and grow side by side in an imitation of the biodiversity of a wild forest and plain, a system which, as any agroecology farmer today can tell you, is much more resilient in the face of weather changes and disease.

The monocrop system of farming came much later, along with colonialism and feudalism, when these reciprocal land practices were largely jettisoned in favor of a warped sense of efficiency, temporary productivity, and financial gain. Later, rampant capitalism and urbanization further erased the forest gardens of old in most non-Indigenous lands.

What inspires hope today in the midst of destructive industrial agriculture is the existence of extensive movements of reciprocal land practices and their growing momentum. These are taking new form in permaculture, agroecology, urban farming, seed sharing, and many other methods of sustainable, Earth-enhancing farming. For example, the Black farmers who make up the Southeastern African American Farmers' Organic Network (SAAFON) respectfully tend the land using ancestral knowledge—in this case, practices passed down from the farmers' African forebears along with techniques learned from the soil of the American South. The farmers of SAAFON believe in "subsistence as resistance," caring for the Earth through time-honored practices of reciprocity, fostering a deep connection to the land, and uplifting the prosperity, independence, and dignity of Black communities through agriculture.[29]

Chinese American filmmaker John Dennis Liu has brought encouragement to many through his films on environmental healing and remediation. One of his earliest and most inspirational projects was the documentation of the restoration of China's Loess Plateau starting in 1995, which showed the world that it is indeed possible to restore a large ecosystem within a relatively short time. The soil of the region had been degraded and eroded over generations of over-farming, turning a landscape the size of Texas into a desert, filling the Yellow River with sediment, and trapping the local farmers in a cycle of poverty that made it next to impossible for them to renew the soil.[30] Fortunately, in this case, the farmers received funding and support from the government and financial institutions that allowed them to plant native trees that now capture rainwater so that it no longer runs down the slope and into the river. The farmers also terraced the slopes, an ingenious, old-time technique practiced across many parts of East Asia. Today, the local peoples of the Loess Plateau enjoy not only a thriving landscape restored through their care but also food abundance.[31]

Liu brought the lessons learned in China to the world, demonstrating in his teaching and multiple documentaries that the way to combat climate crises is through Earth restoration, which, he asserts, requires a change in worldview: "If the intention of human society is to extract, to manufacture, to buy and sell things, then we are still going to have a lot of problems. But when we generate an understanding that the natural ecological functions that create air, water, food, and energy are vastly more valuable than anything that has ever been produced or bought and sold, or anything that ever will be produced and bought and sold—this is the point where we turn the corner to a consciousness which is much more sustainable."[32]

Terracing is also a feature of China's stunning Yunnan province, where the Indigenous Hani People cultivate red rice using techniques they have perfected over tens of generations. The tops of the hills are reserved for forests, as the trees absorb the fog that travels over the region and transfer it to the villages below, and ultimately to the crops on the terraces. Each village appoints a guardian to make sure their community does not take too much water so that each neighboring village has enough. Farmers sow and harvest by hand or with the help of buffalos, as they have for over a millennium, and continue to enjoy Mother Nature's abundance through their ancestral traditions of treating the land with gentleness, wisdom, and respect.[33]

La Via Campesina, or "the peasant's way," brings small-scale farmers, peasants, Indigenous Peoples, and women agriculturalists from over eighty countries together under the umbrella of climate justice, food sovereignty, dignity for migrants and peasants, and rights to seeds and land. Their work is in part an act of resistance to the aggressively capitalist world in which giant agricultural corporations patent and control seeds. Due to the diversity of locations and land-based peoples represented by La Via Campesina, the specific practices vary widely and reflect the local place-based and ancestral knowledge of each region, but all recognize the need to care for Mother Earth by fostering biodiversity, renewing the soil, protecting waterways, growing crops according to their natural seasons rather than the whims of the market, and respecting Traditional Ecological Knowledge.[34]

Like the farmers of La Via Campesina, we can learn much about our local ecosystem, and what it needs from us, by looking at how it has changed through time under the hands and tools of humans, under the hooves of grazing herds, and through generations of knowledge and sharing of reciprocal relationships. Looking back in ecological time, we see how dramatically environments changed in response to new groups of people or novel human activities. While some human communities destroyed or damaged their natural surroundings, many others enriched them, resulting in longstanding soil health and resilient crops. A common thread, however, has been the destructive onslaught of colonization, which disrupts or destroys the local ecosystem as well as the cultural survival of Indigenous Peoples whose identity is inextricably woven into the land. The very purpose of colonization is to plunder resources. In contrast, Indigenous Peoples past and present, and the movements for small-scale organic farming in its many forms, show us we can live in mutually beneficial balance with the land, demonstrating that humans do not have to spell disaster for the world's ecosystems—we can be a force for good.

Learning about the reciprocal and ecology-enhancing practices of Indigenous and place-based communities has encouraged me to explore how I can live in an honorable way with the land—even if only beginning steps. How can I support my plant and animal relatives in rural and urban environments, especially when so much of the natural world is out of balance? And the larger question is, how can the dominant society relearn our role as a keystone species and protect and enrich water, land, forest, animals, and the entirety of the web of life?

Part of my work at WECAN is an attempt to draw nearer to these goals of fulfilling my responsibilities to the sacred ecological systems of life. In this work, the insightful words of Alice Walker have resonated with me: "Activism is the rent I pay for living on the planet."[35]

One of our efforts is the WECAN Women for Forests program in which we work with women in different regions of the world who are fighting extractive industries and defending forests, as well as reforesting damaged lands. We work particularly in the Ecuadorian and Brazilian Amazon rainforest, the Itombwe rainforest in the Congo Basin, and the Tongass rainforest in Alaska—running campaigns and on-the-ground projects, and amplifying the powerful voices and tremendous efforts of frontline women leaders.[36]

Working in the Itombwe rainforest in the Democratic Republic of Congo (DRC) since 2014 with the extraordinary leadership of WECAN Coordinator Neema Namadamu has been a great honor and an ongoing lesson in understanding multiple expressions of reciprocal relations. Neema is also the founder and director of the Synergy of Congolese Women's Associations (SAFECO) and Maman Shujaa: Hero Women of the Congo, through which she has established a media center for Congolese women to make their voices heard on the range of issues affecting their country. Disabled at the age of two by polio, Neema became a disability activist at a young age, first advocating for the rights of children with disabilities in high school. After completing her university education, she served in parliament before founding an NGO to support disabled individuals who had experienced sexual violence.

Through a wonderful friend and colleague, Cynthia Jurs, I met and invited Neema to present at a WECAN International climate conference in 2013. Our partnership for forest work began during this gathering as we explored the ways in which women's leadership could be developed in both protecting and reforesting the Itombwe rainforest. Soon after the conference, we initiated the WECAN DR Congo Itombwe forest program, with Neema as the local coordinator. Our goals were (and are) to protect over 1.6 million acres of rainforest, first and foremost for local communities, but also to help mitigate the climate crisis. This work involves defending the rights of Indigenous forest-dwelling communities; elevating women to leadership positions; helping women generate income and secure land ownership through tree-planting; practicing and teaching hands-on reforestation; facilitating education for girls, since they would no longer have to walk long distances to collect wood; and revitalizing Traditional Ecological Knowledge and cultural practices of reciprocity toward the forest.

The Itombwe rainforest, part of the vast Congo Basin rainforest, is home to the world's remaining gorillas; the elusive owl-faced monkey, the yellow-backed duiker, so essential for the dispersion of fruit seeds; the sweet-voiced Butembo greenbul; and the Itombwe Massif clawed frog, among other wonders. Though the Congo rainforest is second in size only to the Amazon, deforestation is happening at such a high rate that it could be gone by 2100. Local communities have long been in exchange with the old-growth forest for timber, medicines, and food, harvesting according to reciprocal sustaining practices.

But today, that once-balanced system has been disrupted: poachers covet the Itombwe's animals, and mining companies plunder the rich soil for gold and coltan, a mineral used to manufacture electronics. Loggers chop down the trees for the wood and the charcoal, and to clear pasture for livestock. Wars have broken out over the forest's precious gifts, with the forest communities caught in the middle.[37] In an effort to protect the forests from illegal timber harvesting, the women in our program, with great courage and determination, have established conservation committees in their own villages to patrol and protect the forests.

Around ninety-five percent of the participants in the WECAN DR Congo program are women. Their leadership and contribution to the local food system and forest regeneration has elevated their status in their communities, important in a country that has a high rate of violence against women, as well as widespread poverty. Since trees are being planted closer to home, local women are also safer—collecting harvests from the trees used to mean walking long distances, placing women and girls at risk of molestation. As these women tend the forest, the trees exponentially give back to them with food and medicines as well as bringing rain to the region—and the women take great pride and satisfaction in their roles as forest guardians.

Through the trainings, the women learn about environmental protection laws, how to cultivate tree nurseries and plant trees on damaged land, and how their old-growth rainforest is connected to the global climate as they build community with each other and the land. As Neema said, "The women are so happy to be tree-planting. They value our project as tied to their identity—as an opportunity to demonstrate their ability to work a big diverse project to feed their families and reforest their heartland and protect their ancient forests. They see this project as a symbol of the future they are growing and cultivating."[38]

The WECAN DR Congo nurseries grow more than twenty-five types of trees, which are useful for reforestation, medicine, fuel, and food. The community harvests one quarter of the trees for their own needs, and the rest are nurtured to rewild the landscape. These trees include primarily native tree species to help regenerate the natural forest: cypress, African redwood, grevillea, maracuja, acacia, and mimosa scabrella. The diversity increases soil fertility, and many of these trees are resistant to drought and adapt easily to environmental changes. Importantly, the presence of the new trees means the community will no longer need to cut the old-growth trees, allowing 1.6 million acres of these ancients to continue to sequester carbon and shelter biodiversity.

The project is also helping local women grow gardens near displacement camps. Due to ongoing conflict in the DRC and the surrounding region, there are over five million internally displaced people in the DRC; the country also hosts more than five hundred thousand refugees from neighboring countries.[39] The fighting has cost so many people their livelihood, their homes, and their crops, and these gardens are building much-needed food security in the communities.

The women forest guardians are also enjoying more economic security due to the medicines and food harvested in the new forests, which are on land that they now own. According to the forest code of DRC, if a person or community plants trees or crops on public land, they become the owners of that land, so this program secures land-ownership for women and for Indigenous communities, from whom land had been unjustly taken in the past.[40] This increases their status, as well as their incomes. These Indigenous women are also able to share their Traditional Ecological Knowledge and use it to increase the forest's health and longevity.

Neema Namadamu has spoken at international forums about the progress of the forest program and the challenges and successes of the Indigenous forest peoples. Locally, program participants are also meeting with NGOs, government officials, and military members to discuss their work and call for support and accountability from the state. Intrigued, several officials visited the nurseries and the forest and attended a tree-planting ceremony, and they praised the women's work and their respect for the forest. After the successful ceremony, Neema exclaimed, "The women of WECAN DR Congo are determined to change things in our little world of Itombwe, for ourselves and for the rest of the planet. We are taking our stewardship seriously. We know that the difference we make not only affects our world, but everyone globally. We feel the weight of it all and are doing our part. To our sisters around the world, we say: We are TOGETHER!"[41]

Reciprocity has many facets. My family and I listened to a plea for reciprocity from the often underappreciated pollinators of our world. It was during the time my sister and I were caring for our mother, N'ima, in the last years of her life. She was battling several illnesses, and as I helped to midwife her death, I experienced a paradox of emotions from the excruciatingly painful to the profoundly numinous, as anyone who has been through such experiences will understand.

Before my mother was confined to her bed, N'ima asked me to garden with her, but not in her vegetable garden—she specifically requested we plant a flower garden. From there we decided to design a pollinator garden because we had been learning how the bees, crucial in food production, are disappearing at alarming rates due to diseases, intensive industrial farming practices, habitat loss, excessive use of pesticides, and higher temperatures associated with climate change.

As we researched the flora that would best serve bees, butterflies, and hummingbirds (all pollinators), we imagined how lovely the flowers and herbs would look, but also our deep joy in planting something that would give back to a world in such need of restoration. Many pollinator species are dwindling, and a reliable and nutritious food source, filled with specific and diverse native plants that bloom at different times of the year, gives local hummingbirds, butterflies, moths, and bees the sustenance that may help them survive.

Pollinators themselves exist in a reciprocal relationship. As many of us have learned at a young age, when bees flit from plant to plant, pollen clings to their bodies, shaking loose onto other plants and fertilizing them, causing a flower to drop its petals and grow an ovary. That ovary grows and ripens into a fruit that feeds many animals, including humans, or perhaps a bat, also a pollinator, who will carry those seeds in its belly and deposit them elsewhere, propagating the plant ever further. Pollinator success means that vegetable gardens, farms, forests, and fruit trees will all flourish, and in turn, those plants will nourish animals and humans.

I vividly remember designing the garden with my mother as we sat in her front yard on the outskirts of the small coastal town of Mendocino, California, resting our eyes on the pine trees and listening to the songbirds. We did not speak for a long while because there were so many things we wanted to say as we settled into the knowledge that she was dying. There was no way to express everything that was surfacing from deep within as multilayered streams of emotion and reflection overwhelmed us, so we simply sat, holding hands, and looked at the clouds and the pines in a crystalline moment where our spirits connected, mother to daughter in a long stream of women, daughter to mother to grandmothers and great-grandmothers, on and on through the immense cycles of elliptical time. We remained suspended in this portal to eternity, a light breeze on our cheeks, experiencing the past, present, and future as a continuum of profound love held together by great tenderness and immense grief. Rather than speak of the death that was coming, we chose to enjoy being alive together in that moment, in appreciation of the life we had shared, where quiet knowing is stronger than words. I felt an intense, innate devotion for my mother, which was at once sweet, fierce, and primordial. It was in this moment that the garden design took shape.

N'ima chose a large area in her yard that was rather wild and overgrown with invasive species, and I spent many days pulling unwanted flora and adding compost to prepare the ground for the pollinator garden. Among the twelve or so plants we selected were asters, which bloom around the autumnal equinox, earning them the nickname "Michaelmas daisies." They provide nectar when many other plants have stopped blooming. In Greek, *aster* means "star," referring to their brightly colored starburst, punctuated by a golden center. Tea brewed from the flowers can assuage headaches.

We added nodding onion, a member of the lily family named for its drooping umbrella-shaped bunch of blooms supported on its shepherd's-crook neck. Its oniony scent keeps the deer away as the plant reserves itself for pollinators who can hang upside down to sip its nectar. The plant is also quite vigorous, surviving California's droughts, and is beloved by hummingbirds.

In contrast to the pungent aroma of the nodding onion, the flowers of the mock orange bush exude a delicious fragrance, a blend of orange blossoms and pineapple. The bees love their mass of white blossoms, as I know my mother did. Part of the hydrangea family, mock orange blooms flourish for several weeks each year. The wood of this aromatic plant has been cherished by Indigenous Peoples for fishing rods, arrows, cradles, pipes, and snowshoes, and the leaves can be used for soap.

We planted rosemary, whose white and pink blossoms offer nectar to the earliest bees, butterflies, moths, and hummingbirds. Evergreen rosemary perseveres through our dry climate, diffusing its woody, earthy aroma—a smell that lifts one's mood.

Bee balm, as its name suggests, was a perfect addition to our pollinator garden, its brilliant tubular petals circulating a minty fragrance. It is also called wild bergamot, horsemint, and Oswego tea, as Oswego Indigenous Peoples brew it into a tisane that aids relaxation. Bees and hummingbirds cherish its nectar, and finches carry away the seeds in winter.

My mother was an incredible painter, and I witnessed her artistry at work as she selected the plants for the garden, both to sustain our winged relatives and to "paint" the living landscape with flower hues. We grew some plants from seed and bought others as starts from a local nursery. Far more than I had imagined, springtime brought a stunning canvas of vermillion, purple, golden yellow, ivory white, and fuchsia—vibrant hues that pleasantly overwhelmed the eye as ambrosial fragrances drifted through the air. And yes, the butterflies and the bees and the hummingbirds all flocked to the garden.

What I had not anticipated was that N'ima would become bedridden the same spring the garden fully bloomed, and that the location she had chosen for this splendid mosaic of color, this mandala of reciprocity to the pollinators, would lie directly in view of her large bedroom window. As the time approached for her journey to the stars, she could look at this vibrant life, the gift that would live beyond her. I know the special garden dedicated to our pollinator kin brought her great pleasure. As another facet of reciprocity, when we care and give, we also receive—as so many sages have eloquently expressed throughout the ages. Yet we did not plant the pollinator garden with a view to receive. The return gift of joy came because reciprocity inherently creates ripples in kind.

At the sacred time of my mother's passing, I contemplated the way our human bodies, after our spirits depart, become part of the soil, giving back to the living Earth in the cycle of reciprocity as regeneration begins anew, but I also realized reciprocity means becoming a good ancestor, one who teaches responsibility for our Mother Earth to the next generations. This is reciprocity to the spirit of the world and the spirit of humanity: to remember who we are, to transform destructive behaviors, to mentor the next generation, and to do our part to navigate human injustices and harms to the Earth that we have inherited or in which we are complicit.

There is also the importance of listening to the wisdom of our long-ago ancestors. Before her passing, I painted an egg with traditional patterns and planted it in my mother's garden—and I still carry this practice of offering to the land today.

As I worked in my own garden early one spring, I recalled a dream in which an older woman dressed in leaf and herbal foliage emerged from the redwood forests and said to me, "The trees, the creeks and springs, the mountains, the birds, and all the animals—we miss when the people spoke to us and prayed with us every day. Where are our human relatives? You walk among us, but you do not see us or remember us."

Learning to cultivate a deeply intimate connection with the undomesticated natural world has been essential to my well-being, and this has meant spending time in magical lands—from high mountain peaks, to unpeopled beaches, dense forest thickets, and hidden caves. I spent much of my youth backpacking across the Pacific Northwest, especially in Northern California, in an attempt to forge a fully alive relationship to the wild. I was fortunate to have been introduced to wilderness survival skills when I was a young woman and was able to travel with other like-minded friends into what we called "wilderness areas," which I only later learned were most often lands that had been tended by Indigenous Peoples for millennia.

Wilderness walking was an activity my high school—and later, college—friends and I could partake in at little cost, buying mainly used equipment, and hitchhiking in groups for safety or catching a ride with farmer friends to different trailheads. We were not walking for sport. The journeys were not about conquering the highest mountain or counting the miles; we hiked out of pure enthusiasm to delve deeper into the wonder and beauty of the living mysteries of nature, immersing ourselves in a different world populated by wildflower colonies of foxglove and mariposa lilies, tracking black bear and bobcat footprints, breathing in the sunrise, marveling at the tapestry of cloud patterns, and lingering for hours in the quiet shadows and aromas of ponderosa pines and cedars. Developing a relationship with a place takes time, seasons of burrowing into rhythms of the Earth's reverberations, learning the plants, the trees, the animals, the way the sun reflects off the leaves and needles, the dance the bees perform around the trillium. Listening to the stories rising from every footfall helped me locate myself, including my interior world, in the mythical lands of the California back country.

Far away from the cacophony of cars, phones, and industrial modernity overall, I would walk for days, endeavoring in the motion to find a truer part of myself, untethered to harmful cultural norms and expectations. Between work and school obligations, I strove to stay out on backcountry land as long as possible to immerse myself in the heartbeat of the Mother World, staying there long enough that the wild might have a chance to seep in, past the protective shields and cultural conformities I had devised to survive in mainstream society. As a young woman seeking to find my way in our dominant culture, I found myself unable to squeeze into most social norms; my spirit revolted at every attempt. To be in the wild allowed me to be centered and grounded as a human being on Mother Earth, to be my "secret self" that nobody knew about, the one that was unencumbered—and there also to find a soothing balm for my acute loneliness, or perhaps, it was more a feeling of unbelonging. When I think back on it, I sense I was walking long and hard, wandering in the landscape trying to find my people and my long-lost Earth-centered village life that was somehow elusively remembered as if in a dream, while my companions and I traversed our little backpack team across the northern coast. I wonder also if part of me harkened back to my nomadic ancestors, wayfaring through diverse terrains with the changing seasons, following the stars among grasslands and valleys.

During these treks, I learned about the environmental struggles of the lands on which I had been honored to walk, and felt the responsibility to ensure these lands remain protected from exploitation and development—in this manner, my love of nature evolved into a form of activist reciprocity. It was not until later in life that I realized the necessity of learning about and from the Indigenous Peoples whose lands I traversed, many of whom are endlessly fighting to protect their ancestral territories.

On one occasion, I spent a week alone in a Northern Californian madrone, oak, and fir forest, a place that knew me well as I had been there many times. On that assemblage of gentle rolling hills, I found an open meadow in the forest thicket where I set up camp and quietened inside to become a part of that place as much as possible. I listened to melodic birdsongs and wind gently susurrating high in the treetops, watched the wheeling of the stars at night. For the first two afternoons, I noticed a small group of California mule deer grazing in the distance. It was a doe and three of her young; given their size and the time of year, I inferred that two were twin fawns who had recently lost their protective camouflage spots, and the third was a yearling. The mother periodically looked in my direction, well aware of my presence in her forest. I have had a lifelong inexplicable relationship with deer—they have appeared from their wild dwellings at key junctures, and this was surely one of those times.

The following dusk, the doe appeared with only her yearling. They came closer as I sat as still as possible, the mother looking at me the entire time. I soon noticed that the yearling was not well. He did not appear to be injured, only very weak and frail. The next evening, the two approached me again, and the deer mother with her nose seemed to push the yearling even nearer toward me. I sat quiet as the doe nudged the young one closer and closer, until I could see large growths, like cancerous tumors, all over the little buck's face and flank. The yearling made soft crying noises and appeared to be in pain. They departed as soon as the light left the sky. The next day, late afternoon, they came once again, and the mother deer peered at me as the weak buck stumbled along by her side, appearing to be fully exhausted. The doe once again pushed her little one toward me. I wondered if he had eaten some poison, which would explain why in his weakened state he had not been consumed by a mountain lion or bobcat—those animals may have sensed the toxins. I did not know what caused the deer's ailment, but I understood that before me was a dying creature who was suffering.

A knowing surfaced, one which I had been pushing down and trying to ignore for two days, an intense intuition that the doe wanted me to help her—she wanted me to kill her offspring. I sensed that the deer mother was requesting, in her own way, a mercy killing. I stood slowly to not frighten off the deer and reached for my pistol, which I carried not because of wild creatures—a gunshot would only invite a bear to attack harder, not effectively take it down—but to protect myself from men who prey on lone women hikers. I composed a quick prayer as I lifted the pistol. The yearling looked up at me just then with his beguiling black velvet eyes, stepped closer, and turned sideways, making it impossible for me to miss. The precious creature, still beautiful though bone-thin and covered with masses, glanced at me again and bowed his head. Doubt dissolved in these gestures. I shot the deer.

The thunderous shock of gunfire terrified the doe, and she turned and leaped, running into the forest cover. I slowly approached and sat down next to the little buck and sang to him as he journeyed to the spirit world. I cried and apologized for whatever had poisoned him, as I assumed human hands were responsible. I examined his boil-riddled body and found that his teeth seemed soft—it appeared he had literally been starving to death. I carried him to a tree so his body could become one with his forest home once again. Naturally, I questioned myself over and again wondering if I had done the right thing. Was it my place to take this wild life? I reflected on our human role in the natural world, that we have responsibilities toward the beings in the ecosystem we are a part of, and I truly felt that my deer relative had asked me to help her yearling in a way that she could not—that the harm had come from the human family and had to be resolved by the hands of a human. I had been imagining acts of reciprocity with nature—yet, never had I envisioned this would entail a mercy killing. The responsibility in taking a life, and carrying the anguish and anger for what had caused the yearling to become so ill, has now become a part of my learning that is still unfolding.

The deer mother came one last time the next day, this time at dawn, with her twins. She stood in the distance and gazed at me, and I looked back, hoping she could sense the deep compassion and love I felt for her. It is tempting for humans to project our own feelings onto animals, so I must say, I have no idea what the deer was experiencing. I can only speak for myself that I felt humbled and honored, and very much a part of her world. I was in her forest, in her home, as a visitor, and I truly hoped that I had conducted myself in a manner that the doe had wanted.

When we do not live in relationship and reciprocity with the land, we are inevitably a destructive force to ourselves and the sacred systems of life, as countless lived experiences are showing us today, from the escalating climate crises to massive environmental degradation. Mother Earth is overtly blazoning her warning message in blistering drought-stricken landscapes, raging forest fires, tumultuous flooding rivers, and rapidly melting ice caps. We have been warned from ancient times, through thousands of stories passed down from our diverse lineages, to current times from scientists and lived experience, that we must keep our sacred responsibilities and protocols with Nature.

I am reminded of the old-time story of the Fisher King, whose body and kingdom lie wasting away because he has failed to respect the land. There are many versions of this mythic tale from different regions and peoples, but most recently known across the British Isles and France, with each rendering having facets to learn from. Yet, within the various story threads, in any telling, there is a common warning and lesson: When humans abandon their sentient reciprocal relationship to the land and disrespect the Feminine Principle of life, we end up with a wasteland.

The sacred protocol of olden days entailing that a king must marry the Goddess of the Land (or the Goddess of Sovereignty as she is sometimes called) is rooted in Celtic Europe, requiring that the king must be approved by the Goddess for his rule to be legitimate. In old Irish cosmology, as with other Celtic traditions throughout Ireland, Britain, and France, the Goddess *was* (and *is*) the land—and in the case of this tale, was Ireland herself. All fertility and prosperity came from her, symbolized by the cornucopia the Goddess often carries in the stone carvings found in parts of the United Kingdom, France, and Ireland.[42] The holy union of the leader of humans and Earth Goddess ensures that the kingdom will prosper. One such Goddess, Mebd of Connacht from the Ulster Cycle (a collection of medieval Irish legends), was a ruler and warrior herself. As with the other Earth Goddesses of the region, her sexuality ensured fertility but was also linked to sovereignty, warfare, and healing.[43] Mebd legitimized nine kings through a symbolic sacred marriage. This type of marriage with the land was performed in many parts of the world (including Europe, India, the Near East, and Africa), in a variety of expressions, and through varying symbolisms. In most cases, including in Old Ireland, the marriage was enacted sexually, where the king would consummate the marriage with a representative of the Goddess, often a priestess or, occasionally, the queen.[44] Irish kings would also drink a beverage from a sacred chalice given to him by the Goddess's representative. These acts, and the king's ongoing devotion to the Goddess, would ensure the fertility of the land, the health of the king, and the welfare of the kingdom. If the king fails in this, his people may suffer famine and war, and other extreme hardships.[45]

Clearly the Fisher King is a transition story in which peoples and tribes are already moving further into patriarchal social constructs (with kings and kingdoms), yet the times of magic and the sacred and necessary communion and reciprocity with the land and the Goddess are still strongly understood and practiced by the people.

The tale of the Fisher King comes from the older canon of Arthurian legends, but many scholars, including the nineteenth- and early twentieth-century folklorists Jessie Weston and Alfred Nutt, have pointed out that these stories and the symbolism they contain are much older, springing from the rich soil of pre-Christian Celtic mythology.[46] In the story's later patriarchy-influenced renditions, it describes an old king who is suffering from an injury to his thigh or groin—early versions of this tale made it apparent it was a phallic wound—leaving him unable to procreate. The wound has been inflicted by special powers and refuses to heal. As the king declines, so the lands of his kingdom deteriorate into a wasteland. He spends his time fishing to take his mind off the pain. In some popular versions, part of his healing requires the retrieval of the lost Holy Grail, which his lineage was in charge of protecting.

Here it is important to remember the symbolism of the marriage chalice given to the king by the Goddess—it still exists in this story but has been Christianized. Instead of being a symbol of the sacred union between a ruler and the Goddess of the land, the cup has become associated with the sacrifice of Christ. In the older context, the quest for the Grail is the seeking of the renewal and veneration of the Earth Mother that has been lost. Without this honoring, the land and king will continue to deteriorate.

In my reading, the Fisher King story contains a distinct message: Whoever governs the people must pay homage to the land, to the Goddess, and keep the responsibilities and customs of sacred land agreements and reciprocity in order.

Today we see many governing leaders neglecting respectful relationships with the Earth, as corporations ravage the land and water for resource extraction, while our natural world and many frontline communities wither into wastelands. The story of our time is that those with the most political power have forsaken what should be their highest allegiances—to the sacred systems of life, to Mother Earth, to all living beings, to generations to come.

But this is not the end of the story. Community leaders from around the world, representing many struggles, are standing up stronger than ever and demanding that the sacred relationship with the land be restored. It is far past time for all humans to be life-giving enhancers and pay full homage to the Lady of the Land.

Chapter 10:

Composting the Cultural Toxins of Colonization and Capitalism

I MARVEL AT THE TEXTURE and composition of the soil in my garden. The divine marriage between soil and seed is one that farmers and gardeners witness each spring. I dig my cultivating fork into the moist earth, sprinkle some kitchen compost, and work it in. Key elements of healthy soil include rock minerals decomposed over generations, creating nitrogen, carbon, manganese, and phosphorus; decaying plant and animal parts; the life-giving power of water; air; and the presence of worm and insect kin to maintain an ideal balance.

In order for the vegetables I grow to be healthy plants, I am dependent on the soil to be healthy—and the opposite is of course true. Casey Camp-Horinek of the Ponca Nation in Oklahoma tells me her community cannot grow their own food, even though they live under food apartheid and would like to do so, because the hydraulic fracking on their lands is so extensive that the soil is thoroughly polluted—any foods grown there are too toxic to be eaten safely. This is yet another tragic and heartbreaking demonstration of environmental racism in the US where poisonous extraction activities are foisted upon Indigenous lands and Black, Brown, and low-income white communities.

Soil has been ubiquitously neglected and exhausted through modern agricultural processes such as monoculture; synthetic fertilizers, fungicides, herbicides, and pesticides; depriving Mother Earth of rest between crops; and over-tilling with farm vehicles, which exposes the topsoil to wind erosion. While these activities can increase crop yields in the short term, these soils no longer hold nutrients and vital minerals, meaning these farmlands' long-term ability to grow food is in serious jeopardy.

As we think about regenerating a healthy society that lives in reciprocity with the Earth once again, we need to consider the soil in which the seeds of regeneration and reciprocity are being planted. What is the composition of our cultural soil?

Our current interlocking socio-ecological crises demand systemic transformation, and for this, healthy cultural soil is essential; otherwise, we will continue to grow non-nutritious, colonial, patriarchal, ecologically degrading toxic harvests. Consequently, we need to work the cultural loam to bring more awareness to unacknowledged harmful assumptions and practices, or in other cases, to directly confront and change detrimental ideologies and policies that continue to drive dangerous beliefs and fears that lie at the foundation of our societal ills. It is time to gather our excavation tools—not those of extraction, but those of inquiry, exploration, compassionate truth-telling, healing, and action.

We can readily see, through historical and scientific evidence and lived experiences, that burying personal or cultural traumas, or furthering erasures of injustices and never speaking of them, only continues patterns of abuse and oppression. To avoid discussing and understanding the horrors of slavery, witch burnings, land grabs, white supremacy, and genocide in the name of "moving on" and "uniting" is a false approach that we cannot afford. Instead, to bring about real reckonings, reparations, reimaginings, and restorations, it is essential to surface traumas into the sunlight. Absent of sunlighting, we will be unable to recognize and transform the oppressive structures. Yet this endeavor is not about unproductively lingering in harmful past configurations, but instead generating spaces for true liberation based on holistic knowledge, unoppressed histories, and the full intent to repair relations. As Carl Jung wisely said, "One does not become enlightened by imagining figures of light, but by making the darkness conscious."[1]

As I work in my garden, the soil speaks to me about composting the toxins and injustices our society has accumulated in order to build dreamed-about, Earth-centered communities for all. I run my hands through the fertile compost pile; gardeners call it "black gold" for the way it enriches the soil, retains water (vital in my California garden), sequesters carbon, and nourishes plants. But this compost was once a pile of fruit and vegetable peels and cores, eggshells, and coffee grounds from my kitchen—presumably otherwise undesirable things that are thrown away. I added chicken manure from a friend's farm, and turned the pile regularly to expose the contents to the air. I allowed time for the center of the heap to heat up and decompose with the help of bacteria, insects, and earthworms. Gradually my kitchen "waste" and the manure transformed into the deep chocolate-brown, crumbly, earthy-smelling, highly useful life-force elixir I now add to my garden. For soils suffering from chemical toxicity, much more work is in order, such as remediation through soil washing, biochar, beneficial bacteria that break down the contaminants, or bioremediation by fungi.[2]

Here in my garden, I consider the cultural toxins that fuel bigotries and most prevent society from fulfilling our responsibilities toward Mother Earth and practicing a reciprocal relationship with the land. For there to be real regeneration, we need to decompose and transform malignant social constructs and enterprises in the compost pile. Because we are inherently part of daily life as it exists now, we cannot magically disappear patriarchy or pretend to purify ourselves or our world, so let's compost what we have to metamorphize it into something useful, something that builds the world we seek. What needs to be composted to remedy the deleterious consequences of the Anthropocene?

I come back to two forces that are fully entwined and have been hugely destructive: capitalism and colonialism. Capitalism is not only an economic system, but a worldview that commodifies everything, including humans and natural resources (or, more accurately, our plant, animal, and mineral relatives). That same avaricious appetite is the foundation of colonialism. If you look back through history and dig down to discover the reasons for colonial invasions and violence, you will see that nearly every instance of exploitation concerned access to resources or the desire for free or cheap forced labor. It is tragic and infuriating that a great many egregious oppressions of the world can be traced to some people wanting to possess and exploit other people and lands—often with white supremacy as a central lever.

Capitalism has always depended on colonization, exploitation, slavery, cheap labor, disposable peoples, sacrifice zones, plundering of resources, and the bullying of certain countries by wealthy countries and corporations.

The fuse of colonial capitalism was lit nearly five centuries ago, when Europeans began looting the Americas for gold and plundering the Indian Ocean region for cloves, nutmeg, and other spices. The colonists built military forts to ensure their control over the lands where the coveted plants grew, and murdered anyone who objected to their presence.[3]

Marxist political theorist Ellen Meiksins Wood points out that the roots of capitalism and colonialism reach back to the English enclosure movement, in which everyday people were ousted from meadows and forests as privileged landowners—and, later, the British Parliament—walled off public lands.[4] People were now forbidden from foraging in the woodlands, planting gardens, and openly grazing their cows and sheep in what had been shared public land, the commons. This left many no choice but to leave their land-based way of life and beg or work in factories for low wages under horrible conditions. Because the land had been privatized, peasants were forced to pay rent. Because leases were given based on how much—and how quickly—people produced, they had to compete with their peers in order to keep their homes and feed their families. This exploitative dynamic caused more resources and crops to be extracted from the land, resulting in excess products and labor that only benefited the wealthy—a paradigm carried into the Industrial Age.[5] The implementation of privatization relied on force: violently removed from land and the gifts of the harvest, countryfolk had to unavoidably fall in line to survive in this system.

As they did to the peasants in their own countries, European elites created false scarcity by privatizing land abroad. As economic anthropologist Jason Hickel explains, the land was still there as always (not scarce), but it was now unavailable under new notions of "private property," indispensable to modern capitalism.[6] Colonial powers from Britain, France, the Netherlands, Belgium, Spain, and Portugal used this strategy to force Indigenous Peoples the world over to work for them. In South Africa beginning in the seventeenth century, Dutch and British colonizers, though a tiny minority in the region, compelled Africans living on the land to pay taxes in European currency, which Africans could only obtain by working for Europeans, particularly in the hellish diamond and gold mines—a cruel industry that still kills dozens of workers each year.[7]

Meanwhile, as women were tortured, burned, and hanged as so-called witches back home, European men on their colonial pursuits asserted their supremacy over women in every land they invaded. And even the constant exploitation of women and Mother Earth did not content them—they stripped land-based peoples of their communal, reciprocal, and Earth-centered worldviews through forced Christianization and Eurocentric schooling. They targeted the very identity and core of each community whose land they invaded, trying to rob them of anything that gave them the power to resist colonist hegemony and control. Instigating false scarcity was a primary tool throughout these exploits.

False scarcity is integral to colonialism and capitalism in order for the elites to maintain power and wealth at the expense of everyone else. This is especially deceitful because plentifulness and generosity are the living language of Mother Earth, a language readily understood by anyone who takes the time to observe the thousands of acorns one oak tree produces annually, the life teeming in the soil, or explosions of wildflowers and herbal medicinal plants each spring. Cutting open a butternut squash from my garden, I find dozens of seeds inside, freely offered, each one of them capable of creating four or five more squash. Nature showers us with abundance, as she herself is a seat of fertility and power. However, this ability to replicate does not mean we can live in an endless economic growth model of extractivism, but that we need to live within the natural laws of the Earth—both her abundance and her ecological boundaries.

The creation of scarcity through privatization has been expressed in the "Lauderdale Paradox," which tells us that public and private wealth are inversely correlated: to increase private wealth, public wealth must decrease, creating artificial scarcity and making people pay for access to land, food, and water, using money they could only earn from the elites.[8]

Central to composting colonial capitalism is ending false scarcity, and instead embracing a worldview that when we share land, water, seeds, food, forests, and wealth; respect nature's balanced ecological systems; and practice reciprocity, there can be enough for all.

The First Industrial Revolution was galvanized by cotton, planted and harvested primarily through the labor of enslaved Africans, as well as various inventions, including steam power and iron-ore smelting in the eighteenth century. While many products of the Industrial Revolution made life easier for some people, because of the capitalist mindset it also resulted in overcrowded cities, factories with horrible and often deadly working conditions, and pollution. Coal, the cheap energy used to heat homes, fuel factories, and power transport, became more and more prevalent and led to increasing pollution as well as the deaths of thousands of coal mine workers worldwide. Industry leaders eventually discovered the myriad uses of crude oil and natural gas, and with this, fossil fuel consumption continued to expand and revved into a frenzied growth after World War II that spelled disaster for the natural world and communities in extraction zones and territories. Had those who first began processing fossil fuels known what it would do to the Earth, would they have stopped? I suspect early oil capitalists with dollar signs flashing before their eyes and caught in the dominant extractivst worldview would have said, "Drill away," just as we know that fossil fuel corporations have long been aware of the dangers of carbon emissions into the atmosphere, concealed them from the public, and ravaged our world in the name of profit.

By the 1980s, the causes and effects of climate change were known. Many fossil fuel companies were fully aware of the risks—Exxon corporation hired scientists to investigate any connection between fossil fuels and climate change in the 1970s and 1980s. In 1977, their scientist James Black informed the company that "the most likely manner in which mankind is influencing the global climate is through carbon dioxide release from the burning of fossil fuels."[9] Instead of acting, however, Exxon buried this data and spent millions of dollars promoting climate disinformation, even after James Hansen, director of NASA's Goddard Institute for Space Studies, delivered Senate testimony in 1988 declaring that he was ninety-nine percent certain that the Earth was warming, that this was caused by the "greenhouse effect," that warming would result in extreme weather events, and that "draconian emission cuts" could mitigate the greenhouse effect by the year 2000.[10]

The urgent words of these and other scientists gave corporations and governments a chance to assume the mantle of responsibility, but they did not—the lure of profit was too great, and any consequences for the wealthy seemed far away or not a threat to them at all.

Unfortunately, the widespread realization that the climate had entered an exponentially increasing human-made crisis came at the worst possible time. The late 1980s saw a world on the cusp of corporate globalization with the Canada-US Free Trade Agreement—the embryo of the North American

Free Trade Agreement (NAFTA)—the fall of Communist governments and the privatization of their economies, so-called trickle-down economics, further deregulations of corporate activities, and public-spending slashes.[11] Hence, the force of wildly unregulated capitalism, now virtually global, steamrollered the emerging climate consciousness. If we had acted back then, we would have prevented much of the climate-related disasters we see today and will see in the future, but it would have required government regulation of polluting enterprises—the very corporations that were making rich people richer and ensuring government cooperation in their enterprises. As one example, the free trade deals signed under the watch of President Reagan in the US, Prime Minister Thatcher in the UK, and their successors made many prudent climate initiatives impossible.[12]

Naomi Klein, in her seminal book *This Changes Everything: Capitalism vs. The Climate,* summarizes the conflict this way:

> We have not done the things that are necessary to lower emissions because those things fundamentally conflict with deregulated capitalism, the reigning ideology for the entire period we have been struggling to find a way out of this crisis. We are stuck because the actions that would give us the best chance of averting catastrophe—and would benefit the vast majority—are extremely threatening to an elite minority that has a stranglehold over our economy, our political process, and most of our major media outlets. That problem might not have been insurmountable had it presented itself at another point in our history. But it is our great collective misfortune that the scientific community made its decisive diagnosis of the climate threat at the precise moment when those elites were enjoying more unfettered political, cultural, and intellectual power than at any point since the 1920s. Indeed, governments and scientists began talking seriously about radical cuts to greenhouse gas emissions in 1988—the exact year that marked the dawning of what came to be called "globalization."[13]

Klein also points out that authoritarian socialist nations, in which economic power lies in the hands of the state, have been no less guilty than capitalist nations in perpetuating the myth of scarcity, profiting from forced labor, and enthusiastically backing extractive and destructive industries.[14] In this sense, capitalism could be described as one dominant form of extractivism, which is older and has varied forms, including colonial extractivism, state-socialist extractivism, totalitarian extractivism, and so on.

The avaricious colonial-capitalist system of the mainstream culture coupled with its ingrained dominion-over-nature worldview have been central catalysts of our climate, racism, and environmental emergencies. This is a toxic paradigm that must be deconstructed and composted into a wholly new trajectory. As Taily Terena, of the Indigenous Terena Nation in Brazil and WECAN Brazilian Amazon Coordinator, stated in the Indigenous Peoples' Caucus Opening Statement for COP26 in 2021, "Colonialism caused climate change. Our Rights and Traditional Knowledge are the solution."[15] Some have rebuffed this assertion about colonialism as an overstated claim, but it is not—in April 2022, even the conservative IPCC deemed colonialism "not only as a driver of the climate crisis but also as an ongoing issue that is exacerbating communities' vulnerability to it."[16]

Finishing my garden work one afternoon, I sit on the soft earth to revel in the variety of sweet, sharp, and bright scents. As the day's effort ebbs from my limbs, the hum of a honeybee captures my attention. The bee hovers over the brilliant yellow zucchini flowers, gathering the nectar that will make golden honey. I reflect that this tiny creature does not need to rip the flower apart to take what it needs—the plant offers the nectar willingly, courting the pollinator's attentions with radiant colors and enticing fragrances, and the bee, in turn, facilitates the plant's reproductive cycle.

Countless mutually enhancing acts are the way of Nature: deep beneath the surface of the Pacific Ocean, a clownfish is cleaning a stationary anemone of harmful parasites and nourishing it with her excretions while benefiting from the anemone's shelter and protection, even laying her eggs near the guarding embrace of her neighbor. Beavers, too, enhance the environment for their fellow species by restoring nurseries for salmon, creating wetlands that refresh groundwater and purify the water of pollutants, providing safe homes for nesting birds, and helping farmers irrigate their crops more efficiently.[17]

My more-than-human kin, including the bee, the clownfish, and the beaver, impart their lessons, teaching me how to be a beneficial relative who lives in reciprocity with the web of life. They show me how to approach Nature not as a one-way extractor who seeks to plunder but as a family member who seeks communion and mutual support. Like these relatives, we can learn to take only what we need, resist the impulse to hoard, and reciprocate as a life-enhancing species. We have so much to learn from Mother Earth's wisdom, design, and balance, if we take the time to listen, watch, respect, and learn.

In this context of a dynamic Earth populated with myriad examples of reciprocity and abundance, it becomes even more relevant and timely that we recognize and remember worldviews and cosmologies in which the Earth is understood to be animate, whose perennial gifts make all life possible; worldviews free of colonialism, predatory greed, racism, and patriarchy; cosmologies of harmony and reciprocity. How would such understandings translate into alternative economic frameworks for our world?

Right now, most governments and countries of the world base their grand marker of success on Gross Domestic Product (GDP), which perfectly reflects the colonial-capitalist worldview by measuring only the monetary value of products and services offered on the market. Yet, GDP will be irrelevant when major cities and entire island nations are under water, clean water is unavailable, fires and droughts make regions uninhabitable, billions of people become climate refugees, and a pandemic perhaps deadlier than COVID-19 rips through the population. As we say in our climate justice movement spaces, there is no business on a dead planet.

According to a group of scientists supported by the European Union's Horizon 2020 research and innovation program, the modern capitalist system is incapable of transition to clean energy. And another study argues that such a transition will only work if the overall level of consumption in industrial societies is reduced, which of course goes against the maxims of capitalism.[18] To avert the worst catastrophes from the climate crisis, scientists, social and environmental movements, and many others are calling on leaders, particularly those from wealthy countries, to respond quickly by ending the extraction of fossil fuels, mending the Earth's vital and balancing ecosystems, decreasing the production and consumption of meat, and implementing a transition away from GDP growth.[19]

In any viable scenario, we must end the dominant culture's obsession with GDP and endless economic growth and instead live within the natural laws of Mother Earth—inclusive of all peoples in their diverse cultures, kinship formations, ways of knowing and being. This will entail a vigorous and ongoing process of decolonization, degrowth, and alternative economies—all of which are intentional approaches that work to avoid socioeconomic and ecological collapse.

One way or another, we *will* have to transition from a predatory capitalist economy based on extractivism to an economy that relies on renewable, regenerative energy and the boundaries set by natural systems, and this transition *will* be forced upon us by Nature, even if unjustly, if we do not enter upon it voluntarily, equitably, and with planning and organization. The old power structures will remain in place unless we ourselves ensure the transition is an equitable one. This is why what is called a Just Transition becomes critical, because it seeks reparations—financial, structural, societal, and more—to remedy the injustices of the past, while providing a framework for a thriving future for all. A Just Transition dismantles white-supremacist, heteropatriarchal, and extractivist paradigms and systems; it grants agency to all and supports workers in a new energy economy; it heals the Earth of extractive wounds; and it centers the needs of localized communities, keeping their traditions and cultures intact.[20]

Degrowth is core to this process because it questions the dominant culture's existing mindset that GDP equals positive progress and development. We need to define and choose other metrics of societal well-being such as health, peace, community building, ecological balance, happiness, creativity, and connection with the web of life.

The point of degrowth, in part, is to demonstrate that constant GDP growth is not only impossible but unnecessary for human thriving and happiness. Imagine a life where you and those around you have enough of nature's gifts to sustain you without taking too much; adapting your diet to the cycles of nature in your local area and eating less refined products; using decentralized, regenerative renewable energy and doing so consciously and efficiently; ensuring everyone is guaranteed housing, education, and healthcare; supporting a localized economy rather than large global corporations; bartering with neighbors for needs; working together in community gardens; and using our collective creativity to navigate the conversion of old infrastructure into sustainable infrastructure. Communities the world over are already successfully engaging in these practices through models such as ecovillages, defined by Global Ecovillage Network (GEN) as "an intentional, traditional or urban community that is consciously designed through locally owned participatory processes in all four dimensions of sustainability (social, culture, ecology, and economy) to regenerate social and natural environments."[21] On GEN's website is a map displaying locations of active ecovillages on six continents as well as islands—including communities such as Earthlands Farm in Massachusetts and a network of two thousand Sri Lankan villages. At scale, however, this will require a shift in the dominant culture's worldview. Not an easy task, but it is possible if we keep envisioning it and speaking it awake. Some say that we cannot live this way because humans are too greedy and selfish by nature; my approach is that our neoliberal capitalist indoctrination is the root cause—a root we can compost and transform.

Reducing the economy's material production, within a Just Transition framework, is a central aim because that will have a significant effect on the climate, ecological integrity, and the health of our communities, but as degrowth experts point out, this will likely result in a GDP decrease.[22] And yet, this is precisely the point because, as many researchers have shown, GDP does not equal happiness or success as is so commonly lauded.[23] Additionally, Jason Hickel states that central to degrowth economics is that "by restoring public services and expanding the commons, people will be able to access the goods that they need to live well without needing high levels of income."[24] In contrast, many nations impose austerity on their citizens to generate a false sense of scarcity, as mentioned earlier, which then induces GDP growth that only benefits the wealthy. Degrowth recognizes and uplifts the actual abundance that exists when resources are not enclosed or hoarded.

British economist Kate Raworth proposes a type of degrowth using an ingenious set of indicators that form the shape of a hoop, which is why she calls her concept Doughnut Economics. Raworth's Doughnut features an inner circle delineating the social foundation that would ensure well-being for everyone, and an outer circle marking the ecological ceiling. Between them lies the balance, a "safe and just space for humanity."[25] Within the hole of the Doughnut are twelve indicators that show when our economic system is failing people, including in the areas of food, gender equality, housing, and justice. Outside the Doughnut lie nine planetary boundaries—including ocean acidification and biodiversity loss—many of which humanity is currently overshooting. The goal is to stay inside the "sweet spot" of the Doughnut, even though that means GDP will decrease. Importantly, Doughnut Economics explores the mindset that calls for distribution and regeneration of Mother Earth's gifts—a mindset which is diametrically opposed to capitalism.

Unfortunately, a move away from GDP growth is exactly the pathway the wealthy elite minority and most world governments have been unwilling to consider despite pleas and warnings from scientists, progressive politicians, advocates, and activists. When speaking of a transition to clean and renewable energy, CEOs and heads of governments use the word "sustainable," but what they often mean is "without changing the status quo" and "without changing the market-based economic structures"—the structures that got us into this mess in the first place and that only benefit the wealthiest. There is also a misguided belief, even among some progressives, that the market will inevitably shift, with no need for regulation or interrogation, as green technology evolves, and more green jobs become available.

How do we generate and demand polices that are really needed to meet this moment if only a narrow path of incrementalism and dangerous compromise are offered? Often policies are measured by what is deemed politically possible, which confines us to the old dominant cultural and economic paradigms. Any allusion to material growth decline can weaken a policy proposal given the political dynamics of our time, yet the fact remains that GDP growth cannot go on forever, and attempting to prolong its life is futile and deadly. The systemic shift called for here is the need to proclaim a global climate emergency so that our politics have the potential to break business as usual and render a proper response to the crises we face. Embracing that we are in a climate emergency, as well as a socio-ecological emergency overall, might just be what cracks open the political and economic space for a radical transformation at the scale needed.

Likewise, the mitigation pathways offered in IPCC reports assume scenarios of continuing GDP growth, and most of the action plans prescribed involve more use of technologies to continue this endless growth model, including some technologies that do not yet even exist and/or cannot be produced at speed or scale to date, such as carbon capture and storage.[26]

These action plans offer anything but centering equity and considering how to change our way of life so that we do not destroy ourselves and life on the planet, challenging as this proposition is. In this context, politicians and corporations make promises to achieve "net zero" emissions, which are commitments that claim to counter current emission levels with carbon removal projects in order to balance out the global carbon budget. In reality, however, these commitments often greenwash efforts meant to appear to address the crisis while countries and corporations continue to pollute and expand fossil fuel infrastructure.[27]

Given the climate emergency, and the fact that these deficient plans set us on track for well over the 1.5-degree increase in global temperatures needed to preserve a habitable Earth for most species, we need *real* zero now, not an incremental or false net-zero solution that kicks the can down the road. Analysis shows we are on track to warm by at least 2.4 degrees by 2100, a deadly and unacceptable threshold.[28] This means we must drastically slash emissions and transition to democratized, decentralized, and diversified renewable energy, while also reducing our consumption, increasing energy efficiency, stopping all new fossil fuel expansion, and protecting and expanding forests and biodiverse regions.

Within the net-zero framework, governments are planning to advance "nature-based" solutions, some of which have the potential to offer a path forward, but unfortunately often refer to land-based carbon offset schemes—such as appropriating land in the Global South to plant trees—when studies already prove that the Earth simply lacks the amount of available land needed to plant enough trees to offset all of the emissions that human activity continues to produce.[29] Nature-based solutions are primarily part of wider market-based mechanisms that include carbon pricing, mega-dams, geo-engineering, bioenergy, forest offsets, carbon trading schemes, and carbon capture and storage, all of which slap a price tag on nature and continue unsustainable business-as-usual practices.[30] They are "dangerous distractions" that serve to buffer corporate and political narratives of altruism while keeping the specter of accountability and change at bay.[31]

Market-based mechanisms in essence are a false solution to curb catastrophic climate change and deforestation. As an example, these mechanisms allow big polluters to continue to poison communities at sites of extraction and at points of distribution and processing by buying up pollution permits from forests around the world and simultaneously continuing dirty pollution practices at home or abroad.[32] In addition, pollution permits or offsets in forest areas often lead to land theft from and dispossession of Indigenous and local communities. These "solutions" enable polluters to keep polluting, while Indigenous and frontline communities suffer the consequences.[33] Furthermore, these carbon credit schemes are often a scam: a 2023 analysis found that 90% of the forest carbon-offsets credits claimed by the world's most significant offsets company, Verra, were worthless.[34]

From the view of climate justice, it is the responsibility of wealthy nations to reduce their consumption and provide material assistance to the countries that are bearing the brunt of the climate crisis and the historical extractive practices of wealthy countries. This accountability and funding is very possible (just take a look at the outsized military budgets of wealthy countries like the US), but it will take a great deal of pressure from both people's movements and innovation to popularize the idea. At the UN climate talks at COP27 a Loss and Damage financial mechanism was adopted to hold wealthy countries accountable to pay for climate-caused damages to vulnerable countries. While this policy is a historic and significant victory for the climate justice movement, it remains to be seen if sufficient funds will be committed by wealthy countries for distribution to vulnerable countries. Again, this points to the need for deep systemic change in our economic frameworks that can only be accomplished by a speedy embrace of a worldview transition that prioritizes the web of life and people over profit. As Waorani Indigenous leader Nemonte Nenquimo said in an interview that was translated from Spanish, "If we keep fighting collectively, the Earth will remain as such, full of animals, fish. If we don't protect it, the forces of capitalism will come in and destroy us."[35]

Many well-known economists, including John Maynard Keynes and John Stuart Mill, realized long ago that we would one day have to move past a growth-focused economic model, and a recent study from multiple major educational institutions around the world confirms that it is in fact possible to move beyond capitalism.[36] Our ancestors lived very well without capitalism for millennia, and many Indigenous Peoples, local communities, and place-based peoples around the world still do, because nature provides amply for all when ecosystems and local communities are respected. The obstacle is not that we do not have viable alternatives to capitalism. Instead, the problem is the insistent structural interference that the wealthy elite are exercising in response to people who are taking real action from the bottom up.

Moreover, a future without additional fossil fuels is possible. The sun provides us three hundred times the amount of solar energy we need to power the world, and the winds that traverse the Earth provide seven times the energy we need.[37] Of course, solar panels, wind turbines, and batteries for electric cars require minerals that are mined. This continues the extraction model, and safeguards must be put in place to ensure frontline communities and fragile ecosystems are not harmed by these activities. Some argue that the overall effects of mining might be counterbalanced by the closure of oil wells, tar sands, and fracking fields, yet there exists an extraction-free way forward that others have proposed: valuable minerals already lie aboveground in the form of mining waste, and many materials can be recycled—as the oldest solar panels reach the end of their lives, scientists are researching how to reuse their materials and components.[38] In any case, the goal in a transition away from fossil fuels cannot be to continue the capitalist extractivism agenda and ceaseless consumerism model. A Just Transition, degrowth, less consumption—particularly by the wealthy elite—and healing our relationship with the Earth must remain core principles and practices in this effort.

Scientists and planners have already developed numerous road maps guiding us toward eighty percent renewable energy by 2030 and one hundred percent by 2050.[39] And other research, conducted a decade ago but just coming to light, tells us that if we act quickly and decisively to reduce greenhouse gas emissions to zero, the planet's temperature will stabilize within three or four years, as opposed to the thirty or forty years previously assumed.[40] Reducing emissions will likely not, however, stop glaciers, sea ice, and permafrost from melting and thawing. That said, stabilizing the temperature still provides us with a more viable path forward.

Despite the climate emergency, fossil fuel production continues to expand, with governments enabling expansion in the face of their climate commitments. As of the end of 2022, sixty of the largest banks in the world have collectively financed 5.5 trillion dollars toward fossil fuel companies since the Paris Agreement.[41] The UN Environment Program's *2021 Production Gap Report* states that "the world's governments plan to produce around 110% more fossil fuels in 2030 than would be consistent with limiting warming to 1.5°C, and 45% more than would be consistent with 2°C."[42] This is reckless and inexcusable.

Additionally, a treacherous relationship that cannot be overlooked exists between the US government, fossil fuel corporations, and the US military industrial complex—an enormous and increasingly privatized institution whose growing influence President Eisenhower warned us about years ago in 1961.[43] The Pentagon emits and consumes more fossil fuels than any other global institution, and we must never forget the wars generated by efforts to dominate these fuels.[44]

As one important counter to this negative momentum, WECAN is honored to be on the Steering Committee for the fossil fuel Non-Proliferation Treaty. This is an initiative rooted in crucial efforts by Pacific Island nations and officially launched in its current form in 2020 by an international coalition of activists, scholars, and movement leaders chaired by Canadian activist Tzeporah Berman. Inspired by the Nuclear Non-Proliferation Treaty, this campaign was designed to phase out fossil fuels and fast-track climate solutions.[45] It is based on three pillars: non-proliferation of coal, oil, and gas by ending exploration and production; global disarmament of fossil fuels, or phasing out existing stockpiles and production in line with the Paris 1.5°C goal; and peaceful transition, meaning fast-tracking solutions and a Just Transition for every worker, community, and country. These pillars, in contrast to the "net zero" promises that Berman calls "delusional," are realistic and practical.[46] We are advocating for governments and financial institutions to stop financing and subsidizing fossil fuels, choosing instead to invest in and protect our climate and communities. The Treaty is gaining global momentum, and as of early 2023, the countries of Vanuatu and Tuvalu formally called for the Treaty in UN forums, and 84 cities and subnational governments have endorsed the Treaty along with thousands of other groups and institutions globally.

The need to end the fossil fuel era and to decolonize our climate solutions was eloquently and precisely stated at a COP26 event in Glasgow, Scotland, by Eriel Tchekwie Deranger who is the Executive Director of the Indigenous Climate Action organization and is a member of the Athabasca Chipewyan First Nation in Canada. She is also a dear friend and colleague, and one of the most brilliant leaders I am honored to know. She said:

> In our communities, we have full economies centered around our relationship to the land and all species, and the rivers, and the oceans and the grasses to the plants and trees, and to the rocks and the mountains—every living being in those spaces, we build our economies and our systems around that.
>
> Those solutions look like language revitalization and cultural revitalization. They have proven when those things are intact, biodiversity thrives, the health and livelihoods of people thrive, and we are no longer dependent on inputting for food systems.
>
> Just to give you one more fact on this…in my territory before fossil fuels took off in the 1990s, we were eighty-percent subsistence, and now we don't have that luxury. Now we are reliant on grocery stores where things have to be imported in.
>
> Imagine if we were able to return to our traditional food systems. That would decrease our footprint, and there are so many ways in which Indigenous lifeways and ways of being can help us decrease our reliance on fossil fuels and bring us to a place of harmony. But instead, we have to fight, and Indigenous resistance on the frontlines, civil disobedience, and assertion of our self-determination and traditional government systems has been criminalized and vilified. But this resistance is also responsible for keeping 1,800,000,000 tons of carbon in the ground in North America alone. That was documented in a 2021 report called *Indigenous Resistance Against Carbon.*[47] . . . Our communities are demanding that we not only phase out of fossil fuels, but that we decolonize the very systems that allow these industries to proliferate and be leading negotiations and spaces like the UNFCCC.[48]

There is much more to say about alternative healthy economic systems, but to embrace and operationalize them, we need to do some further composting. I bend close to the ground and stir the soil further.

In my exploration of capitalism and colonialism, I was set on a critical trail by Eriel Tchekwie Deranger. She encouraged me to read the book *Columbus and Other Cannibals* by Jack Forbes, which delves into a powerful analysis and narrative of what he calls the wétiko disease. This is crucial terrain that takes us into our interior world as we search for deep systemic societal change.

Wétiko comes from the knowledge stories of many North American Indigenous Peoples, in particular the Algonquian communities. As I researched the work of Forbes and other scholars and cultural knowledge keepers, I learned that there are various interpretations, stories, and names for the wétiko—commonly spelled as wendigo, and sometimes as windego, wentigo, or wentiko—which is often depicted as a creature who was once a human but, due to extreme greed or overconsumption, betrays the laws of nature and becomes insatiably avaricious. The wétiko would even develop an insatiable appetite for human flesh, leading the creature to ceaselessly prowl the frozen winter nights, looking for more victims.[49] However, wétiko is not only a corrupted human but also an overall cultural "disease of aggression against other living things and, more precisely, the disease of the consuming of other creatures' lives and possessions."[50]

This is a difficult topic to discuss, but if we cannot name and diligently address these deleterious, violent aspects of our society, we will be incapable of transformative change and unable to generate the well-being we strive to manifest in our own inner lives as well as in our communities. Healthy and purposeful discomfort, guided by love and compassion, is a needed friend to us now, and a sign that we are growing.

In many versions of the wétiko teaching story, an exceptionally rapacious person could become a wétiko; thus, these important tales served as warning stories that enjoined moderation and reciprocity among community members—which was, in fact, the only way to survive the dead of winter, when famine was a real possibility. A greedy individual could put the survival of the community at risk, and that person might, indeed, be cast out of the group if they refused to change their ways. The stories taught that one's responsibility to the community comes first, and that a self-serving attempt to enrich oneself at the expense of the group will erode one's humanity, almost as if one has actually eaten another person.

The wétiko is not always described as a physical monster. Sometimes, it is "a power that is fueled by the greed, fear, and pain that exists in the hearts of humanity."[51] This rendering of the wétiko concept is important (albeit disturbing), as it informs us that this force can be inside all of us, and we must continually reflect and practice self-care to ensure normal life experiences that include pain, fear, uncertainty, and more do not overcome us. The wétiko can lurk in our psyches, whispering that we can take whatever we want; we deserve more even if it harms others, just this once, no one will notice; we're not responsible for the group anyway, and why shouldn't we take more. It is an imbalance that destroys us and our relationships to each other and our Earth, leaving us alone and unappeasable at many levels. It erodes the very roots of what makes us human, because we have betrayed our community to benefit ourselves alone. This point is particularly salient because the dominant culture of today promotes an ideal of hyper-individualism, which feeds predatory capitalism. Naming and bringing awareness to wétiko through teaching stories is a powerful way to address this human imbalance, so that we can overcome this disease and thrive as mutually enhancing communities.

As I am learning from Indigenous colleagues, wétiko is also used by Indigenous Peoples to describe catastrophic events and the efforts of a group to oppress others. Thus, the concept also describes colonialism, genocide, forced assimilation, rape, land theft, and all of the other atrocities the Europeans committed toward the Indigenous Peoples of the Americas. Jack Forbes provides a complete description of the wétiko impulse behind colonialism in his writing and work as an activist and scholar of Lenape and Powhatan Renapé descent, a founder of the tribal college Deganawidah-Quetzalcoatl University, and the creator of the Native American Studies program at University of California, Davis.

In his analysis, Forbes explains how the universe starts with language and love, but what happens to mentally ill persons is that they become exploiters. I was struck by Forbes's articulation of wétiko as a mental disease because, as mentioned, several of my Indigenous friends and colleagues have told me that their Elders said that when Europeans first came to the shores of Turtle Island, they were assumed to be mentally ill. This was not expressed with negative judgement, but rather as a logical conclusion based on observing the values, religious stories, and actions of the settlers. The Elders did not believe that individual Europeans were evil, but rather that the very foundation of European cultural mentality was inherently ill.

Forbes translates *wétiko* as "cannibalism" and explains that raping a woman, polluting and destroying the land, oppressing a community, and abusing animals are all manifestations of wétiko. Those who commit such acts are insane, and this insanity has been spreading throughout the Earth like a plague for thousands of years. Enter Columbus and his men, who enslaved thousands of Taino, the original peoples of Puerto Rico and many other Caribbean islands, and sent them to Spain to be sold; forced Indigenous People to find gold and cut off their hands if they failed to bring enough; raped Native women; and knifed Indigenous People, including children, for fun. When Indigenous communities dared to fight back, Columbus and his men slaughtered them.[52]

Forbes argues that Columbus was not simply a rapacious person but a wétiko whose avarice-fueled culture had turned him into a mentally insane "cannibal" who carried an infectious psychological disease.[53] Forbes equates imperialism and exploitation to particularly diabolical forms of cannibalism, which he defines as "the consuming of another's life for one's own private purpose or profit."[54] Certainly, the Europeans consumed Indigenous Nations, land, and resources to enrich themselves, extinguishing millions of lives and destroying landscapes and cultures under the banner of "civilization."

I am not saying this to disparage all Europeans (as I have made clear in other chapters in the book), and I do not believe Forbes did either, but there were certainly powerful elements of European culture that propelled a major subset of them to colonize and plunder lands around the globe or to be complicit in these activities. The same wétiko element accompanied Europeans to Africa and South America, where they stole Black and Brown humans and sold them to other wétikos who forced them to work on sugar and cotton plantations, a move that would enrich the elite whites of Europe and North America and pave the way for centuries of white supremacy and structural racism that continue today. This is the same mindset that leads temporary workers at mines and oil-extraction sites in the Americas to ravage Indigenous bodies by night and the body of our Mother Earth by day to feed what renowned Anishinaabe activist Winona LaDuke calls the "Wendigo economy," which "destroys its mother."[55]

It is crucial that we name and recognize the wétiko in our past and present so that we can best navigate this deadly impulse. The wétiko denies the reality of our place in an interdependent community of humans and the web of nature and corrupts the human mind and spirit. Needless to say, colonialism and capitalism are wétiko systems. They glorify continuous consumption and domination, regardless of who suffers.

A significant challenge we face as we work to liberate ourselves and communities from a wétiko culture is that overconsumption is a success symbol in the dominant culture, euphemized as "having a good quality of life." The sheer volume of accumulated wétiko deeds continues to consume the lives of frontline communities and is rapidly swallowing up forests, clean air and water, the polar ice caps, glaciers worldwide, and millions of endangered species—in short, the Earth, our only home, is being devoured by wétiko psychosis. The primary way to halt the disease's progress is to change our individual and societal worldviews and actions. In a 2008 article, Forbes states that his intention in writing *Columbus and Other Cannibals* was to inspire readers to take personal responsibility and commit to living a better life, and he points out a path to doing so, saying, "It has been my intention that CANNIBALS, instead of dwelling solely upon the evils of violence, aggression and one-sided economics, should offer insight into the ancient spiritual philosophies of Native Americans and contrast the core of Indigenous beliefs and actions with the 'empire-building,' 'getting bigger,' 'more and more' notions which have come to dominate many churches, religions, nation-states, corporations, groups, and even individuals and their families."[56]

Most of us are complicit in some aspect of the wétiko mindset, and we are thus responsible for curing ourselves and our communities of this deeply rooted illness. What I have witnessed, and what brings me hope, is that as people consciously refuse to conform to the wétiko malady, and instead educate themselves and take responsibility, they become revolutionized in their personal and political lives, working to reverse and repair the violence and destruction of centuries. Yet, let us be clear, this is certainly difficult work and requires the strength of unwavering honesty and action born from radical love of ourselves, our communities, our planet, and the generations to come.

I am inspired by the perspective shared by Aboriginal art critic and professor Tyson Yunkaporta of the Apalech Clan in his book *Sand Talk*, in which he provides alternate paradigms for our world. Just as the Algonquian and other Indigenous Peoples from Turtle Island share cautionary tales of the wétiko, Indigenous Peoples of Australia share their own warning stories of people who have committed the "original sin" of placing themselves and their own needs over those of the community, and over their responsibilities toward the land.[57] These cautionary tales are embedded in the landscape of Australia, as the participants in these teaching stories have been memorialized as rock formations—eternally visible reminders of the essentialness of reciprocity.

These stone formations were once people who acted outside of good behavior to their neighbors, damaged relations of trust within their community, or broke their reciprocal relationships with nature. While they forever stand as warnings about misconduct, Yunkaporta explains that Aboriginal Peoples do not revile these figures—rather, they revere and respect them. This is because, in their culture, while punishment for a wrongdoing is implemented, it is immediately followed by forgiveness and reintegration into the community. Because of the promise of reintegration, offenders are motivated to acknowledge their accountability to the community and the land, accept their punishment, and strive to change their behavior, becoming examples of keepers of the Law.

Thus, the rock formations convey not only a teaching story but an illustration of community, a promise that anyone can return to a milieu of love, respect, and reciprocity after making a mistake and learning from it.[58]

Here, I would be remiss to leave the gendered aspects of colonization and capitalism unturned in the composting process. Capitalism depends on women's unpaid labor and has further perpetuated patriarchy as a system that subjugates women and their bodies, devalues their work, and leaves many of them vulnerable to increased violence. There is also an intrinsic connection between capitalist extractivism and violence against women—the same worldview that promotes human dominance over the Earth promotes male dominance over women. The increasing destruction of our Mother Earth parallels an insidious violence against women's bodies, livelihoods, and voices.

The intersection of unfathomable acts of violence against the land and violence against women is unfortunately something that I witness as part of my work at a women and environment organization.

In 2015, I was invited to North Dakota by Indigenous leaders to participate with other movement organizers in a conference to stop extractive energy activities.[59] Many of us are experiencing more and more stark reminders of how ruptured our mainstream society's relationship with nature has become. As I stood on the rolling golden prairie lands, looking in every direction at the gas flares from the Bakken fracking fields, I underwent one of these heart-sinking moments. The ominous steel structures digging relentlessly into the Earth for gas and the noxious flames shooting pollution into the air were a clear symbol and fact of entitlement and violence toward the land and the original Indigenous Peoples of these territories.

The fossil fuel industry in the region took a temporary hit in 2020 since demand dropped during the coronavirus pandemic, but in 2015, North Dakota was the second-largest oil-producing state in the US after Texas.[60] The boom destroyed much of the rich, fertile land that had been farmed by the Three Affiliated Tribes (the Mandan, Hidatsa, and Arikara) before they were displaced in 1947. The government had promised the Tribes twelve million acres in an 1851 treaty, but only one million were delivered, in the Fort Berthold Reservation. With the fracking boom, cities in the region, such as Williston, swelled with new residents, mainly men who were there to work in fossil fuel production, overburdening police and social services. Tragically, Indigenous women and girls experience unique threats and dangers when extractive industry workers arrive in their communities, and that was the case in North Dakota. One of the most well-documented and devastating gendered implications of the oil and gas industry is the increase in rates of sexual assault and violence against Indigenous women and girls in areas where temporary man camps are erected for transient male workers.[61]

The presence of man camps, anti-Indigenous racism, a pretense of domination over the land and over female bodies, the lack of police protection, and legal negligence all combine to form a nightmare maelstrom for Indigenous women and girls. In the "Wild West" atmosphere surrounding the man camps, substance abuse meets a history of exploitation and genocide against Native Peoples, and many men think nothing of assaulting, raping, and sometimes murdering an Indigenous woman or girl. Native females exist not as human beings with very real lives but as bodies to be violated. They are "others," viewed as recreation after a long day at work. In the US, over half of Native women experience sexual violence in their lifetime, and they are twice as likely as white women to experience rape; further, most of their rapists are non-Native, which means they cannot be prosecuted by tribal authorities—and the cases are usually ignored by non-Native law enforcement.[62] Sex trafficking during the fracking boom was rampant in the Bakken region, so much so that two very small towns in the area (along with four major cities—Boston, Houston, Atlanta, and Oakland) had been named as places in the US most in need of assistance to combat these abuses.

Meanwhile, the land across the Bakken fracking fields has been desecrated by pollution. Reported and unreported incidents—including spills, fires, leaks, and blowouts—occurred up to two thousand times per year in North Dakota at the height of the boom. The fracking and extraction processes generated seventy-five tons of oil waste each day—a third of it radioactive brine that sickens workers—as well as leaks from disposal trucks, contaminating land, water, animals, and humans all along the routes taken by the trucks. Again, this is allowed to happen, because in the rush to fill corporate pockets, the people who live in these lands are seen as disposable.

As part of the conference in North Dakota, Kandi White, who is Mandan, Hidatsa, and Arikara and a climate campaign organizer with the Indigenous Environment Network, led us on a "toxic tour" of the Fort Berthold Reservation and surrounding areas. Devastation and rage can barely begin to express what many of us felt as we saw endless open flares, man camps, oil derricks and drills, and contaminated sites. We saw a radioactive storage facility meters away from an elementary school, and in one town near the reservation, children were found playing with radioactive waste that had not been properly removed from an extraction site.

Then, with tears streaming down her cheeks, Kandi recounted the story of an Indigenous girl, less than five years old, found running, naked, away from a man camp after having been sexually violated by a worker. Upon hearing this, I physically bent over with horror, grief, and outrage—my emotions piling chaotically on top of each other—and I too wept for the little girl and for the many who have become victims of the epidemic of Missing and Murdered Indigenous Women and Girls across North America. There is no avoiding the fact: capitalism is driving rape of the land and rape of Indigenous women. This predatory, extractive, and insane behavior must stop. It is egregiously violating women and the Earth and driving the climate crisis. Yes, this is excruciating to hear—and even more so to have witnessed or suffered as a lived experience—but we must not avert our gaze or it will not stop.

I was relieved to learn about local groups that are fighting back, like Fort Berthold Protectors of Water and Earth Rights (POWER), which is a grassroots organization that works to conserve and protect the land, water, and air.[63] In 2020, Fort Berthold POWER filed a lawsuit against the Environmental Protection Agency regarding their discarded methane emissions standards for fossil fuel corporations, citing that the resulting increase in emissions threatened the health of Tribal residents and their children as well as the land, all to enrich already-wealthy extractive industries. The lawsuit paid off, and in 2021, under the Biden administration, the rollback on emissions standards was rescinded.[64] The Three Affiliated Tribes are also using the revenues from gas and oil to pay for litigation for economic and environmental sovereignty, an administration building, new schools, and a greenhouse to feed their community. These hard-won victories are tremendous in the ongoing fight against colonial capitalism and patriarchy.[65]

I ponder the lessons I learned in North Dakota as I sit in my garden, all at once feeling heavy grief but also incredible inspiration from the leaders I met there. With soil in my hands that is unpolluted by fracking waste, I also feel tremendous gratitude and am aware of my privilege in having it. Tiny green tomatoes burgeon from their fragrant vines, cascading over thick mulch. I gently cradle one warm fruit, marveling at Mother Nature's mystery in manifesting what was not there just a few days ago. These daily occurrences of life magic, imperative for our existence, can easily be overlooked as ordinary and expected. And yet, in paying attention, we see that Nature's creative power and infinite imagination is truly awe-inspiring. In my garden, I am surrounded by a veritable cornucopia of color, aromas, and taste. I run my fingers along the soft, fuzzy filament covering the tomato vines, which protects the plant from insects, balances its temperature, and helps it resist drought. The carrot sprouts flutter in the breeze as the bright orange treasures grow beneath them, still hidden. I close my eyes and envision myself physically rooted, connected to the creative rhythms and impulse of the life-force, deeply appreciating the immeasurable genius of Mother Earth where a pileated woodpecker's extra-long tongue has evolved to reach insects deep inside the bodies of trees, where the regenerative cells at the end of its beak maintain its continued sharpness after years of pecking, where the bird strikes elaborate patterns on the trees to accompany a whistle-like song to serenade a potential mate. I imagine the genius that evolved into the gentle giants of the sea, the wondrous humpback whales who swim the California coast on their long migrations from Mexico to the Aleutian Islands, clicking and whistling to one another with their unique songs, and who excrete nutrients into the ocean to nourish the phytoplankton, who in turn exhale more than half of the Earth's oxygen and sequester up to fifty percent of anthropogenic carbon, carrying it to the ocean floor—these tiny algae are essential in fighting the climate crisis but are themselves threatened by warming seas. In seeing this panoply of true wonder, I contemplate how these times call for the same evolutionary and world-shaking imagination that Mother Earth is undeniably displaying.

The late award-winning American novelist Ursula LeGuin gifted many of us with her extraordinary imagination deftly spun into speculative fiction, in which she explored alternative political and social structures, fluid gender identities, racial equality, and the quest for cosmic equilibrium. In her acceptance speech for the Medal for Distinguished Contribution to American Letters, she famously said, "We live in capitalism, its power seems inescapable—but then, so did the divine right of kings. Any human power can be resisted and changed by human beings."[66]

The late British philosopher and cultural theorist Mark Fisher invites people to imagine a world without capitalism in his book *Capitalist Realism: Is There No Alternative?* Fisher defines the repurposed term "capitalist realism" as "the widespread sense that not only is capitalism the only viable political and economic system, but also that it is now impossible even to imagine a coherent alternative to it."[67] Capitalism has indeed come to be seen and enforced as the only possible economic and social system in most regions of the world, especially since the end of the Cold War, permeating every aspect of culture, including entertainment, education, methods of resistance, and politics until "it is easier to imagine the end of the world than it is to imagine the end of capitalism," a saying Fisher attributes to both Slavoj Žižek and Fredric Jameson.[68]

Capitalism generated by the dominant culture is propagandistic and exhausting, ceaselessly working to prevent people from questioning the system or seeing around it to imagine and discuss new social structures. Even after the 2008 financial crisis and the coronavirus-related lockdowns of 2020, when the current flawed economic system was laid bare, the wealthy elite ensured neoliberal economics remained in place, and the visible crack in the system was summarily sealed back up. But it inevitably will shatter open again because we are not living sustainably or justly—not even close.

Even universities, whose job it is to prepare mostly young people for an uncertain future, have been unwilling to teach economic systems other than capitalism and socialism, and many students are not even aware of the existing critiques of capitalism or the movements to explore alternative systems. In 2011, this lack prompted a group of economics students in Manchester, UK, to create an international network that is now known as Rethinking Economics, which campaigns for pluralism in the curricula of economics departments and in the social sciences; in particular, they are demanding to be taught economic models that are applicable to the real world of their futures.[69]

A core component in returning to a reciprocal relationship with nature requires an entirely different economic system that moves beyond the extractive colonial economy and instead is based on social and ecological justice and people's needs within the healthy boundaries of Nature. What brings hope is that we see these alternative economies already in practice in communities and countries globally—some are based on frameworks that are quite old and have been rehydrated into modern formations, and others are new explorations and experiments. All these initiatives have their fits and starts, to be sure, but they are vital undertakings, such as the

Schumacher Center for a New Economics that is working toward a just local economy in their Massachusetts community, with democratized money and a natural, public commons; the country of Bhutan, known for its policy of Gross National Happiness, that values metrics such as well-being, health, ecological and cultural resilience, and education instead of GDP; the Transition Network working in communities around the world on localizing economies; the New Economy Coalition, which envisions a "solidarity economy ecosystem" in which community members govern all institutions in their region and historically marginalized groups lead the way; Detroit-based groups in the US who practice bartering; the Zapatistas in Mexico who value economies based on mutual aid; the Economy for the Common Good, which centers dignity, solidarity, and ecological sustainability; the Mondragon community, which is a corporation and federation of worker cooperatives based in Spain; and the Commons Economy, which focuses on communal management of resources. These models lean toward an economy of reciprocal relations and include degrowth, post-growth, cooperatives, feminist economics, land-based economics, various Indigenous economic systems, and circular economies.[70]

Imagination and education are vital to this effort, and for this we need intentional spaces of regeneration and collective council as we re-vision an economy with social, racial, and ecological justice at its core. In any of the new economic formations, it is critical that the voices, needs, wisdom, and lived experiences of the frontline communities are centered and leading the way.

In my garden, among the tomatoes and the carrots, grounded by a woodpecker's staccato rhythm, I reflect again on the compost nourishing the fertile and fragrant soil. A regenerative economy, one that has transformed artificial scarcity and the accumulated inequalities of colonialism and white supremacy, should be like this compost—what has decomposed can circle back to serve the web of life once more.

Chapter 11:

Reciprocal Relationships with People and Land

THERE IS THE SIMPLE but profound truth that for all humans to live, we must take from the Earth. We kill plants or animals every day in order to live, but the question here is how can we do so in a courteous and coherent way that leads to an existence of respect and mutual well-being? In my youth, when I was first seeking to live differently from the over-consumptive, detached from nature, despiritualized mainstream culture, this inquiry led me on a journey into the wild.

I brought this question of what and how I consume to the Yolla Bolly–Middle Eel Wilderness, which rises from the California coastal mountain range and takes its name from the tumbling headwaters of the Eel River's Middle Fork and the Yolla Bolly mountains. This terrain is part of traditional Wintun lands, and in their language *Yo-la* means "snow-covered" and *Bo-li* means "high peak." This stunning, rugged region adorned with white- and Douglas-firs, ponderosa and sugar pines, and incense cedar is backcountry unknown even to most Californians.

Just when I turned twenty, I hiked into this remote landscape with Logan, my sweetheart at the time. Watching our steps with care on the uneven trail, we made our way by sunset to Long Lake in the Yolla Bolly. We passed through dense forest and pockets of glades and meadows and inhaled the intoxicating fragrance of forest loam diffused by the fallen branches and needles in the summer heat. Though in awe of the extraordinary wonder surrounding me, my mind also remained focused on the purpose of the venture. It was my first time fishing and—should I be fortunate enough to catch a fish—my first time killing an animal to eat, a common experience for many still living close to their own subsistence, but not for me. I had requested Logan, an experienced fisher, to guide me, and as we trekked up toward the lake, I replayed in my mind how I had arrived at this decision to kill a fish.

When I was ten years old, my mother moved my sister and me from an urban life in San Francisco to the small country town of Mendocino to raise us as a single mother. Though the radical shift from city to country life was one that I had originally feared and loathed, it was in fact the best gift my mother could have given me. One of the transformative facets of this new life was an immersion into relationships with plants and animals—one not possible in the same way in my urban experience—and this is when my love for gardening was initiated.

Plants fascinated me, and I loved growing them and learning to harvest them from the wild. Behind our house was a small apple orchard that had not been tended in years. It consisted of maybe fifteen trees, bearing a delicious apple variety called King, or Tompkins King, which I have never seen in grocery stores. I remember cutting the apples in half and sharing them with family or friends since they were too big for most anyone to eat by themselves. The orchard was also home to an abundance of persistent blackberry bushes, which needed constant pruning so they would not take over the entire field. In autumn, I would pick blackberries and apples, and my sister made fabulous pies from these delicious fruits. We also dried sliced apples by the wood-burning stove, which was the primary source of heat in the house. These experiences with the land brought me closer to understanding where my food came from, breaking the cycle of only seeing food in a market, processed and wrapped in plastic.

My resolve to kill a fish arose within me as a deepening inquiry about how I source my food. It came to me during an intense experience I had on a solo backpacking journey before my trip with Logan. Like other young people, I had been questioning the philosophical and ideological premises that formed the foundation of our destructive dominant society, which led me to explore diverse spiritual texts. It was in this venture that I learned that fasting was a common thread in many spiritual practices. In Asian, Middle Eastern, Nordic, Native American, and other traditions, fasting was described as an entrance to profound mystical experiences. Through fasting, perceptions and perspectives were altered, and the spiritual senses were strengthened. Specifically, I understood that, as our bodies thirst and hunger, the portal to the deeper universe, in all its wonder and divinity, could open to meet these longings.

At the time of this earlier solo fasting journey, I was also keenly impressed with reading *Siddhartha* by Hermann Hesse, and how Siddhartha practiced physical abstinence in his spiritual quest. I decided to see if fasting would deepen my own spiritual awareness through the experience of "doing without," and teach me self-discipline. I also learned from *Siddhartha* that there was something quintessential about water, specifically rivers, that brought about spiritual development; thus, I decided to fast from food and water for three days near a sacred stream.

I have conveyed the full story of this experience in other writings, but for now I will say that each day of the fast my mind and spirit became acutely clearer and stronger while my physical body became slower and more fragile. I began to understand more vividly than ever that I am an eater, a consumer, that the death of other beings bestows life upon me, that I kill every day to live. I tried to contemplate how much I consumed every day, every month, every year, and eventually imagined mounds piled high with a large variety of plants and animals that had died, given away to me, in order for me to live. And, aside from my wild foraging and vegetable gardening, the fact was, other hands had tended, raised, and killed and processed my food. I understood myself to be a disengaged consumer—distant from farmers, fishers, food pickers, and those who prepared my food, let alone from their well-being and rights in society.

As I fasted, I could feel the life force shifting in my body, generating a passionate impulse to honor evermore the sentient beings who give me life. It is one thing to pull carrots from the garden, which is indeed taking life from the Earth to feed myself, but still another matter to take the life of an animal with a beating heart and eyes that can gaze into mine. After this experience, I asked Logan to take me fishing.

We reached Long Lake and, before it was fully dark, set up camp. The following day, after making prayers to the land and asking permission of the lake and trout to fish there, I cast my rod into the shimmering mirror and waited. As experienced fishers know, I learned that a great deal of this kind of fishing involves sitting quietly in nature, which I deeply enjoyed, but also requires a learned patience—I did not catch any fish that day. Toward evening, Logan and I watched the rainbow trout jump out of the water to feed on insects that were hatching on the surface of the lake. There was certainly no lack of fish there. As we warmed ourselves next to the campfire that night, Logan told me how to listen for the fish, how to become more aware of what is happening under the water, and how to become more sensitive to my instincts as well as the undercurrent, which all inform and indicate where to drop the fishing line.

The next day, I aspired to listen for the fish and to my intuition, which led me to a shady spot along the lakeshore. There were a few nibbles on my line throughout the day from fish who cleverly devoured my bait and avoided the hook, but then toward early evening, when the trout became more active, I felt a real bite, much stronger than I had anticipated. The being at the other end of the line was no longer the hypothetical fish from Logan's lessons, but a wild living creature fiercely fighting for its life—while the trout pulled against my hook and line, it was pulling hard at my internal conflict about taking its life! At this moment, I suddenly remembered Logan's instructions to swiftly commit once I felt the trout's strong jolt, since timidity does the fish no favors.

Adrenaline and determination surged inside me, and instinct kicked in; my body somehow knew what to do and how to enter this sacred space between myself and the life of the trout. I yanked the rod to further set the hook and began reeling in the fish. There was no longer room for conflict, only precise action. Logan came near and encouraged me without interfering as I pulled the magnificent and unusually large rainbow trout from the water. The glow of the sunset reflected off the wet metallic body, revealing the colors of the frantically flopping fish. I swiftly pulled the trout close, lifted a large round rock, and struck the fish behind the eyes, immediately killing it.

The trout stopped its wild dance as the life left its slippery body. I gazed at the wondrous creature who, seconds ago, had been struggling for its life, now given to me, dead by my own hands. Hugely humbled, I offered the fish my gratitude, and softly said, "Thank you for giving away to me that I might live." Logan then helped me catch my inner calm and instructed me how to gut the fish, which we cooked and ate right away. The experience of eating a being that I myself had killed brought me closer to the gifts given daily by the animals and plants. I also learned that day that we cry tender grief tears when we remember our very existence necessitates killing beauty. When we go to the grocery store and see the packaged food, it is easy to forget that other beings, whether plant or animal relatives, give their lives so that we may live.

It is truly humbling to contemplate the enormity of Mother Earth's generosity to us. The Earth freely provides life-giving offerings of water, fruits, vegetables, animals, herbal medicines, shelters, and so much more. As we work to revive and renew a reciprocal relationship with the Earth that enhances all ecological systems and is socially just, we need to uplift a paradigmatic economic shift that respects and honors the gifts from the web of life and a relationship with the community all around us—what I like to call an Earth Community Economy.[1] At this time of multiple crises, we have an urgent need and opportunity to remake our systems of exchange and re-evaluate what is actually valuable, and how we approach our relationship to the web of life. As LeGuin expressed, any human power can be resisted, and we can make other choices.

Our extractivist, colonial economic model demands that humans endlessly take more and more from the Earth without long-term considerations. By design, it requires that we exploit Mother Earth and turn her gifts into a commodity. This exploitive economy also takes activities that people used to share and exchange with each other without money, and runs them through this same predatory economic system.

Yet when we take a deep breath and slow down to really think about it, the things that actually make our lives possible and joyous very often are not the things that require extractivism or items to purchase. Photosynthesis, the hydrologic cycle, love, friendship, walking in the beauty of nature, mutual aid, sharing stories and meals—these are all things that can be done without financial exchange. In the current dominant culture worldview, which commodifies everything and sees things only for their deemed financial worth, we become further detached from what is given freely—wildflowers and glaciers are exiled to the realm of the superfluous since there is no monetary value assigned to them. We are taught in the commodification culture to lose respect and appreciation for many nonfinancially related treasures.

To counter this rapacious economic worldview, we can work to restore what many have called a gift economy—a system that many of our ancestors participated in, some communities still practice today, and has engaged an ever-growing network of thought leaders.[2] One of the central principles is not to hoard wealth; rather, the gift economy framework is about understanding that the only way the entire breathing, living ecology of place and community survives and thrives is through mutual aid. By making and giving gifts and moving those gifts through the community, we ensure our human and nonhuman relatives are taken care of. The essence of this exchange model is about aligning our economies with the natural laws of the Earth and our neighbors.

In communities that have practiced a gift economy, valuables are not exchanged for money or for other goods but are instead given with no outward agreement that anything will be immediately returned. There is an implicit expectation that the person who received the gift will also give at some point in the future, whether back to the giver, to another person, to the land, or to the whole community, and until that happens, one remains in a state of positive debt. While debt is uncomfortable for those of us who live in a market economy, in a gift economy, debt is what binds one to the gift-giver and to the whole community, intentionally creating an atmosphere of ongoing reciprocity.

A gift economy was, and is today, often practiced by hunter-gatherer communities: hunters would share their kill with the whole community; gatherers would share their plants, fruits, and nuts; and artisans and healers would share their skills. Individuals within the community were not paid for what they hunted, foraged, created, or offered, but they could rely on the reciprocal relationship between themselves and their community to care for them. When a hunter from Brazil's Indigenous Piraha Tribe was asked why he shared the substance of a particularly successful hunt with his guests at a large celebration rather than drying and saving the meat, he explained, "I store meat in the belly of my brother."[3]

An important facet of a true gift economy is how it is integrally wedded to upholding practices of taking and gifting with ecological well-being and future generations in mind. Plant ecologist and author Robin Wall Kimmerer of the Citizen Potawatomi Nation tells us, "Collectively, the indigenous canon of principles and practices that govern the exchange of life for life is known as the Honorable Harvest. They are rules of sorts that govern our taking, shape our relationships with the natural world, and rein in our tendency to consume—that the world might be as rich for the seventh generation as it is for our own. The details are highly specific to different cultures and ecosystems, but the fundamental principles are nearly universal among peoples who live close to the land."[4] The Honorable Harvest is not a set of rules that can be enforced, but an agreement humans make with the land and with those who provide for us.

I gather the weeds I have plucked from my garden and yard and fold them into the compost pile. Maintaining a temperature of about 140 degrees Fahrenheit in the mound is essential to transform weeds into beneficial soil; if the compost pile is not turned often enough to rotate the heat through, the seeds can survive, and our gardens can end up with more weeds or invasive species. Similarly, as we imagine a worldview that embraces reciprocity, we need to ensure we do not reintroduce old paradigms and ideologies that have led to injustices or harms to communities or natural systems.

As I remember my great-grandmothers and their mothers extending back through the years, I know that much subsistence work, such as foraging and gardening, was often done by the women of the community. This is still true the world over for land-based communities—women are often the primary seed savers and farmers for household food. Before the encroachment of patriarchal capitalism, such work was valued as much as that of hunters, metalworkers, and traders, as all work contributed to a community's well-being. With capitalism, however, subsistence work and care work were devalued, effectively erasing many women from the economy. The subjugation of women occurs when the care economy and the market economy collide.

The care economy, the maternal giving economy, and feminist economics are central to a gift economy, with their vital focus on removing patriarchal ideologies and practices from economics and instead centering care work as the foundation upon which all economic activities are possible. Care is the economy we first internalize as infants, as participants in the parent-child dynamic and as children of Mother Earth herself, receiving her life-sustaining gifts; thus, our first step in economic revolution should be to deeply respect, value, and appreciate these gifts. A care economy focuses on ensuring all humans, all life, and the planet are tended to and protected.

The COVID-19 pandemic has exposed longstanding social inequities and uncaring economic practices worldwide. In the US, which does not guarantee healthcare to all, the poor are more likely to get sick because of their jobs and housing situations, more likely to die from a disease (preexisting health conditions are linked to poverty), and more likely to face exponentially increased financial hardship after an illness. This is particularly true of Black, Brown, and Indigenous communities, who are more susceptible due to environmental racism and preexisting health conditions.[5]

Along with clearly showing care work as essential, the pandemic has offered an opportunity to center the people who perform it. Many care workers come from marginalized communities, and most of them are underpaid and lack benefits.[6] Moreover, much care work is unpaid, such as caring for one's children and elderly or ill family members, cooking, and cleaning. Most of the time, it is taken for granted that the women of the family or community willingly perform this work. In contrast, a care economy would value these people and their essential work.

It should be noted that actually we already do depend heavily upon a care and gift economy, but the "gifts" are devoid of principles of equity. As American semiotician and feminist Genevieve Vaughan explains, "The capitalist mode of production is built on top of the gift economy and functions by surreptitiously taking the free gifts of all and making them into profit."[7]

Capitalism portrays economies based on gift-giving, turn-taking, and care as primitive, backward, and unrealistic in a modern society. But devaluing daily labor and incentivizing people to compete for resources that would not, under socially just circumstances, be scarce, is the truly uncivilized worldview. And it is untrue to assert that equitable gift and care economies cannot succeed when they have flourished in various models for thousands of years.[8]

Feminist economics can help show a way forward. This economic system advocates for an intersectional approach to policy that takes into account social and environmental justice, racial and gender equality, and attention to work currently ignored by market economies. Lidy Nacpil, activist and convener of the Philippine Movement for Climate Justice, summarizes feminist economics in this way: "A new order in which economic systems give primacy to the provisioning for a dignified and empowering life for all people in ways that are compatible with the capacity and health of the planet, values social reproduction as equally important as production, and recognizes the often invisible role and contributions of women. Beyond simply recognition, what has traditionally been considered women's work should be socialized and redistributed."[9]

One recent arena for activating a feminist economy arose through the idea of the Green New Deal (GND), which was conceived over a decade ago but gained greater recognition, publicity, and acceptance in February 2019 when United States Representative Alexandria Ocasio-Cortez and Senator Edward Markey introduced a bill that attracted support from many other senators and representatives, as well as expected criticism.[10] Soon after their announcement, a group of women's rights and climate justice organizations—including WECAN—assembled to create the Coalition for a Feminist GND, seeking to ensure that essential issues of gender equity were included in the discourse around the GND.[11]

Because the climate crisis has been caused by so many interlocking systems, diverse policy sectors must be involved in the solutions, including trade, development, and the military, as we advocate for regenerative and care economies. A key component of the Coalition for a Feminist GND's work is to highlight the critical role of a care economy throughout the Just Transition, and as a viable response to the capitalism-rooted climate crisis. Care is the essence of what makes us a valuable keystone species on this planet—as crises unfold around us, we must uplift this dignified aspect of humanity. We are not meant to be cogs in the machine of market GDP—we are meant to create well-being for our human and more-than-human communities in a framework of justice and reciprocity.

Our Coalition is certainly not the only group advocating for feminist economic policies in the US, not to mention the many inspiring feminist groups globally. Climate Justice Alliance is made up of seventy frontline communities and organizations who also advocate for economic policies that center frontline and marginalized communities in forming solutions for a future that is equitable, just, and sustainable. They have laid out their concerns and proposed solutions in *A People's Orientation to a Regenerative Economy,* which includes more than eighty policy ideas presented as fourteen planks, including Indigenous and Tribal sovereignty, justice for Black communities, justice for immigrant communities, healthcare for all, energy democracy, and investing in a feminist economy.[12]

In short, there is no way forward for our societies without racial, economic, climate, immigrant, gender, and global justice. There will be no true forward motion without all of us and our interlinked movements working together in reciprocal relationships. We are indebted to and responsible for each other and to our Mother Earth. This understanding is central to entering the portal of the worldview shift now necessary.

The deeply erudite Indigenous artist, author, and teacher Martín Prechtel, who grew up on a Pueblo Indian reservation in New Mexico, instructively describes a community built on mutual indebtedness. In his book *Long Life, Honey in the Heart,* Prechtel narrates his time in a Mayan Tz'utujil village near Lake Atitlan in Guatemala, where he studied with a village leader and eventually became a community leader himself. Prechtel recounts his experiences of weaving his life into the Mayan community, which operated on a sort of care-based gift economy. He noticed that villagers who had been of great service to the community, often giving of their time and finances to the point of being "financially flattened and exhausted physically," were compensated in plentiful and powerful ways, receiving exactly what they needed at just the right moment, often from anonymous sources: special tomato or corn seeds that grew with extraordinary abundance and enriched the recipients, or large amounts of fabric for launching a new cloth business.[13] When Prechtel found himself in severe financial difficulty after a year's work as a village head, generously offering all that he had, many people, including ambassadors and international consuls, suddenly showed up at his hut to buy his paintings, including some that were not even finished.

With money now in his pocket, Prechtel immediately decided to pay back the village leaders who had kindly helped him when he was in financial need that year, but when he set out to do so, with deep appreciation for the loans they had given him, the Elders became intensely angry, chasing and bitterly insulting him until he sought refuge in the home of one of his village advisors, A Sisay. His bewildered mentor and friend asked Prechtel what he had done to so upset the shouting women outside. When Prechtel told him, his beloved advisor also became angry and asked why he was pushing them away, and whether Prechtel wished to belong to the village or if he preferred to be alone again, an orphan.[14]

His mentor continued, explaining, "*Kas-limaal,* that's what we call it, kas-limaal, mutual indebtedness, mutual insparkedness. Everything comes into this Earth hungry and interdependent on all things, animals, and people, so they can eat, be warmed, not be lonely, and survive. I know you know this, but why do you push it all away now? We don't have a word for that kind of death, that isolation of not belonging to all life."[15]

Prechtel then conveys the importance of trusting that the community he belonged to would always come through for him when he most needed help, and that being mutually responsible to them was key to becoming an adult. He was indeed in debt to those who had helped him, but immediately and fully repaying the debt was an act of separation. A Sisay beautifully expressed this knowledge to Martín, saying:

> The knowledge that every animal, plant, person, wind, and season is indebted to the fruit of everything else is an adult knowledge. To get out of debt means you don't want to be part of life, and you don't want to grow into an adult.... A further-initiated man and woman become struck with spiritual lightning, their tree full of honey cracked open by the sharp tooth of the needs of the village. And then the Honey in their Hearts can run freely into the greater Village Heart. When you owe every chief in the village and do not pay it back, this means that every chief in the village has conspired to be closer to you than a blood relative. ... You can only properly buy your way out of village indebtedness by more service, which alleviates the initial debt but sets you deeper and deeper into debt with the village itself and to the other heavier chiefs and spirits.
>
> The idea is to get so entangled in debt that no normal human can possibly remember who owes whom what, and how much. In our business dealings, we keep close tabs on all exchanges, but in sacred dealings we think just like nature, where all is entangled and deliciously confused, dedicated to making the Earth flower in a bigger plan of spirit beyond our minds and understandings.[16]

These words pierce my very being with inspiration to learn more about reciprocal relationships and mutual indebtedness, and learn to practice them as core principles of not only a care economy, but my daily life and worldview.

The potatoes are ready to be harvested, igniting my imagination as I unearth the golden-brown treasures from the garden. There is no uniform shape to this wildly untamed crop, and as I turn the potatoes in my hands, I am captivated by their unique forms, evoking images of various animal creatures—turtles, dragons, buffalos, and horses—all emerging from deep within the mother soil. The harvest is plentiful, and the potatoes delicious—I appreciate the resilient flora lineage and the composted materials that enriched the soil for this bounty.

In my search for healthy soil to grow reciprocity practices and to encourage radically different economies, I have been introduced to powerful movements such as land rematriation, Land Back, the principles of the commons, and Buen Vivir (Living Well), to name a few that make the connection that our relationship with the land has everything to do with reciprocity principles and decolonized economies.

For instance, we cannot truly talk about a post-colonial economy without heeding the call from Indigenous Peoples for Land Back so they can once again steward their lands without invasion or exploitation. Land Back is a movement that seeks to return the traditional homelands of Indigenous Peoples of Turtle Island (North America) to their stewardship and governance, politically, legally, and economically. Popularized in 2018 by Blackfoot member Arnell Tailfeathers in a series of posts criticizing Canadian Prime Minister Justin Trudeau's support for oil pipelines and refusal to consider reparations, the term "Land Back" quickly became a widespread hashtag among Indigenous activists and allies.[17] The campaign was taken up by the powerful Indigenous-led organization NDN Collective, who define Land Back as "the reclamation of everything stolen from the original peoples."[18] The Collective further clarifies that "LANDBACK is a movement that has existed for generations with a long legacy of organizing and sacrifice to get Indigenous territories back into Indigenous hands. Currently, there are LANDBACK battles being fought all across Turtle Island."[19]

These struggles are also taking place globally under different campaign names as land grabs and colonization accelerate worldwide within the capitalist economy. Adhering to Land Back demands would first and foremost begin to mend atrocities and violations against Indigenous Peoples and respect their rights, treaties, languages, traditions, and sovereignty. It would mean dismantling the white-supremacist colonial structures that continue to uphold the plundering and violation of the land and exclude Indigenous Peoples from decision-making and stewardship roles.[20]

Land Back also means bringing healing to the land itself as reciprocal land relations and knowledge are intrinsically embodied in Indigenous lifeways that we all can learn from and implement. Because ecosystems are in much need of healing due to centuries of mismanagement by colonial governments, several scientists have uplifted the call for Land Back, hoping that Indigenous land stewardship can help to avert the worst of the climate crisis and harmful impacts to communities.[21]

As well as including return of the land, the concept of Land Back involves the broader issues of making reparations and undoing the violence of capitalism. The movement, while unsurprisingly demeaned in mainstream society and often seen as an impossibility, has experienced growing instances of actual Land Back—both in actual gifts and in other cases through land purchases.

With the help of the Seventh Generation Fund and through community fundraisers, the Wiyot community bought back a small portion of Tuluwat Island, which formed part of their ancestral homelands in coastal Northern California. In 2015, Tribal Chairwoman Cheryl Seidner requested that the rest of the island be returned, and her requests were echoed by an ally named Kim Bergel, who ran for Eureka City Council for this purpose. The city of Eureka promised to give the island back, though four years of red tape followed before its formal return in 2019. Since then, the Wiyot community has been working to heal Tuluwat Island and restore its natural state, removing the chemicals and oils that seeped into the earth in the time it was used as a shipyard.[22] While the island now officially belongs to the Tribe, a few descendants of settlers continue to live there, accepted by the Wiyot as neighbors; as Seidner explains, "We know how it feels to be taken away from our land."[23] The Wiyot community, in 2020, then succeeded in reclaiming human remains and funerary items from the University of California Berkeley, and in 2022 they purchased fifty acres of their homelands on Humboldt Bay. This gives the Tribe access to plants significant in their culture, and also allows them to enhance the local ecosystem by weeding out invasive species, improving the water quality, and readying the land for the coming impacts of the climate crisis.[24] As this case demonstrates, returning land to Indigenous hands is an act of justice not only for Indigenous Peoples but for the entire Earth community.

The enduring effects of long-ago Indigenous land stewardship may provide legal grounds for at least one community to reclaim their ancestral homelands. Along the coast of what is now British Columbia, Canada, lie villages once inhabited by members of the Ts'msyen and Coast Salish First Nations communities. Though the Nations were forced to leave many years ago, the forests surrounding the villages still teem with their reciprocal practices as crab apple, hazelnut, blackberries, and healing herbs, as well as pollinators and animals, thrive and proliferate. These forests were tended for generations by Indigenous caretakers, who shaped them using fire, pruning, and coppicing—and the signs of these activities prove the communities used the forests as farms. These agroforestry practices translate into "sufficient occupation" in the language of the Canadian courts, and therefore into a case for the return of these lands, which were never ceded, to Indigenous ownership.[25]

Much Indigenous land that had been left to Indigenous Peoples under treaty in North America has since been seized under the cover of conservation purposes, perpetuating further settler-colonization. Such an occurrence happened in 1908 in Montana, when land was stolen from the Confederated Salish and Kootenai Tribes in order to protect bison—the very animal nearly driven to local extinction by settlers. In 2020, stewardship of the land, which is now the National Bison Range, was finally returned to the Tribes, who will care for the bison themselves using their Traditional Ecological Knowledge. This includes controlled burns of the pastures to stimulate the growth of grasses and keeping bison families together.[26]

At the end of 2021, timber corporation Port Blakely Companies returned over one thousand acres of Washington land to the Squaxin Island Tribe. While this land was given at no cost, the community also purchased 875 more acres of timberland, also part of their ancestors' homeland. Some of this land skirts the shores of Puget Sound, allowing the Squaxin to fish as their forebears did. They also plan to use the shoreline for ceremonial gatherings, according to Kris Peters, the Tribe's chairman, who said, "I can't wait to drum, and sing, and dance out on those beaches, just like our people did hundreds, and thousands of years ago ... To me it is a very spiritual thing; it fills my heart."[27]

Save the Redwoods League announced in early 2022 that they were returning over five hundred acres of coastal and forest land to the InterTribal Sinkyone Wilderness Council, which represents ten Indigenous communities in Northern California. Forced to leave their homelands generations ago ahead of aggressive timber companies, the Peoples of the Sinkyone Council now tend the land called "Fish Run Place," or in Sinkyone, *Tc'ih-Léh-Dûñ*, as a Tribal protected area.[28]

The Land Back movement is a demand for justice that is long overdue. Nickita Longman, a Saulteaux writer from George Gordon First Nation in Saskatchewan, explains her view: "Landback, to me, means a combination of a return to things while also taking into account the ways in which Indigenous people have evolved into the present and how we will continue to evolve into the future. Landback means access to sustainable food from the land, and it means affordable housing in urban settings. It means a return to our languages and incorporating harm-reduction strategies into ceremony. It means a return to matriarchy in a way that my generation has never known. Landback, to me, is this beautiful fusion of the core foundations of our ancestors, with room for growth and expansion in envisioning Indigenous futurisms."[29]

Corrina Gould of the Confederated Villages of Lisjan/Ohlone, and spokesperson for her People, is a co-founder of the intertribal Sogorea Te' Land Trust, along with Johnella LaRose. Based in the San Francisco East Bay area of California, Corrina has been working for decades to bring awareness to people in her region about the Lisjan and their sacred sites and way of life. The co-founders received their first quarter acre of ancestral land through the Land Trust in 2018 from an organic nursery called Planting Justice. The owners of the nursery had been moved by the protests at Standing Rock, and on the advice of Native Elders, they allied with the Sogorea Te' Land Trust and returned a portion of the land the nursery had been using. A redwood arbor was then built on the land for ceremonial use, so that the Indigenous community could hold ceremonies on their ancestral territory without having to seek a permit from a park.[30]

In 2022, the city of Oakland returned stewardship of five acres of Joaquin Miller Park to the local Indigenous Peoples, under the care of the Sogorea Te' Land Trust. The area previously known as Sequoia Point will now be called Rinihmu Pulte'irekne, which means "Above the Red Ochre" in Chochenyo, and will host native vegetation as well as ceremonies. The land transfer was effected through a cultural conservation easement, a relatively simple and straightforward process that will hopefully be followed by other cities.[31]

The East Bay of what is now called San Francisco was once called Huchiun, and the Lisjan community belonged to that land for millennia.[32] When speaking of Land Back, Corrina Gould uses the term "rematriation" because as she explains, it involves more than simply returning land to Indigenous stewardship. Rematriation also means bringing back the ceremonies, the women's songs for the waters and for healing, reclaiming ancestral remains, and following the Original Instructions.[33]

As a guest living on neighboring Coast Miwok lands just across the bay from Corrina, I have been truly honored to have the opportunity to march in the streets with her for climate justice and Indigenous rights, and to learn from her at educational events, as well as invite her to teach and speak at WECAN events and workshops. Her voice and leadership ripple throughout Northern California and across the country.

Corrina proposed the idea of the Shuumi Land Tax as a way for people who live, work, and benefit from Indigenous Territory to give back to their hosts. This is a voluntary recurring contribution that helps the local Indigenous community build ceremonial areas, develop education materials and curricula to pass on traditional knowledge, and create resilience hubs to support the community in times of need. It allows one's local Indigenous community to properly care for those who live on their land. Gould explains, "Our job is to take care of folks here. How do we do that as good hosts if we don't have good guests also participating in the reciprocal relationship?"[34]

At the 2020 Bioneers Conference, Indigenous educator Dr. Cutcha Risling Baldy, of the Hoopa Valley Tribe in Northern California, pointed out that much of the land in the United States, particularly in the West, is federally owned. This land could be returned to Indigenous Peoples rather than sold to corporations, as the government often does. In her talk, which outlined ways in which we can reimagine climate justice to dream better futures together, Baldy reminds us that it takes just one person to speak out and give power to the future that could or should be, and she describes the gifts Indigenous communities in particular bring to the table: "Native peoples have been theorizing and using radical imagination for radical decolonized futures since the beginning of time. We have been given the tools to radically imagine how to build our futures. We dream them, we sing them, we story them into being. We have always done this, and we must begin again."[35]

In addition to the Land Back movement, we in the United States need to make reparations to communities historically excluded from commonwealth. The dominant society has a responsibility to uplift the leadership and demands of Black communities regarding remunerations due to the continuing legacy of slavery and countless economic injustices since the founding of the country. In 2019, US Representative Sheila Jackson Lee introduced a bill, HR 40, to form a committee that would investigate the possibility of reparations and calculate the cost.[36]

Reparations could mean direct payments to Black individuals and families, or they could take other forms, such as scholarships, student loan forgiveness, business grants, down payment grants, subsidized mortgages, land, social programs to create equity, betterment of majority-Black neighborhoods and schools, renaming institutions currently named for enslavers, removal of monuments to enslavers and segregationists, and endowments for museums and exhibits. It is not necessary that all funds for reparations come from the federal government. There are corporations that benefited from slave labor, many churches owned enslaved humans, some white families continue to inherit wealth generated by slave labor, and the military forced Black Americans to serve.[37] All of these entities, as well as local and state governments, should, after openly acknowledging and taking responsibility for what their predecessors have done, do something for the descendants of the people whom they enslaved or injured. This work needs to be conducted with the respected guidance of Black leadership.

One significant area where reparations are needed, especially in the form of land and grants, emerges in the field of agriculture. In 1920, there were nearly one million Black farmers in the United States, determined individuals who had purchased land amid the broken promises of the Reconstruction Era and continuing discrimination. On their own land, they plied their considerable agricultural skills to nourish their families and communities.

But in 2019, there were only a little over forty-five thousand Black farmers.[38] The Great Depression accounts for some of these losses, but

Black farmers were already more vulnerable to weather and market fluctuations because, on average, they owned much less land than white farmers. They also faced racial discrimination at the bank, where they were denied loans that would have allowed them to purchase new equipment and better seeds, and at the market, where they received less money for their crops and wares than white farmers— a discriminatory practice that continued at least through the 1990s. This has led to Black farmers facing exponential difficulties in growing their farms. On top of this, and tragically, successful Black farmers faced great danger due to the envy of their white neighbors, whose assumed supremacism refused to accept the prosperity of Black people; throughout Reconstruction and the Jim Crow era, white mobs murdered Black farmers and vandalized and burned their farms. Major targets included Black figureheads such as Ralph Gray, who led the Croppers' and Farm Workers Union in Camp Hill, Alabama. In July of 1931, the county sheriff sent an armed white mob to invade a union meeting and brutalize the men and women inside. On the next evening, the sheriff himself showed up and shot Gray; he then sent a hundred and fifty white men to Gray's home to murder the union leader, who was convalescing there, and torch his house.[39] Due to these horrifically violent, discriminatory, and dangerous conditions, many Black farmers were forced out of farming and moved to northern cities, which were at least marginally safer.[40]

Rising with great strength and brilliance from this tragic context are new generations of Black farmers such as Leah Penniman, a Black Kreyol woman farmer who is co-founder of Soul Fire Farm in New York. Penniman teaches youth and Black, Indigenous, and Latinx farmers regenerative farming, commercial aspects of agriculture, Afro-Indigenous farming practices, silvopasture, sustainable construction, spiritual farming practices, and wildcrafting, through hands-on immersion and workshops as well as her books, *Uprooting Racism, Seeding Sovereignty,* and *Farming While Black.*

Penniman grew up in a mostly white town where she faced racist bullying, but her refuge lay in the Earth and the forest.[41] Years later, she found her passion when working on an urban farm in Boston; as she explains, "I found an anchor in the elegant simplicity of working the earth and sharing her bounty. What I was doing was good, right, and unconfused. Shoulder-to-shoulder with my peers of all hues, feet planted firmly in the earth, stewarding life-giving crops for Black community—I was home."[42]

But in her North American context, Penniman continued to struggle with the historical connections between agriculture and the enslavement of Black people. She found integration and answers on a visit to Ghana, where she learned of the Indigenous Earth spirituality of Vodun—she is now a Manye, or Queen Mother, of the tradition. In dreams, she saw herself planting maize on Krobo Mountain, where only the Indigenous Krobo farming people were allowed to plant. Her Ghanaian ancestry was later confirmed.[43] Farming was long in her lineage.

Penniman also strives to increase the land stewardship of Black, Indigenous, and Hispanic farmers and to help them rediscover their powerful ancestral relationships with the land. On her website, she offers people the opportunity to give reparations, which will be distributed to the Northeast Farmers of Color Network.[44]

Other manifestations of alternative economics and reciprocal relationships with the land bring to mind longstanding explorations of the land as a commons—not privately owned, but open for all to enter and fish, forage, and perform ceremonial rites under principles of reciprocity and taking only what one needs. To protect the land, the use and care of the commons were usually governed by a set of courtesies known to the local community.

There are many ideations and practices of the commons to explore, but here it is important to emphasize that a separation-from-nature worldview goes hand-in-hand with the idea of land ownership versus people being stewards of the land and sharing her gifts. Indigenous Peoples the world over conceive of themselves as stewards of lands, or rather belonging to a place—not owning it, but owing respect and attending to responsibilities to care for their ancestors and homelands.

The enclosures of the commons, mentioned earlier, beginning in the sixteenth century in the United Kingdom, further perpetuated the idea of dominion over nature and land-based people, as landowners took over all available land in an extension of feudalism—a system in which a small percentage of elites controlled the land and enslaved and exploited everyone else.

Centuries after the elites began enclosing areas previously considered the commons, an economic theory was developed to justify the mentality of land privatization: the tragedy of the commons, a term popularized by a 1968 article of the same name by Garrett Hardin, in which he cites an 1833 pamphlet that painted a picture of herders allowing their animals to overgraze a common parcel of land. Hardin claims that the existence of the commons would lead to ruin.[45] This is a worldview that proponents of capitalism continue to propagandize: resources are scarce, and others will use them up if they get there first, in an endless competitive struggle of individualism.

Many environmental historians and ecological economists have contradicted Hardin's views, demonstrating that he described an incident that might take place in an open access situation, while commons are generally used by a particular community who establishes rules of conduct so that no one person or family consumes a disproportionate amount of the shared resources. In fact, designating land as a commons most often has positive results, argued economist Elinor Ostrom, a position that won her the Nobel Prize in Economic Sciences, which she shared with Oliver E. Williamson in 2009. Ostrom demonstrated that private citizens can manage shared resources in a better way than corporations or the government.[46] In her research, she detailed design principles for the most successful commons arrangements: they include defined boundaries, collective decision-making, procedures to mediate conflicts, and provision agreements that are congruent with local social and environmental conditions.[47]

Some people refer to the process of managing the commons as "commoning," a practice that must adapt to changing resources and circumstances. This is the community aspect, recognizing that sharing nature's gifts establishes mutual relationships and responsibilities. These kinds of collective protocols are precisely the way we will need to govern ourselves in these times of increasing climate chaos and environmental degradation.

As a key example, a communal approach is sorely needed to protect our drinking water from corporations and pollution. Already, half of the world's wetlands have been lost to "development," our global waterways are fouled by two million tons of sewage and waste per day, our groundwater is being sucked up faster than it can replenish itself, the mining industry permanently poisons trillions of gallons each year, and a child dies every three and a half seconds from drinking dirty water in Global South countries.[48] Corporations, countries, and investors are now "land-grabbing" and purchasing (or stealing) water in mainly Global South countries in a particularly sinister form of colonialism that will almost surely result in catastrophe for those living there.[49] As we witness now, wherever Nestlé goes, communities lose their drinking water.[50] Coca-Cola has also famously been under fire for nearly two decades for depleting aquifers in and near their India locations, parching local farmlands and cities. The company was fined by the Indian government and subsequently promised to replenish the groundwater they had stolen, but as recent as early 2022, Coca-Cola was still illegally siphoning that water.[51] Water means life, and this sacred and critical element that humans cannot manufacture is being treated as just another commodity.

Clearly, corporations cannot be trusted to manage water. The most sensible and life-affirming solution is for waterways to be protected by public trust and managed by the humans who live in a shared watershed commons—it is in their interest to care for the water they will drink, fish in, and use for washing. Local water management is already being done in some regions, such as the rural areas of Licto, Ecuador. In the early 1990s, the Indigenous majority in Licto challenged the state's water management system, which had a history of discrimination against the Indigenous population and the poor, and eventually won the right to manage their own irrigation water.[52] And in Porto Alegre, Brazil, the local community's management of the water system enabled poorer residents to access water that had previously been available only to wealthy residents.[53] These are only beginning steps, and they need to be amplified everywhere.

Models of thriving commons are starting to spring up globally, including fishery cooperatives in Japan, Norway, India, and Scotland.[54] As was the case in England before the enclosures, meadows and grasslands have been successfully organized as grazing grounds in commons arrangements in several regions around the world, including Törbel in the Swiss Alps, which inspired Ostrom's theoretical work.[55] Similarly, in Mongolia, research showed that degradation in seasonal grazing pastures was relatively low when shepherds were allowed to guide their herds between them collectively—moving just as nomadic communities have done for millennia, though on a much smaller scale—while grassland degradation remained high in neighboring Russia and China, where the state did not allow the herds to make those journeys.[56] Collective management of land and water is most successful when those who participate espouse a place-based identity and a communal reciprocal relationship with each other and the landscape.

For most of us who have grown up in the capitalist dominant culture, such seemingly radical ideas of sharing in a commons or engaging in reciprocal practices might seem incredibly unrealistic, but that's where revolutionary imagination comes in—especially because the climate crises, environmental degradation, and the limits of the extractive, endless-growth economy will continue to transform our daily lives whether we are prepared or not. This change does not need to be frightening, but instead a welcomed opportunity to build the world we know is possible. The joy and promise of commoning have been proven by many communities, including the city of Atlanta, Georgia, where a seven-acre public park and old pecan farm has been revitalized as a food forest, serving a region that was once void of fresh food. The park is now a commons, abounding with mushrooms, fruit trees, vegetables, and herbs that line the beautiful walking trails. Bees are raised in an apiary to pollinate the organic plants and trees, and locals are invited to participate in composting.[57]

In 2020, the metropolis of Beirut, Lebanon, was struggling with economic and political issues, while also recovering from a devastating explosion (caused by ammonium nitrate that had been stored improperly) that killed over two hundred people and left more than three hundred thousand people without homes. From the ashes arose a beautiful outpouring of love and healing in the form of Tariq El Nahl herbal collective. *Tariq El Nahl* means "Way of the Bees" in Arabic; as collective member Paul Saad explains, bees maintain the balance of an ecosystem, working together and thriving among the beauty of biodiversity.[58] And this is what the collective does, as they work together to create regional botanical gardens throughout Lebanon that teem with native medicinal plants that people can take home to plant and tend themselves. The collective also seeks to revive ancestral traditions of healing with herbs, and they offer baskets of herbal medicine to the larger community, hoping the act of working the soil together and using the traditional local herbs will help to heal the region's multilayered struggles.

The capital of Senegal, Dakar, has opened the School of the Commons to encourage citizens to pursue the collective good through arts and culture, with priority given to youth and women. The project, which has gradually grown since 2011, focuses on the importance of collaborating, sharing, and living together—as artists commune in the collective spaces provided by the school, they teach each other new skills they can use in their jobs. The school has also planted a public garden designed by a local artist, built a playground, begun classes on ecology, and opened a FabLab, where people can attend free electronics workshops, learn to repair computers, and benefit from free software.[59]

In addition, many resources are now available for people who wish to implement commoning in their community, including *The Commoner's Catalog for Changemaking: Tools for the Transition Ahead,* written by scholar and activist David Bollier.[60] Observing that "the world we have inherited is no longer working," Bollier spent years learning about the commons and commoning, and he has filled this book with resources and recommendations for coordinating work in a shared system, solving problems that may arise in such an arrangement, experimenting with new ideas, and more.[61] As he explains, "We have relied for too long on centralized, unaccountable, hierarchical institutions that disempower us. But there is an alternative: Commoners can emulate each other's work and federate their initiatives into larger collaborative systems that constitute a different type of economy."[62]

At WECAN, I have been envisioning nets of resilience or connected circles of resilience, thriving archipelagoes of survivability as we face the times to come. Many others share these same dreams and projects, expressed and practiced in countless and creative ways. WECAN is honored to have co-initiated a facet of this work by partnering with the Okla Hina Ikhish Holo collective, meaning People of the Sacred Medicine Trail, which is an intertribal network of women, femme, and nonbinary Indigenous gardeners and farmers. The collective is working to restore their lands and waters in the ancient and fertile Mississippi River Delta, called Bvlbancha, or Bulbancha, by the Chahta (Choctaw). The network spans the US Gulf South, along old Native trade routes, within the traditional and contemporary territories of the Chahta, the Mvskoke, the Biloxi, the Tunica, the Attakapas-Ishak, the United Houma Nation, the Chitimatcha, the Washa, the Chawasha, the Tchopitoulas, the Bayougoula, and those whose names were erased during the early years of European contact. Additionally, these are the adopted homelands of other Nations who found their way to the southeast in search of sovereignty. Situated downstream from the poisoned region called Cancer Alley, Bvlbancha has been plagued by pollution, oil spills, and catastrophic hurricanes and flooding, which, along with persistent discrimination, means that Indigenous Territories there are disappearing at one of the fastest rates globally.[63]

Now, Okla Hina Ikhish Holo is re-establishing old trade routes while adapting and co-designing new future paths for trade ways that support local biodiversity, food sovereignty, and stewardship of Traditional Territories. The program is securing and growing food and medicinal herbs for local communities and supporting a sustainable path toward community resiliency during cascading crises of climate and colonization.[64] The WECAN Indigenous Food Sovereignty Program Coordinator for the Gulf South is Monique Verdin, of the United Houma Nation, and she explains the importance of the program this way: "We believe that it is crucial for us to protect and reclaim our relationships with our sacred homelands and waters in order to restore the natural balance, so the intelligence of this system can be respected for communities to survive in deep collaboration with nature like our grandmothers before us, preparing for the unknown by putting our hands in the ground and our hearts into the work needed to strengthen decentralized systems of support from mutual aid to intentional investment and building circular economies."[65] This is reciprocity and regeneration in real time based on a worldview that we are all relatives and part of an animate cosmology.

The garden patch I have been working today is ready to receive precious life-giving seed gems. Reaching for one of my glass seed jars, I open the colorful hand-woven textile bag that I use to carry my jars from storage to the garden. The ornate and sturdy bag returned with me from the Ecuadorian Andes. Like many weavings from this region, the many-hued patterns and symbols woven into the textiles are story filled, telling us of Pachamama Earth Mother; of the celestial bodies overhead; of the animal kin who walk, fly, and swim in the Andean regions; of the Kichwa Peoples' history and cosmology; and the web of life that connects all of us. The weaving is a colorful love song to Nature and all beings.

Holding the bag reminds me of the powerful words of Blanca Chancoso, the Kichwa Indigenous leader from the Ecuadorian High Andes who spoke at the Rights of Nature symposium in 2014 and shared with us the concept of Sumak Kawsay, stating: "Part of the foundation of our people is Sumak Kawsay; this means to live well, or good living with Pachamama, taking into account the whole of the natural world and human society."[66]

Sumak Kawsay, as explained by Blanca Chancoso, is similar but also different from the more well-known concept of *Buen Vivir*, in the Spanish language, which also conveys the concept of "living well" or "collective well-being." The Buen Vivir philosophy emerged in the Andes among the Kichwa communities, though it was also connected, like tree roots joined through mycorrhizal networks, to the worldviews of the Aymara in Bolivia and the Mapuche in Argentina and Chile.[67] This powerful tree of Buen Vivir with its strong Indigenous roots grew until its branches reached throughout South America and beyond, where the concept influenced elements of new economic movements, feminism, and political progressivism globally.[68]

Buen Vivir is a call to live more simply, to ensure everyone has enough without taking too much. It is an invitation to renew our relationship of reciprocity with the land and our human family, to revitalize a knowledge that the Earth does not belong to us, and that all of nature holds intrinsic value beyond its apparent usefulness to humans. It opposes colonialism, especially the colonial mindset that has privatized and commodified the land. Instead, Buen Vivir embraces the commons, calls for an end to extractivism, prioritizes community life and the ecosystem over the individual, and invites healing of the Earth and her peoples.

Eduardo Gudynas, Executive Secretary of the Latin American Centre for Social Ecology in Uruguay and an expert on Buen Vivir, clarifies the difference between Buen Vivir and the Western idea of well-being, saying, "With buen vivir, the subject of wellbeing is not [about the] individual, but the individual in the social context of their community and in a unique environmental situation."[69] Gudynas also explains that a Buen Vivir framework favors small-scale production that prioritizes local communities, their unique culture, and the health of their ecosystem, and this would require a change in the global economic framework of today.

Leandro Garcia, a Kichwa government official, emphasizes the need to participate in one's local community and engage in active dialog to find out what their specific needs are. In his Andean home, this process of collective work and pooling of knowledge is called *minga*. As Garcia says,

"Minga work and democratic living teaches us that we live by doing things for others," adding that modern society, in which many people do not even know their neighbors, has lost the value of solidarity.[70] Minga includes care for the environment. The Cotacachi-based group Citizens' Minga for Environmental Education teaches the local community to manage the region's ecological issues, especially the water and its distribution.[71]

On a larger scale, Kichwa Peoples in the Cotacachi Canton of Ecuador champion Buen Vivir as they resist the highly destructive industrial mining in their homeland. In 2000, they succeeded in persuading the local government to deem Cotacachi Canton South America's first ecological canton by municipal ordinance, all because of grassroots social mobilization by people who hold sacred their responsibilities toward nature.[72]

It does not require a crystal ball to foresee that the endless material growth-based centralized economy will continue to fail, and that multiplying climate and socio-ecological crises will force us to change our lives—the fact is, this is already happening. Now is the time to mobilize to create alternative systems that benefit the Earth and all her peoples, working with the inherent laws of nature rather than against them. Like the intertwined and colorful symbols on my woven seed bag, we as diverse peoples are interconnected in responsibility and reciprocity. The inevitable changes underway will require all of our collective imagination and a receptivity to other worldviews and other ways of being. Aside from the obvious ecological benefits, localized economies result in more jobs than centralized economies, and they provide greater security for communities.[73] Land Back, Buen Vivir, degrowth, commoning, BIPOC farms, care and feminist economics, regenerative agroecology, are all part of an Earth Community Economy that centers a re-emergent and thriving reciprocal relationship to the Earth and each other. For further reading on applying these same principles to a more urban context, I would also suggest the work of Transition Towns—there are hubs on all continents, and the Transition Network has information to start practical community projects such as seed swaps, skill-share events, repair cafes, ride-sharing schemes, and green-energy audits.[74] A plethora of other models are underway that are attuned to specific peoples and place as the need for mutual aid increases and as people seek community self-sufficiency and localized interdependence, including community gardens in almost every city.[75]

As springtime burgeons again after a cold winter, I plant corn and squash seeds in my garden where they will begin their sprouted story. Just as the sun is setting, with the seeds all tucked away in their raised beds, I tenderly plant a painted egg into the ground for the season—both an offering to the gifts of the coming harvest and a tribute to my ancestors. It heartens me to know that the multifaceted custom of egg painting and offering has survived the ages, though its origins are often overlooked. Surely, the egg is one of the oldest and most universal symbols of birth and life, and so it is no wonder this precious oval is a symbol for many in welcoming the spring season, when Mother Earth gives birth to so much life after the long winter.

Since ancient times, the egg has been a symbol of the Creatress Mother, and her World Egg contains all of creation. This understanding is represented in the ancient Greek alphabet, in which the great Goddess is symbolized by the design of the capital letter omega, the great Ω. In the Orphic history, the great Ω and the World Snake make love, and the Goddess lays the World Egg: the birth of all births, which then initiates the procreation of all of existence. The lowercase omega, ω, shows the World Egg separated into two halves to create the universe; this happened with the birth of the Sun, whose presence divides the seasons. Every spring, then, a new year "hatches."[76]

Earth-based peoples from old Britain to the North Sea and east to the Black Sea and the Caucasus Mountains have celebrated the sacred egg. From the Indus River to the Indian Ocean and ancient Egypt, people have revered this powerful symbol of life. In many parts of Old Europe, eggs were colored scarlet in honor of the Sun. In ancient Egypt, the Mother Goddess Hathor laid the Golden Egg of the Sun. The ancient Etruscans' symbol of creation was a black egg of polished stone, and upon the stone, an arrow was carved in relief, representing the life-giving marriage and balance of female and male. In some European traditions, young men received painted eggs from young women at the beginning of spring to signal courtship.

The Thracians honored Mother Night, who brought forth the white World Egg they named the Moon, and they taught that this celestial body designed time through the lunar cycles of birth, growth, maturity, and death.

The ancient Britons had a ceremony in which they built an altar honoring the axis of the four directions and gave eggs and apples to all the tribal peoples. The druids of Albion and Ireland described the Holy Egg as the red egg of the sea serpent.

In many Slavic countries, the tradition of egg painting goes back thousands of years. People painted eggs with designs called *znaki*, which had special meanings and were part of meditation and spiritual practices. The Poles also planted decorated eggs in newly sown fields and at the base of fruit trees to ensure abundant growth. Many a home in these Slavic lands always had a basket of colorfully decorated eggs to bring good health and prosperity to the family and community (and some still do).

In Ukraine, an egg adorned with mystical symbols is called *pysanka*, which means "written." Eggs were traditionally painted by women, and they followed various elaborate rituals to do so. First, a woman gathers water from various springs or wells, or from snow she has collected during the month of March, and she uses this water to prepare natural dyes from plants. She then purifies herself through sacred bathing, paints the symbols onto the egg with molten wax, and dips the egg into the dyes, from light to dark, while calling upon different Goddesses, adding more wax between dye baths to preserve the desired colors. She then polishes the egg to remove the wax and buffs it to a brilliant shine with oil. These Mother World Eggs prepared in the spring season have protective powers and can be gifted to a beloved, offered in the fields, placed under a beehive, or displayed on the home altar, where they shield the home and its occupants from calamity.[77]

Try as they might, it was impossible for the newer patriarchal religions to eradicate and erase the many festivals and celebrations of Mother Life during the spring season. Everywhere wildflowers painted the countryside in ribbons of color, and young animals from baby squirrels and bear cubs to fawns peered from the forest. It was impossible to ignore Mother Earth's message of birth and life, so the old traditions were never fully suppressed by the new patriarchal, separation-from-nature religious tenets.

The life force of nature cannot be conquered by humans, nor can humans manipulate the laws of nature. Now is the time for us to liberate ourselves from de-animate, soul-crushing, ecology-shattering systems as we re-establish a reciprocal, mutually enhancing, mutually indebted relationship with all of our magnificent Relatives, and once again become a beneficial force in the thousandfold beauty and brilliance of our living Earth.

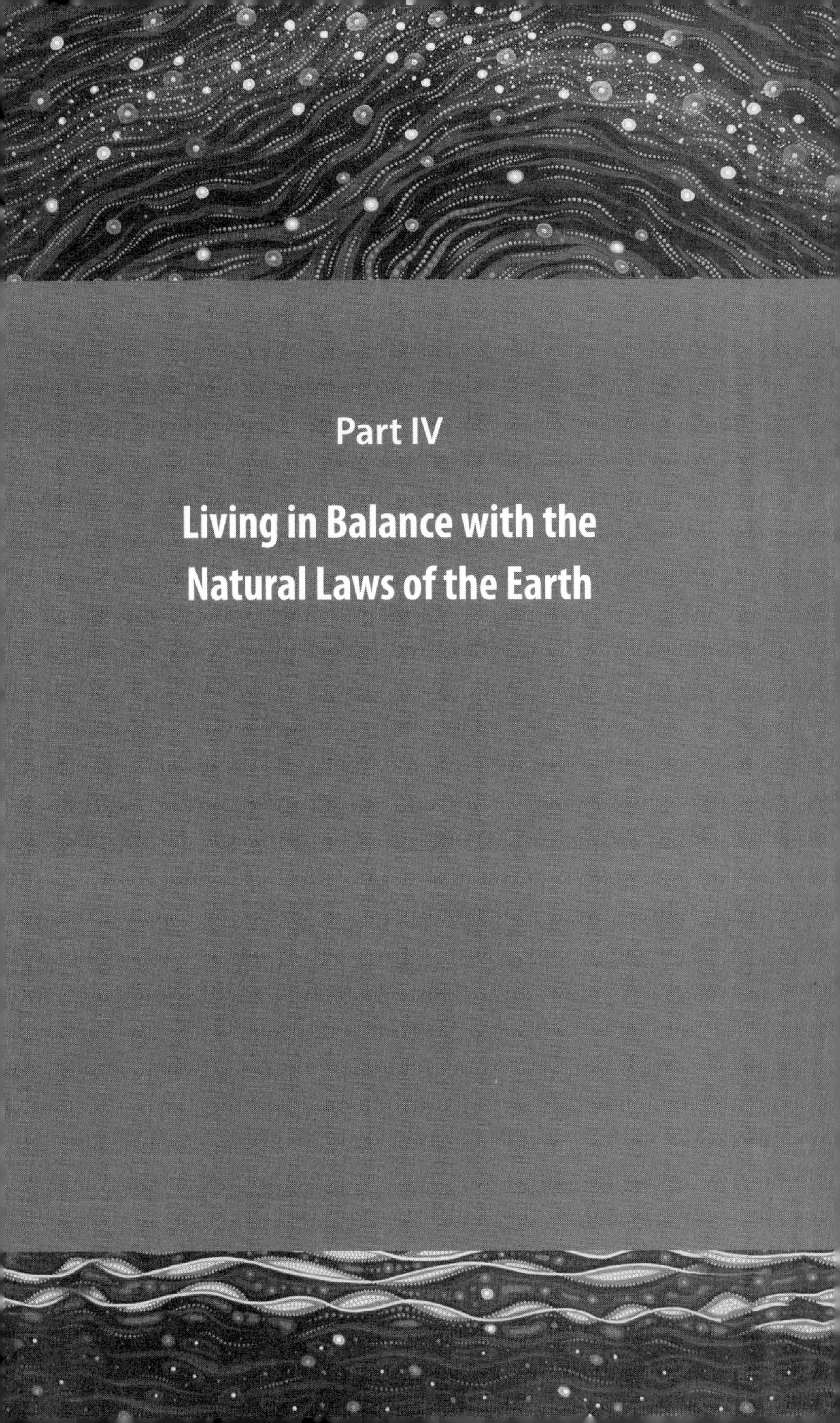

Part IV

Living in Balance with the Natural Laws of the Earth

Chapter 12:

Rights of Nature: A Systemic Solution

THERE IS A FAST-GROWING MOVEMENT called Rights of Nature that aspires to recognize Nature legally and culturally as a rights-bearing entity. It offers long-term systemic solutions to interlocking crises. At its core, Rights of Nature presents a worldview shift from an ideology of human dominion over Nature to understanding and practicing that humans in all their diversity are but one part of the sacred ecological systems of life.

To describe what Rights of Nature is and to explore its implications and implementations, I would like to start with a few experiences that have shaped and informed my understanding of this worldview and legal philosophy. The first involves a call from Indigenous Peoples to respect and implement Indigenous rights and Traditional Ecological Knowledge because these rights and knowledge systems reflect the ancient headwaters from which the Rights of Nature movement was born.

Ena Santi, a Kichwa leader from the Sarayaku community in the Ecuadorian Amazon and a council member in charge of women's issues for her people, stood on the stage, speaking at one of the most unique and legendary press conferences I have ever attended. Ena was a representative of one of several delegations of Indigenous Peoples from the Americas advocating at the historic United Nations Framework Convention on Climate Change (UNFCCC) Conference of the Parties (COP) 21 where officials negotiated the Paris Climate Agreement in 2015. She stood before an audience of journalists, global advocates, policymakers, and civil society members, floating on the Péniche Antipode barge along the Bassin de la Villette, a lake that connects to the Seine, the iconic river that winds through Paris. As she spoke, her presence was magnified by intricate obsidian-black markings across her forehead and cheeks. The ink from her traditional *wituk* face painting is made from an Amazonian jungle fruit, also called *wituk*, and is used to render ornate and delicate geometric patterns of great meaning to the person who wears them. Ena later told me her designs represented the jaguar and the sun.

The press conference was hosted by various Indigenous organizations, communities, and movements, including Indigenous Environmental Network, Idle No More, and the Kichwa community of Sarayaku. Attendees and organizers filled the long red-and-black barge, which had been decorated with banners featuring powerful Indigenous figures and proclaiming messages in English and French such as "Indigenous Women of the Americas for the Defense of Mother Earth," "Honor Indigenous Treaties," and "Respect the Sacred System of Life."

It was standing room only on the lower deck of the barge, where Indigenous Peoples had called the press conference to demand real climate solutions, including grassroots initiatives originating in Indigenous knowledge, culture, and spirituality. They specifically were advocating for governments to include Indigenous rights in the Paris Agreement, as many nations, including the US, Australia, and the European Union, had pressured the UN to drop Indigenous rights from the sections of the new accord that were legally binding. In the end, Indigenous rights were only mentioned in the preamble, rendering them aspirational but not legally enforceable, and that is where they remain to this day.

One after another, Indigenous speakers, dressed in their traditional attire, stood on the riser at the front of the audience and delivered the words of courage, strategy, and ferocity demanded by the moment. Most wore red bandanas or other red clothing to symbolically demonstrate the red line not to be crossed to protect communities and Mother Earth.

"We, Indigenous Peoples, are the redline," they stated.[1] With each testimonial offered from Indigenous leaders, those of us in attendance felt the heat and passion of the Earth crying out from all the harms done to our living planet and frontline communities. They explained that because of their knowledge, passed through generations of stewarding still-healthy lands, Indigenous governance and voices are a solution to our current crises.

The barge could accommodate perhaps one hundred people on the lower deck, but it seemed as though thousands occupied the little boat—the press room resonated with the clarion call for resistance and survival over many generations. Indigenous leaders had come to warn the world that our Mother Earth's voice must be heard and that their rights to protect the natural world must not be denied. As Indigenous Peoples and allies bound together at COP21 in common cause, we were collectively committed to fighting ongoing colonial ideology—it must end, or the world as we know it will end. Each speaker moved the audience in ripples of illuminations and emotions, and there was more than one listener, including me, quietly wiping tears as our hearts cracked open with beauty and grief in that visceral encounter.

When it was her turn to speak, Ena Santi focused her penetrating gaze on the women, beckoning us in particular to lead and struggle forward with all our strength and power. She had traveled for some days, crossing her rainforest homeland by foot and canoe before taking a small plane, and then a jet, to Paris. I can still feel her pronouncement ringing deep in my core of being: "We women can do anything. We women run through the jungle at night with our babies on our backs. We women can do anything. We are the descendants of the jaguars. We women can do anything!"

Her transmission stirred within me while a layer of modern-day, dominant-cultural malaise began to fall away as she reminded us all that we originate from nature, we are nature, we are Earth's children. Through the long history of evolution, we are literally the descendants of our plant and animal relatives, and other older elemental ancestors. Ena explained that she and her people are descendants of the magnificent jaguar, and she implored all the women to remember this strength for the work ahead.

In listening to Ena, I thought this is precisely the kind of life force and worldview we need to recognize in our bodies and minds to transform the deleterious systems of domination and oppression that we are up against. Ena reminded us that we as women can rise in this power of the Earth, which we hold in our very cells, which exists as our birthright. We need to act as Mother Earth's immune system, rising up as midwives, warrioresses, leaders, and healers. Ena's voice, thundering from the spiritual depth of the Amazon rainforest, summoned us to think differently and remember who we are and our duty to this moment in time.

The entire day of the press conference was remarkable in every way. It had started with a gathering of primarily Indigenous women from around the world holding a ceremony for the Indigenous Women of the Americas Defending Mother Earth Treaty. In the early afternoon, youth from the Lummi Nation drummed and led songs as a flotilla of more than twenty-five kayaks, paddled by Indigenous leaders, floated along the waterway to make visible the fight for Indigenous rights.[2] The Sarayaku "Canoe of Life," carved from a Canella tree in the form of the rare hummingbird fish, which lives in the lagoons of the Sarayaku homeland, would arrive two days later to partake in the event, concluding its six-thousand-mile journey from the Amazon.

The kayaks were bedecked with banners displaying Indigenous artwork, and the flags of many Indigenous Nations undulated in the breeze. Thousands of people lined the sides of the waterway and the bridges above, affirming their solidarity, unfurling supportive banners, and beating drums. As the kayakers paddled toward the barge for the press conference, the Lummi singers carried them forward with their mesmerizing, undulating song of the waters as many of us called out in encouragement.

During interviews at the flotilla action, Indigenous leaders made their vital message clear. "We cannot negotiate a climate agreement at this critical time without the recognition of the rights of Indigenous Peoples, who are on the front lines of the impacts of climate change and the innovators of solutions we need to stabilize our climate," stated Tom Goldtooth, Executive Director of the Indigenous Environmental Network.[3] "We belong in this treaty, and we have a place in this discussion. Our future and the future of our children are not up for negotiation," stated Crystal Lameman, Treaty Coordinator for the Beaver Lake Cree First Nation, which lies in Canada's tar sands region.[4] This Indigenous-led action was one of many critical events that arose at COP21.

Outside the UNFCCC negotiating halls, environmental and social movement leaders, community organizations, and activists gathered for strategy sessions, conferences, and nonviolent civil disobedience actions on the streets of Paris. Our voices joined in a collective call for bold, transformative change, advocating for centering of frontline leadership, community-led solutions, feminist policies, forest protections, keeping fossil fuels in the ground, and the accountability of polluters. Like ripples in a deep pond, these alternative parallel proceedings, and actions continue to reach outward and broaden into ever-widening circles to this day, connecting one to another and spreading globally.

Just days before Ena Santi delivered her remarks at the press conference, I was honored to meet her at the International Rights of Nature Tribunal, which WECAN and other groups had co-organized under the umbrella of the Global Alliance for the Rights of Nature (GARN). This global alliance had been facilitating Rights of Nature tribunals for the past years, purposely conducting them in parallel to the annual UN climate conferences to specifically demonstrate alternative solutions deriving from a worldview that differs sharply from that of the government officials in the UN climate meetings.

Rights of Nature is a potent tool that could significantly help provide a framework to navigate climate and environmental justice issues, guided by the natural laws of Mother Earth.[5]

While it is vital to take on struggles in sites of immediate harm, injustice, and environmental destruction with all the tools and legal levers available, at the same time we need deep structural reconfigurations for long-term systemic solutions, healing, and transformative strategies—and this is where solutions such as a Rights of Nature framework are essential.

Each year brings new record-breaking temperatures, massive deforestation, pollution of waters, colossal species extinction, and the list goes on. The question is: How is it that we have not stopped these destructive actions that are not only lethal to the web of life but also perfectly *legal* under our current laws?

In her seminal book *Silent Spring*, Rachel Carson proclaims, "The question is whether any civilization can wage relentless war on life without destroying itself, and without losing the right to be called civilized."[6] Today, as we face the destruction of all life as we know it, the burning, polluted, and dying world forces us to reckon with what we deem "civilized." Under the banner of so-called civilization, we have failed our most primal duty to the living world and to each other.

Rights of Nature, also referred to as Rights of Mother Earth, is a step toward becoming truly civilized in the highest sense of the word: living civilly with other humans, the Earth, and the more-than-human members of the web of life; respecting the laws of nature and the boundaries of the ecosystem; and recognizing the world as a living being hosting myriad other animate beings that enrich all aspects of existence.

The question, as I have asked throughout these pages, is how can we address root causes and enact systemic change in order to reorient a human presence that is Earth-enhancing and reciprocal with the natural world so that we live in harmony with each other and with Mother Earth. A primary impediment lies within our current legal frameworks, which need to be challenged and changed. Most countries have environmental laws, but they are clearly not working. If those laws functioned as a guide to proper human activity, we would not be careening full speed off a cliff. But at their core, these laws are an expression of the values and worldview of the exploitive, dominant culture.

In this context, Rights of Nature has the potential to not only help prevent some of the worst effects of climate change and ecological devastation but also help restore the role of humans as a beneficial species on the planet.

For the past three decades, environmental lawyers, visionary thinkers, Indigenous leaders, and activists around the globe have been developing a new theory of jurisprudence, the Rights of Nature approach, which promotes a structure of law that recognizes that our living planet has rights of its own. When a Rights of Nature legal framework is implemented, environmentally disruptive activities that harm the ability of ecosystems and natural communities to thrive and naturally restore themselves are deemed legal violations of Nature's rights.

The Rights of Nature framework recognizes the inherent meaning, sacredness, and value of the natural world: not reducing it to mere economic value. Respecting these rights, along with human rights, encompasses what it means to be civil.

The erosion of reciprocity practices and rootedness in nature-oriented cultures has unfortunately informed our legal systems in the dominant culture. To illustrate this point further, at a WECAN event hosted in New York, Aura Tegría Cristancho, a young leader and representative of the U'wa People of Colombia, who live in the Cloud Forests, reflected upon the Original Instructions and stories her grandparents passed on to her. In the beginning, Aura told us, the Creator designed the Earth in equilibrium and, in the heart of the U'wa People, left an original message—instructions to care for and protect Mother Earth. Protecting the Earth thus became the mission and center of the U'wa mind, body, and spirit.

When oil companies invaded their homeland in the 1990s, the U'wa knew they had to speak out in defense of the land. Having seen the violent devastation that their nearby neighbors endured at the hands of industry, the U'wa perfectly understood the perils of extraction. The U'wa call oil *ruiria,* and Aura explained that this is the source of all plant and animal life, as well as the human spirit. She stated, "For the U'wa, the oil is like the blood of Mother Earth; it is vital to life—if you take out the blood, the whole planet will die, including human beings."[7] She continued, saying, "In this sense, we must gather our strength and unite in defense of the Earth, in defense of life. All of us together: Black, White, Indigenous, old, young, all of us."[8]

Aura's community took a staggeringly brave public stand, avowing that they would rather literally die collectively by suicide than witness the degradation of their ancestral homeland. Support poured in from international groups to bolster the U'wa. The Tribe's fierce dedication compelled the oil company to withdraw in 2002—though the corporation claimed their capitulation was due to economic troubles after digging a dry well.[9]

While Rights of Nature is a developing framework by and for modern societies, the sources of this ideation to protect and defend the sacred living systems of life are firmly grounded in Indigenous worldviews, such as those of the U'wa. Thus, protecting Indigenous rights, sovereignty, and knowledge should be an inherent priority of Rights of Nature work, as the Indigenous Day of Action demonstrated in Paris.

Establishing Rights of Nature globally is one of the most effective ways we can simultaneously and coherently address our dysfunctional environmental, legal, colonial, and economic frameworks as well as injustices to land defenders, and begin to swim upstream to renew reciprocal relations, and healthy agreements and practices with Mother Earth. We need to be like the salmon who follow their innate impulse to return to their ancestral streams with an undaunting dedication to future generations and the life-enhancing practices that come with this duty.

In this moment, the urgency of the climate crisis and the vast destruction of ecosystems demand that we completely redesign our economic and legal systems.[10] To truly live sustainably and in harmony with the Earth, we need to change the very foundations, the very structural DNA, of our legal frameworks to respect the natural laws of the Earth. By implementing Rights of Nature, we have the opportunity to not only change our laws but also begin to establish a culture of reciprocity. In our global movements for justice, we are looking for revolutionary and evolutionary transformations for society, and Rights of Nature can deliver an important model for structural metamorphosis.

The past decades of environmental protection laws have surely achieved some notable successes and are essential in the immediate term, but in the larger scheme of human-nature relationships, our legal systems have failed to prevent the increasingly grave threats of global warming, the degradation of Earth's ecological systems, and the growing displacement and death of humans and other species.

Today, the majority of the world's legal scaffoldings are based on treating nature as a treasure trove of resources for the shrewdest capitalists. Our life-giving rivers, forests, and mountains are treated as property, which is a main crux of the problem. By treating them as property, the law allows these living sources of life to be bought, sold, and consumed, with no process to recognize these natural communities and ecosystems as rights-bearing entities, rendering them invisible to courts. By adhering to the current structure of law and economy, we are furthering a dangerous relationship of ownership and exploitation with the natural world—a way of life that is highly uncivilized.

We need an entirely different legal configuration that recognizes that Earth's living systems are not the enslaved property of humans. Just as it is wrong for men to consider women property, and for white people to enslave non-white peoples, it is wrong for humans to claim dominion over the Earth. It is also important to be clear here: humans cannot arrogantly grant rights to Mother Earth; rather, the aim is to recognize the rights that Nature inherently possesses, or in other terms, to respect the natural laws of the Earth and our ancient agreement with and responsibilities to the web of life. When endorsing Rights of Nature, Sami parliament member Marie Persson Njajta put it this way: "We, the Sami people, believe that we belong to the land, not the other way around. Today we see how a colonial perspective, exploitation, and climate change threaten our culture. And it is not just us; it is a global issue."[11]

Christopher Stone, a professor at the University of Southern California law school, was an early proponent of Rights of Nature, outlining the concept at length in his 1972 essay "Should Trees Have Standing?" in which he asserted that nature possesses value far beyond its ability to benefit us, and it should thus be protected on its own accord. Stone also anticipated the concept's reception in the legal realm: "The fact is, that each time there is a movement to confer rights onto some new 'entity,' the proposal is bound to sound odd or frightening or laughable. This is partly because until the rightless thing receives its rights, we cannot see it as anything but a thing for the use of 'us'—those who are holding rights at the time."[12] White supremacy was not the familiar talk of Stone's day, but the sentiment in his article seems to also aim at this infringement of the dominant culture upon Indigenous Peoples, People of Color, women, and Nature—a connection examined more explicitly by environmental studies professor Roderick Nash in his 1989 book, *The Rights of Nature: A History of Environmental Ethics*.[13] Stone also pointed out the need for lawyers to speak for natural systems, as they already do for humans and corporations. If corporations can have a legal voice, then the ocean should too.

Cultural historian Thomas Berry introduced the term Earth Jurisprudence to describe a legal philosophy that prioritizes natural laws over human interest and recognizes that every part of nature possesses the intrinsic right to exist and evolve. Particularly in his 1999 book, *The Great Work: Our Way into the Future*, he called for a shift from laws that center humans to those that center the Earth.[14] Berry emphasized that Western nations had much to learn from Indigenous worldviews and Indigenous leadership. He also pointed out that earlier environmental laws had been written for a natural world composed of distinct parts rather than an interconnected, interdependent, animate, cosmic, multivalent, and mystical realm. These exclusively human-centric and compartmentalized laws prioritize economic interests over the well-being of Earth's living systems.

A primary premise of the Rights of Nature movement is that, to ensure a thriving future, humans must reorient themselves away from an exploitative and ultimately self-destructive relationship with nature to one that honors the deep interrelation of all life and contributes to the health and integrity of the natural world. And this Earth-centered worldview must be reflected in our jurisprudence. At GARN, we advocate that breaking out of the human-centered limitations of our current legal systems is one of the most transformative, highly leveraged actions that humanity can take in this time of crisis.

Unfortunately, most of our current laws simply call for environmental regulations, and these regulatory laws do just that—they regulate how much pollution or destruction of nature can occur under the law. They do not prevent pollution and environmental destruction but specify how much can be done. Many of us are engaged in these regulatory processes because emergency measures are needed to protect and defend communities and threatened areas of nature under immediate attack, yet we are aware that these efforts aim to extinguish the proverbial urgent fires while leaving the gas turned on to spark future infernos. At best, the current laws may help us to slow the rate of environmental degradation, but they cannot stop it or address systemic change.

And the need is now. The world saw an eleven percent decrease in tree cover between 2001 and 2021; one million species are in danger of going extinct; global temperatures have been rising by 0.18 degrees Celsius each decade since 1981; only two thirds of large rivers worldwide are permitted to flow uninterrupted; and we have lost seventy percent of our wetlands since 1900.[15]

It is clear we cannot solve the climate and environmental crises by further subjecting the Earth to the same destructive legal system and human-centric worldview that caused them. These systems are irrevocably flawed because they are far removed from the fundamental natural laws and sacred codes of conduct with nature. As in the example of capitalism, which is based upon a legalized ownership of the Earth, they encourage, and even require, the exploitation of humans, forests, waters, minerals, and animals.

If Rights of Nature legal frameworks were implemented, activities that harm the ability of ecosystems and natural communities to thrive and naturally restore themselves would be fully illegal.

And for the nay-sayers, just to be clear, Rights of Nature focuses on protecting natural systems so that they can thrive and evolve. No one is saying a person cannot go out and catch a fish, for example; what Rights of Nature prevents are practices such as overfishing or using fishing methods that cause mass damage to the ecosystem or marine species. Mother Nature offers us gifts, and Rights of Nature laws are reciprocal practices at the legal level: if we care for the oceans, forests, and soil, Mother Nature can continue to provide her generous abundance. Rights of Nature is a way to fulfill our responsibilities to the web of life through legal means.

Around the world, social and ecological movements are increasingly denouncing the commodification and financialization of Mother Earth. Our sacred earth, water, air, and forests should not be in the marketplace, nor should we accept market-based carbon-trading schemes to mitigate the climate crisis. Instead, pollution must be stopped at its source, and all humans, governments, and corporations must rather learn to live within planetary boundaries and reciprocity with Pachamama, our Mother Earth.

This work involves changing local laws to recognize that natural communities and ecosystems possess a fundamental right to exist and flourish, and that residents of those communities possess the legal authority to enforce those rights on behalf of those ecosystems.

We *can* change our laws—think civil rights, suffrage, and the end of apartheid. This movement is about taking back the right to environmental decision-making in our communities. We have seen communities and governments from Chile to Uganda to Switzerland pushing forward legal protections for nature systems, clearly proclaiming that Rights of Nature is an idea whose time has come.

In 2008, Ecuador became the first country in the world to recognize Rights of Nature in its constitution, which states that nature has the right to exist, to be cared for according to its natural life cycles and ecosystems, and to be restored if it is harmed.[16] Although there have been significant challenges to the implementation of this legislation, Right of Nature efforts bore fruit as early as 2011, when defenders of the Vilcabamba River won a lawsuit filed in protest of a project to widen a road. The project, which had resulted in debris and contaminants falling into the water, was stopped because it violated the rights of the river.[17] In December 2021, Ecuador's highest court ruled that plans to mine for copper and gold in Los Cedros, a protected cloud forest, are unconstitutional and violate the Rights of Nature protected in Ecuador's constitution.[18]

Communities in the US have been successful in Rights of Nature cases, including the borough of Tamaqua, Pennsylvania, in 2006. Corporations and farmers had been filling abandoned mine pits with toxic sludge consisting of sewage and chemicals, and this sludge had been leaching into the river, sickening residents and wildlife who depended on its water. Now, anyone living in Tamaqua can sue a corporation for violating Rights of Nature, and any damages won in court go toward healing the local ecosystem.[19]

Indigenous Peoples in North America are also implementing Rights of Nature legislation. The Ponca Nation made history as the first Tribe in the US to implement Rights of Nature laws to protect their territory from fossil fuel extraction, which is causing asthma, cancer, and earthquakes in their region of Oklahoma. In July 2022, they expanded this legislation to include the "immutable Rights of Rivers," most significantly the Ní'skà (the Arkansas River) and Ni'ží'dè (the Salt Fork River).[20] The council of the Ho-Chunk Nation has also voted to include a Rights of Nature amendment in their Tribal constitution. In addition, the US has passed more than three dozen Rights of Nature ordinances to protect communities from harmful practices such as shale-gas drilling and fracking.

In 2018, the Chippewa Nation's White Earth community in Minnesota implemented a measure to protect the rights of manomin, or wild rice, and the rights of the freshwater habitats in which it grows. Wild rice is a nourishing and delicious staple food of the Ojibwe and other communities, and harvesting it—traversing the waterway by canoe and tapping the stalks to release the grains—is a tradition that has been passed down for generations, a spiritual practice and cultural relationship with the land. As Anishinaabe activist Winona LaDuke expresses it, "Wild rice is our life. Where there's Anishinaabe there's rice. Where there's rice there's Anishinaabe. It's our most sacred food ... It's who we are."[21] For the Indigenous Tribes of the area, recognizing wild rice as "an independent entity with the right 'to exist, flourish, regenerate and evolve'" is an act of sovereignty and reciprocity, fulfilling a responsibility to their sacred land.[22]

A treaty signed by the federal government and the Ojibwe in 1837 granted the community access to the manoomin on ceded territory. Wild rice is the only grain guaranteed in a treaty, and the 2018 law is the first to grant legal rights to a specific species of plant. But industrial development is threatening the waterways, destroying this food that grows on water. Since 2021, the Enbridge Line 3 oil pipeline has torn through parts of Minnesota, where the sacred manoomin grows. And the Ojibwe land defenders' concerns are justified because Enbridge projects have already resulted in many oil spills.[23] One of the powerful and main leaders of the Line 3 resistance, Tara Houska, who is Couchiching First Nation Anishinaabe and Founder of Giniw Collective, states, "This is the only place in the world where wild rice grows, and a threat to that is a threat to our survival as a people. It's not only a violation of our treaty rights and the expansion of the fossil fuel industry, and climate change and all of that, it is a threat to our ability to survive and to pass on our practices of wild rice to the next generation and to the generations to come."[24]

The White Earth Nation has sued the Minnesota Department of Natural Resources for allowing the pipeline to be built without Indigenous consent, and because manoomin has legal personhood, the grain was the lead plaintiff in the case.[25] Unfortunately, the case was dismissed because the Nation lacks jurisdiction outside the borders of their reservation, but, as of mid-2022, White Earth Nation members are still fighting to reverse the court's order.[26]

Over eight thousand miles away, the High Court of Bangladesh deemed the Turag River, along with all rivers in Bangladesh, a living entity with legal rights under the guardianship of the National River Protection Commission. This resolution was adopted in the wake of a 2019 case dealing with illegal development and pollution of the river.

In 2018, the Colombian Supreme Court of Justice issued a ruling to halt deforestation in the Colombian Amazon, granting the forest status as a rights-bearing entity deserving of protection and restoration. This action was prompted by a lawsuit filed by twenty-five children and youth, demanding that the government protect their rights to life, health, food, water, and a healthy environment. Deforestation, they stated, increased greenhouse gas emissions, causing climate change and endangering their future and that of their rainforest.

Lake Erie, the twelfth-largest lake in the world, has also achieved legal standing, with the ability to sue polluters. In 2019, residents of Toledo, Ohio, voted for the Lake Erie Bill of Rights, which recognizes the waterway's right to thrive and evolve, free from pollution. Advocates of the bill asserted that the lake has been damaged by industry and agriculture, citing toxic algae blooms caused by fertilizer runoff—a tragedy seen in waterways across North America—which not only turned the picturesque glacier-carved lake a sickly green but also contaminated Toledo residents' drinking water.

Opponents of the bill worry that minor actions of farmers, corporations, and coast-dwellers will be blown out of proportion and render them liable for fines, and they have challenged the law in courts. Those who defend the bill acknowledge that it is a new idea that will take time to get used to, and that deciding how to implement and enforce the new legislation will be challenging, but they deem the opponents' supposed worries as simply a ploy to distract voters.[27]

Another facet of the work to create new legal standing for nature was developed by the late Polly Higgins, who in 2009 began campaigning for the inclusion of ecocide as a crime in the final Rome Statute of the International Court. Ecocide refers to destruction of ecosystems and systemic severe damage to nature, and there were efforts in 1996 to classify it as an international crime against peace. Higgins, a barrister from Scotland, joined the fight after experiencing an epiphany during a long court case. Her client had been injured at work, and Higgins realized that the Earth was also being injured and needed a good lawyer. She left her job and campaigned tirelessly for international laws that would protect the Earth and hold nations and corporations accountable for any damage they caused. There is now a global network across many disciplines specifically dedicated to making ecocide an international crime.

Bolivia established eleven new Rights of Nature laws after hosting the World People's Conference on Climate Change and the Rights of Mother Earth in April 2010, which also produced the Universal Declaration of the Rights of Mother Earth, a key document in the movement.

Bolivia decided to hold the conference because the government parties at the 2009 UNFCCC in Copenhagen (COP15) were unable to come to an effective agreement on climate legislation. Unlike COP15, where civil society was barricaded from key negotiating spaces, the World People's Conference in Cochabamba, Bolivia, was open to anyone who wished to attend, and over thirty-five thousand people came—even more had planned to participate, but flight cancellations due to ash from Iceland's Eyjafjallajökull volcano kept most Asians and Europeans at home.

At COP15, parties argued over exactly how much damage Mother Earth could withstand before *humans* experienced catastrophic change, debated how much of the climate budget should be given to which countries, and negotiated how little support wealthy countries, who have historically polluted the most, could get away with providing to vulnerable nations who are mainly affected. In contrast, attendees at the World People's Conference openly shared their lived experiences in seventeen working groups, discussing how climate change had affected their communities and nature, and what strategies could be devised by looking through a climate justice lens.

Among other outcomes, the World People's Conference produced the Universal Declaration of the Rights of Mother Earth (UDRME). The document was born from drafts which had been posted online, where individuals could participate in the collective statement, and many of the elements were provided by Indigenous groups that had traveled to Cochabamba.

The UDRME is a companion piece to the Universal Declaration of Human Rights (UDHR) in that people, nations, and corporations that perpetrate harm are held accountable under the law, and they are responsible for rectifying any damage they have done. The document also states that people and institutions must be empowered to defend Mother Earth's rights.[28]

Efforts have been made to prioritize the adoption of the UDRME by the United Nations, but that has not yet occurred, though there are ongoing discussions at the UN under the umbrella of the Harmony with Nature program. Nevertheless, many people and organizations have been inspired by the declaration and have drafted their own legislation. In 2012, a Universal Declaration of the Rights of Nature was proposed in a resolution by the International Union for Conservation of Nature. At their 2016 conference in Hawai'i, the incorporation of the Rights of Nature was formally adopted.

Five months after the World People's Conference in Cochabamba, the Global Alliance for the Rights of Nature was formed to promote the values in the UDRME. On a personal note, I am honored to serve on the Executive Committee of GARN as we continue the fight for our Mother Earth's rights.

Implementing Rights of Nature laws would also require fundamental changes in the world's economy, since extractive activities such as mountaintop removal would become illegal, as would industrial agriculture practices that genetically alter natural species, poison soil and water, and negatively impact biodiversity. Corporations, industries, governments, and individuals would need to keep their activities well within the healthy and finite boundaries of the Earth—within the Earth Community Economy that I have mentioned in Chapter 11, based on respecting healthy ecosystems, equity for all, and reciprocal relationships as the guiding principles for any economic system.

Earth laws can, in part, help to restore the ancient and life-sustaining concepts that have been damaged in our current times, and bring together the entirety of the Earth community—humans, rivers, mountains, forests, deserts, oceans, animals—to prosper and thrive.

Of course, a Rights of Nature framework will not solve all our mounting problems, but it is an important tool, and it does offer a foundation upon which healthy and equitable legal, social, ecological, and economic principles may be built. As we look to transform the dominant culture's responsibilities to and relationship with the natural world, Rights of Nature offers an elegant and functioning model at the local, national, and global level as more and more communities and countries engage with this breakthrough development of Earth jurisprudence.

Given the prevailing power structures, it is prudent to recognize that the changes so necessary for our survival will not happen without impassioned advocacy from change-makers, whether they are activists, artists, farmers, journalists, scientists, land defenders, students, politicians, parents, or teachers. Consequently, a vital component of the struggle for ecological, social, racial, and economic justice—as well as cultural transformation—is, in fact, what happens in the streets, the schools, the fields, the forests, the marketplace, and the halls of decision-makers. True transformative policy changes will only occur when we have strong social movements pushing for them.

With this in mind, one way that GARN is building a movement is by demonstrating Rights of Nature through International Rights of Nature Tribunals, which illuminate how the legal framework can be implemented in communities worldwide. The formation of the tribunals was inspired by the International Criminal Court and the Permanent Peoples' Tribunal, during which citizens investigate violations of human rights. The Rights of Nature Tribunals provide a vehicle for reframing prominent environmental and social justice cases, and they also build awareness, educate people, facilitate movement building, spotlight struggles on the ground, and have led to action in active cases in communities.

GARN has organized five international tribunals globally since 2014; and our third major tribunal was held during the UNFCCC at COP21 in Paris in order to provide a different vision from those of the world governments.

In the Maison des Métallos, a building where musical instruments were once made, over six hundred people gathered for the two-day tribunal, in which sixty-five people from around the world participated as judges, witnesses, and land defenders. There was such a keen interest in Rights of Nature work and new approaches to our current dilemma that the auditorium was filled to capacity, and we unfortunately had to turn people away.

Many of us from GARN were honored to not only organize the event but also participate in various roles. Along with twelve remarkable leaders and colleagues, I was humbled to act as a judge during the tribunal, during which we heard and questioned case presenters from different global hotspots. We sat on the main stage, which had been transformed into a simulated courtroom with a judge's table and podiums for case presenters, witnesses, and prosecutors for Nature's rights.

Approaching the human-nature relationship is not a narrowly intellectual matter when we collectively stand at the threshold of apocalypse. Rather, in this particular moment, we confront a complex circumstance requiring us to bring our full selves to the discourse and deliberation. In light of this, each tribunal begins with Indigenous-led ceremonies, providing an opportunity to listen to our deepest inner wisdom and the voice of Mother Earth.

At the tribunal in Paris, Indigenous Peoples from the Global South and North conducted these important ceremonial protocols: from the South, Ena Santi, Nina Gualinga, and Miriam Cisneros, who are all Kichwa women from the Amazon; and from the Turtle Island Indigenous territories of the North, Casey Camp-Horinek, Tom Goldtooth, Cherri Antea, and Tantoo Cardinal.

When the ceremonial singing began, we no longer existed only in the confines of the building in Paris; our spirits were transported from the Amazonian jungle to vast prairie lands and beyond, through prayer, song, and the delicate smoke and sweet fragrances emanating from the burning of herbal plants from each of those lands. The ritual proceeding was a manifestation of the participants' deep reverence for the Earth, and a necessary preparation for what we were about to do.

The ceremonial weaving together of Indigenous Nations of the Americas from the South and North was especially significant, as was explained to us on the opening day of the tribunal, because in that moment we were experiencing a manifestation of the ancient Indigenous prophecy of the Condor and the Eagle. I later learned more about the prophecy as elucidated by Chief Phil Lane, Jr., of the Ihanktonwan Dakota and Chickasaw Nations:

> Prior to the arrival of the first European explorers an Indigenous kinship network based in spiritual and cultural relationships and Protocols existed for centuries across the Western Hemisphere. Through this network of many networks, this Hoop of Many Hoops, Indigenous Peoples shared knowledge, capacity and resources for mutual aid, trade and development. This ancient network, based in life-preserving and life-enhancing spiritual values and principles, known by many Indigenous Peoples across the Hemisphere as the Union of the Condor and Eagle, was shattered by European colonization and the subsequent decimation of Indigenous Nations and Peoples across the Americas.
>
> The prophecy of the Reunion of the Condor and the Eagle is retold by many Indigenous Nations and Peoples. The Q'ero People's prophecies, for instance, say that when the Eagle of the North and the Condor of the South fly together, the Peoples of Mother Earth will awaken. They say that the Eagles of the North cannot be free without the Condors of the South.
>
> All of these ancient prophecies predict, after a long and bitter spiritual wintertime, that the "Reunion of the Condor (Indigenous Peoples of the South) and the Eagle (Indigenous

Peoples of the North)" will be realized. It is promised that when the Reunion of the Condor and Eagle is fully realized, a great era of peace, well-being and prosperity will follow for all members of the Human Family.

The Reunion of the Condor and Eagle is intimately connected to other Indigenous prophecies across the Americas. These include the Time of the Eighth Council Fire, the Hopi Prophecies, the Slaying and Return of the White Buffalo, the Mayan 2012 Prophecy and the Emergence of the Fifth Sun, the Return of Quetzalcoatl, the Prophecies of Chilam-Balam, the Prophecies of Sweet Medicine, and the fulfillment of Black Elk's Daybreak Star Prophesy.

So strong is the belief in this prophecy among some Indigenous Peoples, that the Otomi Peoples in the State of Mexico have built a vast ceremonial amphitheater dedicated to the "Reunion of the Condor and the Eagle." The focal point of this amazing construction is a gigantic stone carving of a Condor and an Eagle joined in a loving embrace. It was built largely by the volunteer labor of thousands of materially poor Indigenous Peoples out of love and faith in this prophecy.

The Movement of the Condor and the Eagle continues to gain strength and momentum through the dedicated action of many different Indigenous networks and their allies across the Americas despite on-going opposition from adverse political and economic forces, especially natural resource extraction interests.

This opposition is easy to understand. Collectively the Indigenous Peoples of the Americas have the social capital, land base, natural resources, including water, petroleum, natural gas, timber, rare minerals and gems, fishing and hunting rights and those rights and resources that are still to be justly acquired, that once the Reunion of the Condor and Eagle is fully realized, Indigenous Peoples and their allies will become a decisive and major economic and spiritual force not only in the Americas, but around the world. Indigenous Peoples will then be in a position to mandate the wise and harmonious

> ways Mother Earth's gifts should be safely and sustainability developed, as well as, when development is not appropriate, no matter how much profit is to be made.
>
> Now, following the end of the long-count of the Mayan Calendar, December 21, 2012 and the beginning of the 5th Sun, the rising of Indigenous Peoples and their Allies is accelerating everywhere on Mother Earth. Indigenous women and their sisters of the Human Family are destined to play a great role in this global transformation process![29]

And with that historic opening between Indigenous leaders of the South and North, the Paris Tribunal began as we judges listened and responded over two days of testimony. At a tribunal, case presenters and impacted persons testify before an audience and the judges, and prosecutors for Mother Earth guide them through questions. While we do focus on specific cases, we are also fundamentally placing on trial the totality of our current legal and economic systems that legalize and monetize the destruction of nature. We are denouncing the commodification and financialization of nature, which stems from the pervasive misconception that humans are above the rest of life and that the Earth is here purely for resource extraction.

A goal of the Rights of Nature tribunals is to examine and judge rights violations that have remained outside the consideration of formal governmental institutions. The worldview reflected in the Universal Declaration of the Rights of Mother Earth provides a framework for our investigation, as does our combined knowledge of natural systems. The tribunals also allow people to collectively experience and imagine what it might look like if our courts of law truly and justly served people and the planet. The tribunals serve as examples and educational tools so that legal experts can fine-tune the constructs needed to implement Rights of Nature in active environmental cases.

During the adjudication process, participants analyze various cases using the framework of Rights of Nature. If an action is deemed a violation of the UDRME, presenters will explain why, further educating everyone present. The expert presenters on the tribunal will then lay out what should be done to repair the damage the violation has caused, as well as any measures that should be taken to prevent further harm.

Even though the verdicts of the tribunals are not legally binding, there is the possibility to have real-world impacts on communities. At the fourth GARN international tribunal in Bonn, Germany, a high-profile case was presented involving a road that had been planned in Bolivia. The proposed route would divide the Isiboro Sécure Indigenous Territory and National Park (TIPNIS) in half and end its protected status. This tribunal case was widely covered by the media, particularly in Bolivia, and the publicity played a key role in Bolivian President Evo Morales' decision to pause the project.

By holding these tribunals, GARN wishes to also demonstrate a commitment to justice for the more-than-human world. The survival of many species, including our own, depends on a dramatic shift in practice and worldview at all levels of society. As Cormac Cullinan, environmental lawyer and author of *Wild Law*, explains, "The way forward simply isn't visible from within the existing worldview. Both the international governance system based on national sovereignty over 'natural resources' and virtually all national legal systems and institutions have been designed to serve the imperial project of extending human control over every aspect of the planet so that we can wring the maximum out of it for our exclusive benefit."[30]

In the Maison des Métallos, after the opening ceremony, Casey Camp-Horinek's introduction countered the idea of a world with human dominion over nature by recognizing her kin, including "my relatives with wings, my relatives that crawl, my relatives with fins, those in the spirit world, and the elements."[31]

A rapt audience then listened to the first case, Climate Crimes Against Nature, presented by Pablo Solón, former Bolivian Ambassador to the United Nations and chief negotiator on climate. Solón emphatically described systemic violations against the very functions of nature, affecting all of Earth's rights through global climate change. Half of global emissions are caused by the world's richest ten percent, with the poorest fifty percent of people responsible for only ten percent of emissions.[32]

When Prosecutor Ramiro Ávila asked Solón whether each of us, as human consumers, bore some responsibility for this crime, Solón clarified that, while we all bear some responsibility, it is not equally distributed. A person living in rural India, for example, will consume far less than an average person in the United States. Solón stressed that governments, especially, are guilty of allowing the crimes against the Earth to continue when they fully understand the results of their policies. Referring to the national delegates at COP21, he said they only recognized the logic of power and the logic of money and thus neglected to address and eliminate the root causes of these continuing crimes: fossil fuel extraction and greenhouse gas emissions, deforestation, and agribusiness.[33]

Solón's presentation was followed by testimony from experts and witnesses on the topics of fossil fuels, deforestation, water and climate, market mechanisms, climate-smart agriculture, and land use. In her presentation as an expert on water and climate, Maude Barlow, co-founder of the Blue Planet Project and founding member of the Council of Canadians, stated that current laws only protect water as property, not as a living entity with inherent value. Current agricultural and industrial practices are voraciously swallowing the water of current and future generations, depleting groundwater, forever altering climate patterns, causing whole rivers to disappear, and irreversibly polluting water in much of the world with toxic waste. The demand for water already outstrips the supply in much of the world—often leading to war—and this imbalance is only getting worse. She named the abuse of water by governments and corporations as a leading cause of climate change, and a crime against humanity and against nature.[34]

When Avila invited Barlow to suggest solutions, she called for a "new water ethic," in which policymakers would need to consider the impact on water when making any decision.[35] She highlighted that governments and international institutions, such as the World Bank, are privatizing water and allowing it to be commodified and polluted and emphasized the need for taking water out of the market system.

On the same topic, environmentalist and hydrologist Michal Kravčík acted as a witness, illustrating the connection between water and climate change with an example from his native Slovakia. Kravčík explaining that the country has lost a huge amount of water, due to damaged landscapes and changes in agriculture and forestry practices. The issue, however, has been ignored by national and international institutions.[36]

Nnimmo Bassey, Nigerian architect and Director of the Health of Mother Earth Foundation, and I were judges for the Climate Case. We listened and jotted down notes as the experts and witnesses shared their stories of the heartbreaking ways in which the ecosystem is being destroyed by human avarice. Bassey and I agreed that the rights of Mother Earth were being systemically violated through human-caused global warming, most of which has been allowed to continue because of the failure of governments and corporate actors to prioritize life over profits and trade deals, and because of the false climate solutions that have been proposed that avoid real action.

We called for a halt on new fossil fuel expansion projects and for governments to step up and take real measures to reduce emissions and remove water from the market system. We called for an examination of the World Trade Organization, the World Bank, and the International Monetary Fund, who have collectively failed the world's social and ecological systems, and pointed out how false solutions proposed by governments result in further violations to Mother Earth's rights while ignoring the actual causes of climate change. Geoengineering was also listed as a false solution because it interferes with Mother Earth's ability to balance herself according to her own perfect calculations.[37] We proposed ten actions from a climate justice framework that governments, corporations, financial institutions, and the global community should be taking based on Earth's natural laws.[38]

Later that day, Dr. Vandana Shiva, with the force of decades of experience, presented a case on the agro-food industry and genetically modified organisms, stating that GMOs violate the "integrity of life" because they interfere with nature's self-organization, self-determination, and self-directed evolution, and they also violate Nature's right that the genetic structure of beings not be disrupted.[39] Corporations, themselves inventions, have patented seeds as if they have a right to claim the invention of life itself. When judge Atossa Soltani asked about the ongoing crisis of farmer suicides in India, Dr. Shiva answered that the GMO companies had also violated the rights of the farmers by monopolizing the seeds, spreading lies about the farmers' native seeds, sharply raising seed prices, and introducing new pests that consumed their crops, all of which led to destitution and despair throughout the farming community—and ultimately, to farmers tragically taking their lives.[40]

Her account was supported by evidence from witnesses José Bové and Andre Leu and experts Ronnie Cummins and Marie-Monique Robin, who provided more examples of the ways GMOs and their associated pesticides are harming animals, the soil, biodiversity, farmers, and consumers, as well as the rampant climate destruction caused by factory farming. Andre Leu, organic farmer and president of the International Federation of Organic Agriculture Movements (IFOAM), shared shocking pictures of tumor-riddled animals infected with carcinogenic glyphosate, the active ingredient present in commonly used pesticides, to demonstrate the danger inherent in Roundup Ready crops.[41] The tribunal opted to hear additional evidence in upcoming regional tribunals before giving a ruling so that more testimonies could be spotlighted.

The last case of the first day concerned the criminalization of defenders of Mother Earth, with witnesses from Ecuador and Texas. Belén Páez informed the tribunal that the Pachamama Foundation, an NGO which had for seventeen years advocated for Indigenous territorial rights and rainforest protection, was shut down by the Ecuadorian government after organizing peaceful anti-extraction protests and marches, actions that are allowed by the Ecuadorian constitution. Indigenous activists in the area have been criminalized, sent to jail, and physically attacked.[42] Fellow witness Manari Ushigua, a rainforest defender from the Sápara Nation, pleaded with the judges to take detailed notes and support frontline activists.

Bryan Parras, Texas resident and environmental justice advocate, reported that environmental organizations in Texas have been followed and surveilled by the government and private corporations. Yudith Nieto showed a brief video describing her Houston, Texas, neighborhood of Manchester, a primarily Hispanic neighborhood completely surrounded by industrial activity. As members of a low-income community of color, Manchester residents feel powerless to speak up, and the Environmental Protection Agency (EPA) has failed to protect them from environmental racism as community members, including children, suffer from leukemia and other cancers caused by chemical contamination.

Judge Christophe Bonneuil proposed that the tribunal continue to add the protection of Earth defenders to its documents, and the tribunal wholeheartedly agreed. Case judges Atossa Soltani and Ruth Nyambura agreed that Ecuador had violated not only the UDRME but their own constitution. The government of Ecuador was going backward in its protection of the environment, and the tribunal found the government also guilty of violating human rights.[43] The tribunal called upon the government to free the political prisoners and drop the charges against them, to allow the NGOs to carry out their activities, and to respect the consent of Indigenous Peoples regarding decisions affecting their territories.[44]

During this case, many of us were reminded of José Isidro Tendetza Antún, a leader and land defender of the Shuar Nation, who had been slain the year before. Tendetza had been protesting a major mining operation in Ecuador, which would destroy over 24,000 acres of densely biodiverse forest. He had been scheduled to attend the December 2014 Rights of Nature Tribunal in Lima, Peru, as a witness against the deforestation perpetrated for the sake of the Mirador open mine project, but only days before, Tendetza was abducted, tortured, murdered, and buried in an unmarked grave. To this day, no one has been found guilty of this horrific crime.

When Tendetza's colleagues presented the case at the Lima Tribunal, they praised his courage and kindness. The Tribunal found the Ecuadorian government guilty of violating Rights of Nature laws as well as the rights of Indigenous Peoples, but the mining project unfortunately went forward. Judge Alberto Acosta, former president of the Constituent Assembly from Quito, Ecuador, invited the tribunal audience and jury to applaud in Tendetza's memory to demonstrate that his voice had not been silenced. All of us, including the witnesses from Ecuador and Texas at the 2015 tribunal, would continue to fight in his honor.

On the second day of the Paris Tribunal, we heard a case on hydraulic fracking, a common theme at the tribunals. Presenter Shannon Biggs, co-founder of Movement Rights, opened by referring to fracking as a "rape of the Earth" and submitted detailed data to demonstrate that fracking violates the rights of water and the rights of soil.[45] She also reported that during the frequent droughts, California uses radioactive fracking wastewater to irrigate many fruits and vegetables, which are then consumed by the entire United States and exported internationally. Biggs then invited Casey Camp-Horinek to join her at the podium to share her experiences as an Indigenous activist in Oklahoma, a state with thirty-nine federally recognized tribes.

Camp-Horinek informed us and the shocked audience that, in 2015 alone, Oklahoma, a state that had not previously been prone to seismic activity, experienced five thousand earthquakes, many of them severe since they originated near the Earth's surface, a direct result of the thirteen thousand active injection wells in a region controlled by corporate interests. Aquifers on Tribal lands are defiled, fish are poisoned to death in the waterways, and the EPA fails to warn residents of the toxic water.

"We know firsthand that when you grind into the bones of the Mother, that she begins to shake and shudder with pain, and with the reality that she has to find a way to live with her waters being shifted and poisoned and killed," said Camp-Horinek, and she implored the panel and the audience to make sure the world understands the dangers of fracking.[46]

Witness Kandi White supported the statements of Shannon Biggs and Casey Camp-Horinek as she began her heartfelt presentation by referring to Article 2 of the UDRME, which lays out the inherent rights of Mother Earth, including the rights of water, the rights of all beings to live in well-being, free from cruel treatment, and the right to freedom from contamination, all of which are violated by the act of fracking.[47]

In White's home state of North Dakota, which produces much of the nation's wheat and sunflowers, cattle are born with deformities and soil is sterilized by fracking wastewater spills—and when the damage happens on a reservation, it is even more difficult to hold corporations accountable. She described friends who had died because of the industry's activities and the workers who bring drugs and crime, inscribe racial slurs on buildings, leave socks filled with radioactive fracking material in playgrounds, and attack Indigenous women. She spoke out for her sisters and for the future generations who cannot speak for themselves.

Many of us were also in Paris as official delegates to the UNFCCC COP21 negotiations and were outraged and frustrated to see that the words "fossil fuels" were not even mentioned in the Paris Agreement, which is stunning given that coal, oil, and gas are at the core of greenhouse gas emissions (since then, in 2021 at COP26, a phasedown of coal, but not oil and gas, has entered the text of the Paris Agreement, and the demand from civil society for all fossil fuel phaseout is ongoing). This omission tragically reflects government and industry insistence on only focusing on the demand part of the equation without addressing the essential lever of stopping the supply side, the extraction of fossil fuels. In other words, allowing polluters to keep on with business as usual. Our tribunal confirmed the claims that fracking violates the rights of Mother Earth, and that ending the extraction of fossil fuels must be centered as crucial to the climate crisis discourse.

We heard other cases during the tribunal, on mega dams in the Amazon, which have the potential to destroy much of the Amazon and the livelihoods of those who live there; on deforestation, which deprives Mother Earth of one of her key regulatory functions; on the destructive impacts of industrial farming and market mechanisms on farmers and the land; and more on false solutions such as bioenergy and carbon offsets.[48]

We judges, as well as hundreds of audience members, felt our hearts fractured and crushed again and again as we heard testimonies of ecological and human violations, anguish, assault, wreckage, extinctions, erasures, brutality, and disrespect. The death march of the dominant culture was laid out in one story after another. And yet, we were also moved to great heights of respect and awe by the movement leaders, their knowledge, and their continued resistance against all odds. The tribunal organizers also created space for relaying recent victories in Rights of Nature cases and how the movement is growing globally to help balance the hardships all had endured.

The tribunal then concluded with statements by prosecutors Linda Sheehan and Ramiro Ávila, both attorneys, followed by words from each of the judges. Sheehan had viewed a release from the ongoing COP21 in session and expressed her grave concern and disappointment at what she saw. Sheehan confirmed that there was no mention in the Paris Agreement of fossil fuels and added that there was no mention of the need to be accountable to the Earth's natural systems. Of note, she stated, "We will not bargain for the destruction of Mother Earth. We must insist on laws that recognize the inherent Rights of Nature. Any laws or conventions that aim for less must be rejected."[49]

Then Ramiro Ávila, professor of law in the Universidad Andina Simón Bolívar-Sede, seemed to defy gravity as he leaned forward in an acute angle at the podium to deliver his statements, meant to be heard beyond the auditorium, to reach the skies above and all the realms of nature in one of the most fervent orations in defense of Mother Earth I have ever experienced.

Spoken in a passionate tenor that rose and peaked in repeated crescendo-like waves, Ramiro's remarks flowed as a sacred roll call of all the beautiful creatures and magnificent landscapes of Mother Earth, describing in grand and minute detail every exquisite butterfly, songbird, bee, fish, turtle, jaguar, snake, frog, eagle, wildflower, seed, forest, jungle, creek, whale, snowfield, ocean, mountain, deer, valley, grass, river, and glen, calling out to each of these sacred beings, clearly avowing that we who had heard the testimonies stood in defense of all we hold dear, and of the land and water protectors who risk their lives in the name of life every day. He reaffirmed that the violations were unacceptable, the current systems were unjust, and the perpetrators of violence against Mother Earth must be held accountable.

After days of intense absorption of egregious abuses toward our Mother Earth and all sentient beings, Ramiro's oration spoke for all of us, and we breathed a collective and emotional breath of release. These were the words we all ached to hear in courthouses in every nation around the world. Nearly all six hundred of us joined in a standing ovation as we rode every wave of the prosecutor's remarks and cheered him on, tears brimming our eyes from our overflowing hearts, as he named the phenomenal and awe-inspiring species and lands that had been lost or endangered by human folly and greed. Nothing short of this level of allocution and impassioned expression, which went on at length with each naming, would have satisfied any of us as we searched inwardly to integrate and embrace the grief of the violations.

Through this work, I have realized time and again that our Mother Earth is calling out to us, warning us we are on a path of great devastation. Our shattered hearts are devastated for her and also for our fallen sisters and brothers, defenders of the land, whose stories we will never forget. And while it is true that the Earth will live on even if humans do not, it is tragic to think that our mark as a species on this beautiful, numinous planet will be one of destruction and violence instead of beauty-making, dignity, reciprocity, and harmony.

The merging of so many stories from around the world demonstrated that our legal and economic frameworks are at war with the sacred web of life. In order to live in harmony and safeguard a healthy world for present and future generations, we must radically transform the destructive aspects of our society. Uplifting Rights of Nature can help support a transition to a truly thriving future, one based not on endless material growth but on care for each other and Mother Earth. It is our task to imagine and create modern cultures with a new understanding of ourselves, of progress, and of well-being.

Many more tales and vital moments could be conveyed from that critical point of global juncture during the United Nations Paris climate negotiations. This includes the profound work of remarkable women leaders whose collective efforts are what made possible a global climate accord among world governments.[50] And while the Paris Agreement is insufficient and not commensurate to the crisis, achieving it is still highly significant and quite useful in many important advocacy spaces. I and others have shared the stories of these women leaders and a feminist analysis of COP21 through other mediums for quite some time and will continue to do so, but for now I will conclude my account of those who met in 2015 on the ancestral lands of the Gaul, a Celtic people.

Before traveling home from Paris in that biting cold December, I stopped to pay homage to the powerful River Seine, who flows northwesterly through Paris to the Normandy coast. The Seine emerges from the Earth at Source-Seine, in the Burgundy region of France. Now guarded by a serene stone nymph, the place was once a major gathering point and sacred site for the ancestral peoples of Europe.

As I stood on the riverbank, I opened my ears to hear what the ancestors might offer, the flowing river beckoning me to recall how all the sacred waters of the Earth are connected, including those inside my body. In that moment, I saw a flash in my mind's eye of a river I did not know.

It seems it was these sacred Seine waters that carried me a few years later to another land where the emergence of a new and distinctive Rights of Nature law had carved another waterway of Earth jurisprudence—this time along the Whanganui River of the Māori Peoples of Aotearoa, meaning "land of the long white cloud," also known as New Zealand.

"The first thing you must do is meet the river," announced Māori leader Hinewirangi Kohu-Morgan as soon as I set foot on her homeland of Aotearoa in May 2018. I had been traveling for twenty-seven hours straight because of fossil fuel divestment campaign work on the other side of the world in Europe, but clearly this was not the moment to give way to my body's call for sleep. What mattered now was to muster all my energy and arrive in the present moment and respect the protocol of Hinewirangi Kohu-Morgan, a remarkable leader and poet, and our main guide in Aotearoa.

I was honored to travel to New Zealand as part of a Rights of Nature learning and fact-finding delegation, which WECAN co-sponsored and that was organized by powerful Māori leaders and Movement Rights, an organization dedicated to Rights of Nature work founded by Shannon Biggs and Pennie Opal Plant. Two years earlier, Movement Rights had laid the groundwork for our delegation, with the purpose that the Whanganui River's relationship to the Crown Settlement Agreements needed to be better understood and shared in our collective Rights of Nature work. I had arrived a few days after the delegation began and was warmly welcomed by Whanganui Iwi (Tribe) leader Kahurangi Kanau, and as directed, I was kindly and immediately taken to the Whanganui River. Kahurangi informed me that she was guiding me to a special place along the waterway, as they did not want me to begin cultural and political conversations until I had met their Relative, the river.

There are various oral stories about the origin of the Whanganui River, one being that the waterway was born when Maui, a cultural and mystical leader, caught a fish that transformed into the North Island of New Zealand. When Maui prayed to one of the creation deities, the Sky Father, one of his teardrops descended atop a trio of snowcapped, still-active volcanos, eventually becoming the Whanganui River. Fed by nine tributaries, the river flows northwest from these mountains before curving to the south, twisting and turning around bends guarded by spirit guardians called *taniwha*, and finally flowing into the Tasman Sea between the two main islands of the country.

I was lovingly greeted not only by Kahurangi but also by my dear friend and mentor Casey Camp-Horinek, who was part of the delegation. Together we traveled by car to the river and, once we arrived, were ushered down to a gentle sandy bank with several rock outcroppings.

Kahurangi sang a song to the powerful winding river while Casey and I knelt close to the water. The tears from our eyes were now joining the riverway as we were embraced by the ancient melody of Kahurangi's voice blending with the flowing stream and the reverence of the Māori Elder approaching her ancestor, and all of us connecting to the eternal sacred motion of waters around the world. The Whanganui Iwi are known as the River People, who say *Ko au te awa. Ko te awa ko au,* which means "I am the river. The river is me." Their river is their living ancestor.

When Kahurangi indicated the time was right, I offered Big River water from my homeland in Mendocino that I had carried in a small container for just this moment. Then, with a warm hug, I thanked Kahurangi for inviting our delegation to her people's homeland and introducing us to their sacred river. She took my hand, looked me in the eyes, and gently corrected me. "You mean *our* river. We are all a part of the river, this is *our* ancestor, yours too."

This deeply rooted kin relationship to nature is fundamental to the revolutionary and profound Te Awa Tupua Bill (Whanganui River Claims Settlement), which passed in March 2017, making Aotearoa's Whanganui River, from its source in the mountain to its mouth at the sea, the first river in the world to hold legal personhood. The Whanganui is no longer owned by New Zealand's government, the British Crown, or any other entity, but instead holds the same legal rights, responsibilities, and liabilities as a human person. The Māori Peoples fought for one hundred and forty years for this legislation to gain recognition of the river as an ancestor.

The struggle for Rights of Nature, or more specifically in this case, the rights of personhood for their sacred river in Aotearoa, began in the eighteenth century with the arrival of English traders—a violent colonial force that burned crops, killed animals, fought the Indigenous Māori, and spread diseases. In a short time from first colonial contact, fifty percent of the Māori had died.[51]

The Treaty of Waitangi, signed by most Māori leaders by 1840, appeared to be a gesture of peace with the British settlers but was built on an egregious deception, as there were two completely different versions of the Treaty. The version written in a Māori language was a document inviting the Europeans to share the land with them, provided that the settlers granted the Māori freedom to practice their own religion and traditions. However, the version written in English stated that New Zealand would become part of the British Empire and the Māori were now subjects of the Crown, albeit allowed to continue their traditional practices, though this freedom would prove to be limited.[52]

In the name of this deceptive Treaty arrangement, the government exploited and spoiled the carefully tended and protected ecosystems of Aotearoa by "development," taking control of the land's riverbeds in 1903 in the name of coal mining. Other industries dumped sewage and garbage in the waterways. Along the sacred Whanganui River, settler extractivists and opportunists eradicated the rapids with explosives, plundered gravel from its bottom, and disturbed Māori fishing practices with their riverboat tours. The colonists went so far as to divert the headwaters of the Whanganui—effectively cutting off the head of the Māori Peoples' living water Relative—to power a hydroelectric plant.[53] On several occasions, under threats of arrest, the Māori petitioned against these harmful developments, but again and again the cases were callously dismissed.

Finally, in 1975, the Waitangi Tribunal was implemented as an attempt to address the violations of the Treaty of Waitangi, beginning New Zealand's truth and reconciliation journey. The tribunal listened to Māori representatives in their own languages and noted the problematic phrasing of the treaty, stating that the Māori of the time would not have understood the concept of ownership and transfer of land rights: "Land was not a tradable or disposable item. Having passed down through forebears from Papatuanuku, it was entailed to the tribe's future generations"—Papatuanuku being the name of the Earth Mother.[54] The Waitangi Tribunal acknowledged both the river's ancestral relationship to the Māori and the fact that the ownership of land is not a universal concept.

Still, negotiations dragged on for years through complex court proceedings and appeals, culminating with *Ruruku Whakatupua,* a settlement deed recognizing the Whanganui River as a legal entity called Te Awa Tupua, with rights equivalent to those of humans. The legislation was ratified in 2017, ending New Zealand's longest-running litigation, as Tribal representatives celebrated with open-hearted joy, especially those of the Whanganui Iwi, whose name, identity, and livelihood come from their ancestor river.

The Te Awa Tupua Bill legally transformed the river from a piece of property into what the Māori had considered it all along—a rights-bearing entity with sacred value and agency. From the mountain to the sea, encompassing one hundred eighty miles of river, and the entire ecosystem of life in, above, and along the river, the Whanganui is now recognized as a living being to be respected and protected by two appointed custodians, one from the local Iwi and one from the Crown, guided by values held intrinsic to the river by the Māori, including that it is the source of sustenance both physical and spiritual. The bill recognizes the Māori Peoples' responsibility to the river and their right to care for it based on their pre-colonial laws, and it includes a formal apology from the Crown for the harms perpetrated under colonialism to the river and its people. It also includes an intention to fund education and community development based on Māori cosmology. Now, harm to the river is legally treated the same way as harm to the Tribes.

The Whanganui Iwi present at the passing of the law celebrated with a song called a waiata—people of all ages, a few of them cradling babies, stood on the balcony of the courtroom, offering rhythmic chanting and joyful songs accompanied by traditional hand motions.[55] As the head negotiator for the Whanganui Iwi, Gerrard Albert, explained, "Rather than us being masters of the natural world, we are part of it. We want to live like that as our starting point. And that is not an anti-development, or anti-economic use of the river but to begin with the view that it is a living being, and then consider its future from that central belief."[56]

The Whanganui Iwi had been demanding the Crown recognize their interests in the Whanganui River for well over one hundred years, and it was largely sustained activism by the Māori—including significant Indigenous movements rising in the late 1960s and 1970s—that brought a shift in power. Consequently, the victory of the Whanganui River Claims Settlement is an irreplaceable step forward not only for the protection of this beautiful body of water and entire river ecosystem but also for Indigenous rights and sovereignty.

Part of the struggle and resistance to imperialism and colonization for Indigenous Peoples in Aotearoa and globally has been brilliantly expressed by Māori author and scholar Linda Tuhiwai Smith. In her book *Decolonizing Methodologies: Research and Indigenous Peoples,* she states, "To resist is to retrench in the margins, retrieve 'what we were and remake ourselves.' The past, our stories local and global, the present, our communities, cultures, languages and social practices—all may be spaces of marginalization, but they have also become spaces of resistance and hope."[57] While in Aotearoa, our delegation was elated to meet with Linda and learn more.

The Whanganui Settlement Agreement recognizing the rights of the river and removing it from a property-based regime is a powerful example of a "space of resistance and hope" retrieved and revitalized by the Māori People—one that can teach us all.

As Gerrard Albert says, "This river isn't just water and sand. It is an ancestral being with its own integrity. This river is not the river that has been contended by the crown, that exists in compartments, its bed and its waters. It's an indivisible whole that includes iwi so this concept of legal personhood is the nearest legal approximation to the way in which we relate to our people as being inextricably entwined to it and can never be alienated from it."[58]

When a journalist asked Gerrard Albert how Pakeha—non-Māori people—could achieve a connection to the river, Albert related an event that occurred after the death of an esteemed Tribal leader involved in the Whanganui River Claim. As a boat transferred the leader's body down the river toward the burial grounds, non-Māori leaders gathered along the riverbanks, bearing fern fronds with which they adorned the water as the boat passed. By respecting and participating in a Māori practice to honor the river and the Indigenous leader as kin, non-Māori stood in solidarity and also respected the cause the leader had championed.[59]

The Whanganui River was not the only entity to be granted legal personhood. The Tuhoe Settlement Agreement recognized the rights of the Te Urewera Forest, the home of the Tuhoe Iwi, which lies in the eastern portion of New Zealand's North Island and encompasses 820 square miles. The Tuhoe Iwi never signed England's Treaty of Waitangi, the only tribe to refuse. The Crown once owned Te Urewera, holding it as a national park. The Tuhoe Iwi, led by chief negotiator Tamati Kruger, struggled tirelessly to persuade the government to cede ownership of their forest, declining one unacceptable proposal after another, until the Crown agreed to consider the Māori view of the land.

Kruger's team suggested that no one should own the forest, that it should be recognized as a living being with agency. Finally, with the Tuhoe Settlement Agreement, the forest's legal status matches its spiritual reality, and the Tuhoe Iwi are optimistic that the agreement signifies a move toward their own sovereignty, which is an ongoing process.

These Rights of Nature laws came into being through the unflagging efforts of the Māori Tribes, and they are also part of the New Zealand government's recognition that industries are damaging the environment and triggering natural disasters. The recognition of the rights of the Whanganui River and the Te Urewera Forest is a significant breakthrough exemplifying how modern legal systems can shift to protect the Earth and sacred systems of life and uphold the rights and worldviews of Indigenous Peoples.

Upon my return from Aotearoa, I journeyed to my own home waters, Big River, just south of the town of Mendocino. Standing at the water's edge where the river meets the Pacific Ocean, I slowly poured out my offering to the estuary sanctuary—an alchemical mixture of waters I had collected from the Seine River during the adoption of the Paris Climate Agreement, the Whanganui River, and the Missouri and the Cannonball Rivers, whose waters I had collected while at Standing Rock in North Dakota during the Dakota Access Pipeline resistance.

As the waters mingled, I respectfully asked that my offering be part of the retrieval of our original agreement with our living Mother Earth, and that our modern culture lacking in consent and responsibilities be reoriented to one of reciprocity and asking permission of the wild once again—of the wild animate natural world, our sacred unconquered kin.

As American ecofeminist Carolyn Merchant stated in her book *The Death of Nature,* "The image of the earth as a living organism and nurturing mother had served as a cultural constraint restricting the actions of human beings. One does not readily slay a mother, dig into her entrails for gold or mutilate her body, although commercial mining would soon require that. As long as the earth was considered to be alive and sensitive, it could be considered a breach of human ethical behavior to carry out destructive acts against it."[60]

Somewhere deep within our embodied assemblage lie the forgotten story threads of the wisdom of our Earth-loving ancestors, the fibers of the original agreement we upheld with Mother Earth to live with respect and responsibilities in our eternal and necessary exchange to exist and thrive.

I bend low to cup water into my hands from Big River, and then pour the precious liquid over my head, feeling the tingle of the wild water flowing down my body, cooling and alive. What I love about the possibility of Rights of Nature is that it can help the dominant culture and its legal systems realign with Nature, and help mend a sense of severance from our living Earth, which has proven to be not only spiritually heartbreaking but a disaster in the human experiment.

A culture deprived of its connection to the sacred systems of life, to Nature, is simply not tenable or sustainable. My great hope is that the crises we now face will bring us home, back to Earth and back to each other in reciprocal community. The waters in our bodies coursing in rhythm with the rivers, oceans, lakes, and wellsprings of the living Earth exquisitely affirm that we are Nature. The ceaseless, but often hidden, memory of our wild bodies in kinship with the Earth must not only be remembered, but also reimagined and revitalized into our legal and societal frameworks and practices. In this sense, we are Mother Earth's, wild nature's, immune system rising to protect and defend her.

Part V

The Land Is Speaking: Language, Memory, and a Storied Living Landscape

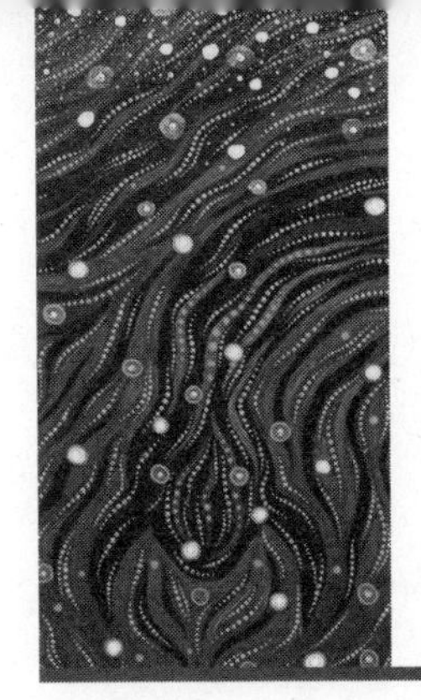

Chapter 13:

Worldviews Conjured by Words

WESTERN RED CEDAR, yellow cedar, Sitka spruce, and western hemlock—these are the primary trees that create the luscious topography of the Tongass National Forest, one of largest intact temperate rainforests left on Earth. Its 16.7 million acres host some of the highest concentrations of wolves, brown and black bears, and bald eagles in North America, with bears outnumbering humans in some areas. Enveloped by these ancient woodlands and wildlife, I slowly make my way through an ecosystem so healthy and abundant that I can hardly believe my eyes. I am left thunderstruck at the staggering beauty.

I come in respect to the Yakwaheiyagu.[1] I say the Tlingit word *Yakwaheiyagu* out loud and in my mind several times to embrace and respect this place as Tlingit leader Kashudoha Wanda Loescher Culp of the Chookeneidi, the Eagle–Brown Bear Clan, and I walk in her forest homeland on the island of Hoonah. Wanda is legendary in this region for her vast knowledge and ceaseless dedication to defend the forest and reclaim Indigenous land management. In the Tlingit language, the *Yakwaheiyagu* or *Yeik* are the spirits, the persons of each living being, including the salmon, the berries, the rivers, the mountains, and the trees. As I listen and learn from Tlingit women, I am deeply moved by their culture, so alive with ebullient language and Oral Traditions that form a wondrous mosaic of life-enhancing practices and sacred knowledge that carry hieroglyphic messages from deep within spiritual realms of understanding and balance with the Tongass.

The Alaskan Tongass is the homelands of the Tlingit, Haida, and Tsimshian Peoples. It rises from an archipelago, comprising fjords, glaciers, and coastal mountains. Cascading from the highest peaks and descending to its lowest tide, life thrives in this unique forest and ocean ecosystem, providing nourishment for Indigenous and local communities. From the beach asparagus that skirts the shoreline to the salmon who journey up the woodland rivers toward their hereditary spawning grounds, the forest feeds, heals, teaches, and endures. Within this vast sylvan landscape, stories, language, and knowledge have been imparted for thousands of years by the original peoples of these northern lands and the forest itself. Viewed from above, the ocean that surrounds this forest-bejeweled archipelago aptly looks like large tree limbs with intricate branches reaching northward.

The ancient forest is the foundation of traditional Tlingit lifeways, from cultural and spiritual practices to the food and medicines Indigenous Peoples depend upon. Yet the forest and its Indigenous residents, whose lives and identities are inextricably woven together, have experienced ongoing assault since Russian and European contact in 1741, and the colonial policies and extractive industries, such as large-scale logging, that came with it.

Since 2016, with the leadership of Indigenous women from the Tongass, WECAN has been honored to engage in ongoing campaigns to protect the forest from further encroachment from extractive industries. One of the most effective strategies involved assembling several delegations of Tlingit and Yupik women in Washington, DC, to meet with lawmakers to advocate for their forest. As the delegates told us, it was the first time Tlingit women had ever traveled to the US capital to protect and defend their Traditional Territory and communities. Since it is the nation's most important forest carbon sink, and thus vital to climate mitigation, the Tongass is heralded as "America's climate forest"; consequently, the delegations were also a mission to stop deforestation as the climate crisis escalates. In coalition with Tribes, environmental groups and local businesses, we had a major victory in early 2023 when the US Forest Service reinstated Roadless Rule protections across the Tongass, which restores federal protection from industrial logging and damaging road-building to just over nine million acres.

Amid vital campaign work such as this, I am learning how critical it is to remain in keen relationship to the lands and peoples WECAN fights for, to stay grounded in the work, or else our efforts can become distant from the rippling creeks, temple woodlands, spawning fish, and the original peoples who live in the places we hold so dear. This has meant, for me, traveling (by kind invitation from our Indigenous partners) to the Tongass to respectfully walk in the rainforest and to experience being present in the lands of the women who lead the WECAN Tongass forest program. Being together supports our relationships as we work for the long-term vision of Indigenous communities here managing their own forests again.

We continue our walk in the forest joined by Kari Ames of the L'uknaxh.ádi, the Coho Salmon Clan, under the Raven moiety from the Frog House. She shares the Tlingit phrase *Aas Kwaani,* meaning Tree People. As we move through the tall thickets, she tells me her people have been strongly rooted to the Tongass for hundreds of generations as stewards of this land. "I am a strong Tlingit woman standing with the forest, speaking for the Aas Kwaani, the Tree People, and I will fight for the Tongass," she states.

An immersive silence has fallen around us in the forest, and without any noticeable signal, we all suddenly stop walking. No one talks or moves—we listen. There is the heartbeat of the Aas Kwaani, the Tree People. Their presence encompasses us. A light breeze stirs the needles in the trees, two eagles fly low enough that we can hear the *whoosh* of their wings, then comes the unexpected awareness of an invisible collective breathing.

Kari describes it so clearly: As we breathe in, the trees give us their breath, and when we breathe out, the trees gather our breath—the Aas Kwaani give us life. We are quiet in the forest; we are breathing together. The inhale and exhale are the embodied conversation that is happening beneath our words; it is the language of the forest, of the tree spirits. Wanda explains to me that the ancestors of her people live in the forest. The Sitka spruce, western hemlock, western red cedar, and yellow cedar are relatives, alive with those who have passed from this realm.

To walk this forest respectfully and with awareness, I feel the need to say these words: Yakwaheiyagu and Aas Kwaani. There are no English words that can truly express the presence of the Tongass living forest, and it requires a shift in my inner world. Saying these words, as I breathe in the verdant woodland fragrances, alters my consciousness from noting single-minded data for climate and forest protection to instead unlocking other ways of knowing. "Yakwaheiyagu" and "Aas Kwaani" contain a metaphysical substrate of meaning of relationship to this place. These are sound and word configurations that have come from this very landscape, from those who come from this place and listened to the story here, and asked the Tree People what they love and how they want to be approached and how they want to be named.

Kari tells me the Tlingit honor each tree, and the spirit of the Tree People. Traditionally, when a tree is cut down, permission is asked, prayers are made, and the tree is addressed with respect as the people describe what the wood will be used for, whether that be a canoe, or paddles, or medicines for the community. Once the tree is cut, an X is carved into the top of the stump and a tree seedling is placed in its center. Moss is placed on top of the seedling and the branches that were cut from the main tree trunk are piled on top of the moss so that the new tree will have nutrients to grow. The surrounding trees will send more nutrients through the roots of the stump to help the new sapling. Finally, the very top of the cut tree is burned with prayers, and as the aromatic smoke rises, prayers are carried upward to the Aas Kwaani of the forest. This practice was shared with me when I had the honor to learn from Kari's grandfather.

Originating from profound antique cosmologies—composed of chirping birds, murmuring springs, thundering skies, and echoing mountain ranges—an assembly of sounds informs human languages with meaning in naming and identifying things in the world around us; describing and defining sensations, relationships, and codes of conduct; understanding and expressing complex realities of time and space; and so much more. All of this is bundled into utterances that strive to verbally convey our very existence and lived experiences.

A significant component of shaping our worldviews resides not only within our cultural stories but also within the very words and the language we hear and speak. Questions stir within me as I make my way through the Tongass forest with my Tlingit friends: Where are the Earth-honoring words in our minds and upon our tongues in the dominant culture? Where is the language that animates the land and inherently connects us to place, a living Earth, and an alive story-filled landscape? Is there any chance that the poetry of our living Earth can still be heard, hidden fugitive-like in the ancestral tongues of the dominant culture? How can we support the leadership of Indigenous Peoples who are working to protect and revive their languages that carry invaluable knowledge, and which are hanging by a thin strand due to the relentless assault of colonization?

I have always been fascinated by words. Not just their meanings, but their etymology—a word's origin and development. But my interest reaches further into a word's actual sound and how it affects our senses and how the word-sound connects us to our antecedents, who formed speech to transmit what is choreographed in the mind.

As many of us learned in high school physics class, sound is produced by waves, passing through air or water in the form of vibrations that most of us can hear with our ears and feel with our bodies. But science alone cannot account for the magical conjuring that transforms sound waves into declarations of poetic "amore" or fiery calls to action that raise our adrenaline; or recitations of wisdom that ring true in our inner knowing; or eloquent prayers and incantations that alter our very consciousness; or the distinct and familiar voices of loved ones that thrill our hearts and make them flutter.

By joining consonants and vowels, we can conjure concepts, actions, realms of existence, mystical imaginings, and a surfeit of emotions into existence. By spelling out words, whether using a written alphabet or a series of phonemes, we create communications that can influence our human experience and perceptions, from the intimacy of a meaningful exchange with a beloved to a visionary oration to the global community. While the ability to use language is a seemingly mundane, everyday experience, it is one of the most puissant agencies we have as humans—one that must be wielded with care, wisdom, and accountability. Language is foundational in influencing and informing our worldview and the way we imagine, experience, and actualize the world around us.

I am particularly interested in how we can redevelop an Earth-loving language that respects nature and the staggering and awe-inspiring reality of existence itself—a language that can hold the multivalent auspicious nature of our living Earth; a decolonized, anti-racist language of equity and care; a language whose very syntax and timbre convey the remarkable and ancient kin relationships within the web of life; a sumptuous, multidimensional language that those of us speaking modern languages can employ to share our histories, cosmologies, and traditions; a language of animacy and enchantment that can knit us to the land, grounding us to a specific place and opening our hearts and minds to the aliveness of the world and our love for nature.

Do we have access to words and languages that are life-enhancing, that awaken our consciousness to an animate Earth and our place in a living cosmology? Do the words we choose, and the order in which we string them together, depict the world as a web of sacred systems with agency, or instead, tragically, as a vault of commodities and dead-matter resources for human use?

Of course, evolving our language is not enough on its own to completely transform our worldviews or our actions toward the sacred land. Using words that convey a living Earth does not necessarily guarantee that speakers will respect the web of nature. That said, the vocabulary, syntax, and word meanings we use every day, out loud and in our minds, do, in the larger cultural context, shape the way we perceive the world, deepening the well-worn paths in our minds that reflect our cosmologies, societal assumptions, and values that ultimately affect our actions. Altering the way we communicate, and exploring new ways of conceptualizing and imagining, can forge new thought patterns and perceptions. Thus, part of making space for diverse worldviews, healing our relationship with nature, and finding new ways of being lies within the realm of language, memory, and a storied living landscape.

It is fortunate that the articulation of a multivalent Nature is brilliantly alive in hundreds of Indigenous languages around the world today, describing the natural world with powerful agency while also transferring invaluable wisdom of ecologies and land relationships through generations of memory.

Tragically, however, Indigenous languages are seriously threatened by the assault of modernity and colonization: in fact, many profound languages have already died, or are dying, due to the ongoing attack on Indigenous Peoples. And in this great offense, we cannot underestimate the truly devastating loss of worldviews and immense knowledge. As philosopher and polyglot George Steiner comments, "When a language dies, a possible world dies with it. ... a language contains within itself the boundless potential of discovery, of re-compositions of reality, of articulate dreams."[2] There has also been a significant loss of words and languages from older land-based times the world over. Noam Chomsky points out, "Some of the most dramatic language loss is in Europe. If you go back a century in Europe, all over the place people were speaking different languages."[3]

Indigenous author and teacher Martín Prechtel recounts one of his many experiences with language in his book *The Disobedience of the Daughter of the Sun*, offering a particularly insightful look at the verbs many Indigenous languages use and the need to keep this knowledge alive. He grew up speaking a refined Indigenous language on a reservation in New Mexico and later spoke Tz'utujil and other intricate Mayan dialects while living in Guatemala. Within this context, he makes the distinct point that none of those ancient complex languages possesses the verb "to be."

For those of us whose first language is English or another Indo-European language, it can be difficult to imagine a conversation without "she is," "they were," or "it is." The primary way to express ourselves without "to be" is through copious use of metaphors. But Prechtel tells us, "The brilliant ingenuity of Indigenous language and what is indigenous in all languages, especially the language of origination, ritual and sacred, though often mounted on rails of metaphor, is the way they zoom way past metaphor into realms of understanding that have metaphor looking rather naïve."[4] Speakers of non-to-be languages do use idioms for other purposes; however, in places where we would use a "to be" construction, they express that connection in ways that reflect their way of thinking about the world and, specifically, the relationships between beings living in the world.

The dependence of Indo-European languages on "to be" constructions also informs the way we interpret a culture's rituals. An Indo-European speaker, for example, often will view an Indigenous ritual as a metaphor: this represents the universe; that represents the sun; this is a fertility symbol, and so on, instead of seeing that the universe *is* versus that it is being represented.[5] Prechtel goes further to describe how the verb "to be" is the way English speakers normally express a connection between two entities. Indigenous and non-to-be languages can express these connections in ways that reflect—and shape—the speakers' understanding of relationships. Prechtel shares an example from the Tz'utujil language, which uses the verbs *ruqan* and *ruxin*. Both are often incorrectly translated as "to be," but in fact they mean "carries" and "belongs to"; for example, *Ruxin wa ja vinaq* literally means "It belongs to those people," but an English speaker would say "That is how those people are." If one were to translate "This is my land; this land is mine" into Tz'utujil, it would become *Javra uleu ruqan cavinnaq joj; ruxin joj ja uleu,* which literally means "This soil carries my people; we belong to this land."[6]

It requires extra attention for a speaker of a "to be" language to create a sense of belonging to a place and the precise nature of the relationship between humans and the land, and between beings and their attributes or states. While English speakers can also misrepresent a relationship using the verb "belong"—for example, "this land belongs to me"—using a verb other than "to be" forces a speaker to stop and think of what they are saying.

On occasion I have experimented with phrasing where I grew up as "The coastal land of Mendocino carries me," or "My family belongs to the land of Mendocino." It might seem strange at first, but if I belong to the land of Mendocino, and its soil carries me, what then is my responsibility to the land, to other beings who live there, and to the original peoples of these lands?

Most Indigenous languages recognize nonhuman beings with agency as "persons" and express these relationships as sentient kin. In contrast, many Indo-European languages such as English are human-centric, and they tend to objectify the land and her natural systems, portraying them grammatically and lexically as nonanimate beings without agency. Internalizing this kind of language and hearing it daily reinforces the dominant culture's worldview of the other-than-human world as separate from and lower than humans, and therefore existing solely for human benefit.

Anishinaabe and Métis scholar Mary Isabelle Young, in her book *Pimatisiwin: Walking in a Good Way,* gives several examples of the way her language treats nonhumans as living beings with a spirit. An Anishinaabe-speaking participant in Young's project shared that the Anishinaabe language uses two suffixes to pluralize a noun: one for animate nouns and one for inanimate nouns. Trees and rocks are animate, living, so they take the same animate suffix used for humans and animals, while the inanimate suffix is reserved for nonliving things such as dishes.[7]

Similarly, in Mundari, a language spoken by the Munda Peoples of northeast India, forces of nature such as wind, rain, and rivers are grammatically "animatized" when they appear as the subject in a sentence with a transitive verb—animacy is indicated by agreement elements affixed to the end of the verb, indicating that their agency is equal to that of humans, animals, deities, and the astral bodies.[8] This is not surprising, given that the Munda Peoples hold an animate cosmology worldview, and the Munda revere their deities, including the Mother Earth Goddess Chalapachho Devi, in sacred groves called *sarna*.[9]

When I consider the kind of human beings we in the dominant culture need to become to live in a post-colonial context and navigate the Anthropocene, I realize our work is not only to support Indigenous-led efforts to revitalize and protect their languages, which is absolutely critical, but also to transform our own tongues into languages of belonging and kinship by learning to bend, circumvent, liberate, and stretch language that currently objectifies nature until we become speakers of languages that affirm a living universe. No doubt, this is a long-arc cultural transformation strategy, but one I think is necessary.

This is one reason why for many years I have kept frayed, yet treasured, notebooks of words that assist me to experience landscapes more intimately. From watery oceanic lexicons to crackled and arid desert locutions, I have attempted to keep my ear attuned to specific words that elicit a full-bodied response in my interior topography and that offer a precise elucidation of place. Noticing and appreciating particulars are the language of relationship. Words such as *cordillera,* a system of parallel mountain ranges together with the intervening plateaus and other features; *atoll,* a ring-shaped reef, island, or chain of islands made of coral; *pelagic,* related to the open sea; *ceja,* which means "eyebrow" in Spanish but also refers to a thin crescent moon, an arc of trees alongside a meadow, or a narrow line of clouds over the mountains; *sastrugi,* from the Russian word *zastrugi,* the peaks, ridges, and vales constructed by the wind-whipped snow.[10]

All these words possess a texture and visual gesture that help to moor my consciousness to the landscape. Seeking and compiling these words is a component of my spiritual practice, part of the effort to come to know and love a place. Cultivating a relationship with a particular land involves coming to know its every facet as much as I can—from respectfully learning about and from the Indigenous Peoples of that land, to learning the names and manners of the plant and animal kin, to studying the geologic layers of the storied soils, to listening to the land and searching for words to define the unique beauty and contours of a region.

For many people, a deep love of place is best expressed through song, as music is an intensely somatic form of language. Perhaps one of the most profound verbal expressions of our living world that I have experienced was a Sami *yoik,* an ancient musical tradition of one of Europe's Indigenous Peoples. Composing and intoning a yoik about a person, the land, or other beings is different from singing a typical song; it is rather a way to evoke and conjure their distinguishing attributes through words and reverberations, a skill that requires dedicated love and acute observation of that which is cherished, both human and more-than-human.[11]

I am grateful for the experience of hearing a yoik in person. Casey Camp-Horinek called for a water ceremony during one of the UN climate conferences. People from all over the world were invited to participate and bring water from their homelands. At the gathering, waters from lakes, rivers, seas, and streams were combined in a central pottery vessel to be given to the local river with our prayers for the healing of the waters, the Earth, and all our relations.

Part of the ceremony included inviting women from various lands and nations to sing water songs from their respective homelands as they poured their waters into the beautifully decorated bowl. All of the songs were entirely stirring, and in particular, I remember a young Sami woman from the far north of Sweden singing a yoik. I closed my eyes to listen with all my senses, and as she sang, I began to visualize the northern lands and hear trickling snowmelt, the rumbling loud *splash* of glaciers calving, the rush of fast-paced rivers tumbling over stones. Her whistle-like guttural tones and trills created the most unusual resonances in my ears, and they invited me to see and sense where her waters had come from—waters that produced tonalities very different from the rivers of my California home. Through her yoik, I was distinctly transported to the contours of her living waterways that rippled through her voice and words, and I will never forget the embodied experience of the Sami woman's heartfelt love song to water. Worldviews can also conjure worlds—in this case, a sentient far-north landscape expressed through music.

Expanding my place-based vocabulary is also part of my quest to move away from linguistic patterns that articulate separation-from-nature narratives, and to keep nature language alive in the ever-increasing thrum and buzz of the modern world, which is a very real concern.

British author Robert Macfarlane has conveyed a similar worry, noting that the 2007 edition of the *Oxford Junior Dictionary* had removed dozens of nature terms—words such as clover, conifer, crocus, cygnet, fern, gerbil, holly, magpie, newt, porcupine, and starling—and instead added words related to technology and popular culture, such as attachment (the file, not the relationship), blog, celebrity, database, MP3 player, and BlackBerry (the device, not the fruit). The publishers defended their decision, explaining that the thin reference volume allowed for a limited number of words, that they should be words that mostly non-rural children up to age seven actually use, and that the more comprehensive *Oxford Primary Dictionary* still included the nature words.[12] This explanation is to be expected—young children spend much more time in front of screens than outside in nature—but it should make us bristle. This is no reason for the erasure of nature words, which only compounds the problem.

In the hopes of countering this trend, Macfarlane compiled a lexicon of unique and pertinent words that describe features of the British landscape and its weather. I was particularly inspired, but not altogether surprised, that much of the content was generated from Macfarlane's practice of collecting nature words from people who live or work intimately with the land and water, such as shepherds, fishers, and scientists. As I see it, the locution collection became Macfarlane's offering to the living Earth; his 2015 book, *Landmarks,* is a celebration of a wild Britain calling to be remembered and cherished.

Macfarlane made sure to research diverse languages and dialects across the British Isles, including words from the Fenland dialect he had learned from an elderly farmer and Gaelic words gleaned from an obscure and remarkable document called "Some Lewis Moorland Terms: A Peat Glossary." The glossary informed Macfarlane that "rionnach maoim" refers to the wandering shadows of clouds that are carried along by the wind on a sunny day, and that "èit" is the custom of leaving quartz stones in waterways at night, where they shimmer under the moon and entice salmon.[13]

When I read *Landmarks*, it became clear to me that the words Macfarlane collected had sprung from a deeply intertwined people and land. Knowledge and love of the sea, forests, heaths, and hills leapt from the page, fostered by intricate and profound reciprocal relationships with the land born from deep listening and observation over generations. The speakers of these words were at one with their homelands.

I was struck by Macfarlane's reflection that learning the word "smeuse," which refers to an opening made by an animal at the bottom of a hedge, caused him to notice more smeuses in his surroundings.[14] When we learn the name of something, a relationship is initiated. The naming can help us crack open the portal of not only seeing but also understanding who and what this sentient being is in the world.

From these same British Isles come other ancestral Irish words, linked not only to naming the natural world but to other ways of knowing, which are woven into broader cosmologies, paradigms, and spiritual understandings of life. Irish author Manchán Magan illuminates otherworldly meanings of Irish words in his book *Thirty-Two Words for Field: Lost Words of the Irish Landscape*—words that recognize the existence of the unseen world, other dimensions, and an animate Earth.

One of these multidimensional words is *cáithnín,* which refers to tiny things such as dust specks, snowflakes, subatomic particles, or eye irritants, but also "the goosebumps you feel in moments when you contemplate how everything is interrelated and how tiny we are in relation to the whole, like that feeling when you realise, or, maybe, remember, that we are all one—all unified."[15] It is a magical word that connects seemingly ordinary occurrences in our own world to the universality and infinity of existence.

Pre-colonial Irish peoples developed and spoke a language that, like most older languages, was not limited to things they could see and touch. *Ceantar* is a word still used in modern Irish, meaning "place," but its opposite, *alltar,* meaning "the other realm," has fallen out of use in our overly demythologized times.[16] The Irish language encompasses the two worlds that exist at once—while our organic matter exists in the ceantar, our minds can wander between the two realms. Likewise, a *púicín* is a "blindfold," but it is also a magical phenomenon that prevents mortals from seeing supernatural beings. *Comhalta* means "hooded," covered in a normal garment of this world, but it can also refer to something that is invisible to us because it is shrouded by a magical fairy robe. *Comhla* can refer to either a normal door or a portal to the other realm, but as you might have guessed, the latter meaning is now almost never the intended one.[17]

Importantly, scientific research on quantum mechanics, especially on virtual particles that disappear and reappear elsewhere like a "quantum foam" of soap bubbles, seems to verify these notions of multiple realms, concepts that many have dismissed as "superstitions."[18] We now know that the world we see and touch is perpetually winking in and out of existence, as are our own bodies. This universe is far from solid and stable, and what we claim as "reality" is based upon narrowly defined perceptions.

We also know that the diverse pre-colonial, pre-Christian European peoples realized that there was more to the story of our material world because the vestiges of language passed down to us reflect a multifaceted, complex universe—these languages are not limited to materiality, and allow for the existence of much broader understandings and relationships with an animate ecology. The etymology of words can assist us in opening spaces into the cosmologies of our antecedents, connecting us with not only them but our living Earth.

As I again reflect on the power of words to conjure worlds and communicate worldviews, to situate us physically and spiritually in the present, living, dynamic web of nature, I am reminded of the legendary Amergin Glúingel. Irish bard, poet, and druid, Amergin was a member of the Milesian People who settled Ireland in approximately 700 BCE. As recounted in various texts, when Amergin set foot on the shores of the Emerald Isle, he summoned the world of nature and his relationship to the land, recited in a poem, now known as the "Song of Amergin":

I am wind on sea
I am ocean wave
I am roar of sea
I am the stag of seven battles
I am an eagle on the cliff
I am a tear of the sun
I am the fairest of plants
I am a wild boar in valour

I am a salmon in the pool
I am a lake in the plain
I am a hill of poetry
I am a word of knowledge
I am the head of the spear in battle
I am the God that puts fire in the head
Who shed light on uncut dolmen?
Who announces the ages of the moon?
Who shows the place where the sun sleeps?
If Not I[19]

The language of this poem proclaims the magic and majesty of nature through the specific conjuring of the landscape, water, flora, fauna, seasons, time, words, and mystical dimensions of the island in an oral incantation revealing identification with these realms. It is said that Amergin reached the island through magical means—an Atlantic Ocean wave, emerging from a storm that had drowned his family, deposited Amergin onto the shore. The evocative invocation is for the land that Amergin approaches to make his home, the land he would belong to and cherish, the land that would sustain and identify his people.

Amergin's conjuration greatly influenced Irish cosmology and poetry, and though the analysis Robert Graves offers in his book *The White Goddess* remains controversial, it is worth noting his stance, as a distinguished poet: "English poetic education should, really, begin not with the *Canterbury Tales,* not with the *Odyssey,* not even with *Genesis,* but with the *Song of Amergin,* an ancient Celtic calendar-alphabet, found in several purposely garbled Irish and Welsh variants, which briefly summarizes the prime poetic myth."[20]

The ancestral homeland of the Sami Indigenous Peoples exists in a wide expanse of land stretching from northern Norway to the Kola Peninsula of Russia. Because it usually snows more than two hundred days per year (though this is not consistent now due to climate disruption), the Sami foster an intimate relationship with the weather of the extreme north, a day-to-day collaboration mirrored in their language.

Many Sami are reindeer herders. To safely guide the reindeer to fresh pastures, and even to traverse the land themselves, the Sami read the landscape carefully, especially the snow and ice, for which their language offers more than two hundred words. While *muohta* is the all-purpose word, *cuokca* refers to ice or snow that shapes a bridge over a waterway. When spring warms the land, dapples of melting snow called *roahtti* form, and *skarta* refers to a tough crust that congeals over snow after the rain.[21]

Prowess in evaluating and discussing different types of snow is essential for Sami livelihood and cultural practices, and their wisdom is helping scientists and environmentalists better understand climate change, since the way the snow accumulates, packs, melts, and refreezes can delineate the story of an entire snow season—and how one season differs from another. At least one group of scientists has already confirmed that the Sami reindeer herders notice the changes in their environment more acutely than the scientists, a fact reflected in the delicate precision of their language.[22] The Sami possess an elaborate hibernal language to convey the story of their changing environment and are, in fact, Indigenous scientists.

Vivid ecological knowledge is also embedded in the Bininj Gunwok languages of northern Australia. In this region, the advent of the rainy season is announced by the emergence of strikingly beautiful blue and orange grasshoppers. They are known as "children of the lightning," and they pay their respects to the Lightning Man, Namarrgon, their mighty father. The grasshoppers are called *alyurr,* but the bushes where they dwell share that name, as does, sometimes, the lightning. Soon after their arrival, the rains will come, along with wild plums and apples. The Bininj Gunwok languages show the complex relationships within the living landscape—in this case, between the grasshoppers, the bush, the seasons, and the growing and ripening fruits.[23]

The kinship between humans and the seasons often is revealed in old-time names of the months of the year. While in the city of Prague working on an art and ecology project, I became intrigued by the Czech language with its unique alveolar trill, *ř,* like rustling leaves in the forest, and its gliding palatals, all combining in pronunciations that twisted my American English tongue.

My friend Milan told me the names of the months in Czech, explaining that they remained close to their old Slavic roots. I loved the way they narrate the physical activities of plants, animals, and humans taking place in the region during a specific time of year. These names beckon the speaker and the listener to internalize the living Earth and seasons in a vibrant, integrated relationship:

January: Leden (Ice)
February: Únor (Hibernation or Ice Lowers)
March: Březen (Birch or Sap, or With Young)
April: Duben (Oak)
May: Květen (Blossom or Flower)
June: Červen (Red)
July: Červenec (Redder or Ripen)
August: Srpen (Sickle)
September: Září (Blazing or Glowing Sun)
October: Říjen (Rutting)
November: Listopad (Leaves Falling)
December: Prosinec (Slaughter of the Pig or, per Milan, Ask for Something)

Many Indigenous communities worldwide name the months of the year according to the cycles of nature and relationships with the land. In the book *The Sixth Grandfather,* there is an account of the monthly names and meanings of the Plains Indians conveyed by Black Elk, the Oglala holy man and educator:

January: The Moon of Frost in the Tipi
February: The Moon of Dark Red Calf
March: The Moon of Snow Blind
April: Red Grass Appearing
May: The Moon When Ponies Shed
June: The Moon of the Blooming Turnip or Making Fat
July: The Moon of Red Cherries
August: The Moon of Black Cherries
September: The Moon When Calf Grows Hair
October: The Moon of Changing Season
November: The Moon of Falling Leaves
December: The Moon of Popping Trees[24]

It is easy to forget that the word "month" means "moon," though the Gregorian calendar obviously does not correspond with the lunar cycles and is calculated by the movement of the sun. Yet moon calendars are quite ancient and have guided practical and ceremonial customs and activities for thousands and thousands of years. A Hawaiian colleague shared that her Elders taught her the names of each phase and day of the lunar cycle, elucidating that within these names there is encoded knowledge about when to fish and for what, when to forage for medicinal plants, and when certain seeds should be planted. She said, "As the traditional Elders have taught me, I know what to do by knowing which month it is and then looking into the face of the moon."

Similarly, the old Japanese names for the months aptly and beautifully describe what is happening during that time from a human-nature relationship:

January: Mutsuki (Month of harmony)
February: Kisaragi (Month of wearing extra layers of clothes)
March: Yayoi (Month of growth)
April: Uzuki (Month of Deutzia—a flowering shrub)
May: Satsuki (Month of planting rice sprouts)
June: Minazuki (Month of no water)
July: Fumizuki (Month of literary)
August: Hazuki (Month of leaves)
September: Nagatsuki (Autumn long month)
October: Kannazuki (Month of no Gods—because they are at the Izumo Shrine)
November: Shimotsuki (Month of frost)
December: Shiwasu (Month of running priests—because of the coming New Year)[25]

July is called the literary month because people once wrote poetry to celebrate the star festival of Tanabata, which commemorates the yearly meeting of Orihime the weaver (the star Vega) and her beloved cow herder, Hikoboshi (the star Altair). Today, there is a tradition of people in Japan conveying their wishes to the stars by writing them on papers and streamers and hanging them on bamboo branches—an artful and meaningful intersection between Earth, the Heavens, language, and time.

Languages can also convey direction in a landscape—how this orientation is shaped by, and shapes in turn, a community's perception of place in the cosmos and upon the land. Like the names for seasons and months, the way we understand our physical place and orientation in the landscape influences how we experience the world.

Manchán Magan shares his observations on how the Irish and English languages have shaped his perception of the world and his position in it, describing that giving directions in Irish requires a knowledge of the cardinal directions. One would not say they were merely going into town or up the road, but rather they were heading northwest or southwards home. Even if a person wanted to indicate a dog in the neighbor's yard, they would say that the dog was in the yard to the north.[26]

As someone attempting to learn from the Indigenous Peoples on whose land I live, I wanted to understand more about how Native peoples of Northern California speak about direction. I explored the work of Professor Leanne Hinton and discovered that many languages in this region rely not on cardinal directions but on landscape features, which makes sense in a location where the terrain is abundant with mountains and etched with rivers.

Particularly in the far north of the state, in the ancestral homelands of the Hupa, Yurok, and Karuk peoples, those geographical features tend to be waterways.[27] In that area, the steep mountains are perilously difficult to traverse, and people commuted alongside or on the river—today, roads run along the same watercourses. Because rivers twist and wind, indicating direction by north, south, east, or west is impractical; instead, one follows the flow of water. In the tongue of the Yurok, *pul* is "downstream," "pets" is "upstream," and *hiko* means "across the stream."[28]

Within this context, some California Indigenous Peoples refer to their own hands in various ways. While some use right and left, many do not, and instead they use the cardinal directions—for example, north and south hands—or water flow, so that a person's hands are upriver or downriver, and they switch positions when they turn around.[29] Hinton points out that left and right are egocentric terms and muses that as human communities lost their connection to nature, the directional terminology shifted away from the landscape and toward one's own self. The right-angled layouts of many modern cities disregard the cardinal directions as well as the mountains and rivers, making it more practical to say "turn left" rather than "north" or "upriver."[30] Again, it makes me ponder, how can modern-day city dwellers remember that each metropolis exists within a much grander terrain, in a living landscape? Many ancient cities and communities were built to reflect the cardinal directions and/or movements of the sun, moon, and stars, to mirror the seasonal transitions and celestial events; thereby, the built world of each community was a living calendar that corresponded to the cosmos that people lived within, a constant reminder that humans are part and parcel of the living Earth.

I also have been intrigued by how many peoples worldwide have integrated the natural world into their early written forms, as in the elaborate and elegant Chinese pictograms and Egyptian hieroglyphs. This also includes Ogham, the script used to write the letters of the early Medieval Irish alphabet, seen traced into standing stones throughout Ireland and Wales.

It is believed, due to kennings (compound expressions with metaphorical meaning) in old manuscripts such as the *Auraicept na n-Éces* (the scholars' primer) and *In Lebor Ogaim* (the Ogham tract), that up to eight of the twenty letters represented in the Ogham script were named after trees—alder, ash, birch, hazel, oak, pine, willow, and yew—though later scholars ascribed trees to the remaining letters.[31] The letters are each built on a vertical line, and varying numbers of horizontal or diagonal lines attach to that central line, crossing it or adhering to the side like branches. On manuscripts, the letters are turned sideways and written right to left.

But I particularly love looking at Ogham carved into rock because the letters are written bottom to top, each sprouting atop the one below so that words grow like trees—each inscription is a storytelling stone forest.

The letters themselves are referred to as *feda* or *nin,* which mean "trees" and "forked branches." A leading scholar of Ogham script, Damian McManus, proposes that the concept of a tree alphabet probably began around the ninth century CE, and he uses old manuscripts to delve into the etymologies of the individual letter names.[32] For example, the letter *duir,* a vertical line with two horizontal lines extending from the left, corresponds with an oak tree. I look out the window and compare the nearest oak with the letter duir, and find it exhilarating to see a letter formed from a tree. This observation pulls me more toward kinship in my thinking and expression, and works to close the gap between my life and the forest around me.

Other confirmed tree letters in Ogham include *fearn,* or alder; *beith,* or birch; *coll,* or hazel; and *saille,* or willow. McManus's research has added *onn,* or ash; *ailm,* or pine; and idad, or yew. I imagine a Medieval Irish scribe relating to the trees around them while carving or writing these letters, inscribing the "branches" to each letter's trunk, embodying this physical manifestation of the ancient Earth-centered language. This is just one more creative facet of the way the languages of place-based peoples can integrate a relationship with the living ecology all around us.

Canadian anthropologist and ethnobotanist Wade Davis states that "a language is not just a body of vocabulary or a set of grammatical rules... Every language is an old-growth forest of the mind, a watershed of thought, an ecosystem of spiritual possibilities."[33] As mentioned, Indigenous languages carry immense realms of knowledge, including insights into the natural world, how nature cares for us, and how we can care for nature. This is directly reflected in the Māori word *kaitiakitanga,* which is the practice of protecting the land in respect for one's ancestors, at the same time preserving natural systems for future generations.

However, many irreplaceable languages are acutely endangered, and it is essential that we particularly support Indigenous-led initiatives to sustain these languages and foster environments where they are valued and protected. Over eighty percent of the world's biodiversity thrives in Indigenous Territories or on lands tended by Indigenous Peoples, which are simultaneously places of great linguistic diversity. This is not a coincidence: protecting Indigenous languages is inseparably interwoven with protecting Earth's biodiversity.

The languages of place-based peoples are teeming with Traditional Ecological Knowledge, including rich and detailed information about soil, seasons, water, plants, insects, and animals of their homelands. This is generational knowledge that does not and cannot exist in the language or mind of someone not from the region—a person who has not lived in a specific place for many generations simply will lack the wisdom and words to describe it. When non-Indigenous scientists "discover" a species of animal or plant, they have most often found something the Indigenous Peoples of that area have named and known for generations. Even more knowledge is passed along through Oral Traditions in the form of song, stories, and ceremonies for a multidimensional understanding. Many oral histories remain consistent over hundreds and thousands of years, and long-ago events described in Indigenous stories are often confirmed by scientific and archaeological findings.

Even though around seven thousand languages remain globally, linguistic imperialism and capitalism have resulted in half of Earth's population speaking only twenty-three languages.[34]

Ultimately, a language dies when people stop speaking it. Peoples forced into a colonial context and immigrants to a dominant-culture country have been harassed, punished, and abused for speaking their languages, with residential schools as one of the most heartbreaking and infuriating examples. Active erasure of language is a calculated technique that seeks to destroy Indigenous Peoples' unique cultures and transform them into a cog in the machine of white supremacy and capitalism. In short, imperialist forces responsible for language death are also inextricably coupled with culture and ecological death.

Most residential schools were begun by Christian missionaries intent on erasing Indigenous cultures and converting Native peoples to Christianity. Indigenous children in North America, Australia, and elsewhere were ripped from their families and imprisoned in schools where they were forced, under threat of severe punishment, to speak only English and dress, pray, and eat like their colonizers, who presented themselves as saviors and the Indigenous students' families and peoples as primitive savages in need of saving. Children were prohibited from speaking their languages or singing their Peoples' songs, and were subjected to horrific physical, mental, emotional, spiritual, and sexual abuse. Thousands of children died at the schools from disease, abuse, neglect, and outright murder—and were thrown into unmarked graves; in many cases, their families were never notified.

Deb Haaland, US Secretary of the Interior and member of the Laguna Pueblo Tribe, initiated an investigation of the abuses and continuing legacies of the boarding schools, and a report released in May 2022 confirmed 408 schools in the US, designed to assimilate Indigenous children into white colonialist culture.[35] "We lost multiple generations of contributions to our communities because they were damaged by their experience—and they were the ones who survived," lamented former Osage Nation Principal Chief Jim Gray.[36]

Most who survived the schools emerged traumatized and demoralized, which was the intended result, since Indigenous languages and cultures were and are a challenge to white-supremacist, capitalist, dead-matter worldviews. Make no mistake—the Christian missionaries behind the schools saw Indigenous languages as a threat. In an institution where Christianity, capitalism, and whiteness demanded allegiance and pursued dominance over Nature, women, Indigenous Peoples, and People of Color, those voices had to be silenced—and the lexicons almost all disappeared. But not entirely.

Fortunately, the tide is slowly turning, as Indigenous educators and leaders and their allies work together to safeguard Indigenous and endangered languages from extinction, and to elevate the dignity of their speakers. Thanks to various groups of Indigenous linguists and teachers, including the Native American Language Issues Institute and the American Indian Language Development Institute, the Native American Languages Act (NALA) was passed in 1990, giving Indigenous Peoples in the US the right to speak and develop their languages. This includes the freedom to use Indigenous languages in schools and in businesses. The bill also urges educational institutions to give academic credit for proficiency in Indigenous languages, as they do for "foreign" languages.[37]

Languages can bounce back under the right circumstances, as we have seen in Hawai'i and New Zealand—and often, individuals and small groups can make a huge impact in that direction, like Professor Larry Kimura, who interviewed Native Hawaiian speakers on a radio show in the 1970s while trying to learn the language himself. His efforts inspired Hawaiian activists and his own students, and together they fought for permission from the Department of Education to create K–12 Hawaiian language immersion schools; there are now twenty-eight of them spread throughout the islands.[38]

So critical is language to one's culture that during the onset of the coronavirus pandemic, Indigenous communities prioritized the speakers of their languages along with the Elders and most vulnerable when distributing vaccines and personal protective equipment (PPE). I was struck to learn about this as I worked with WECAN in two PPE efforts for Indigenous communities in both the Global South and North who have suffered horribly during the pandemic due to government negligence, insufficient healthcare, and the ongoing effects of colonialism, with a death rate twice that of white Americans.[39] When an Indigenous Elder or language speaker dies, the community not only loses a beloved community leader, parent, sibling, and friend, but also a precious fountain of knowledge, embedded in ancient languages. Jason Salsman, of the Muscogee (Creek) Nation, likens the deaths to a "cultural book-burning."[40]

Nola Taken Alive, a member of the Standing Rock Sioux Tribal Council, lost both of her parents to COVID-19. Her father, Jesse Taken Alive, had been a fluent speaker and teacher of Lakota, a language carried by only two thousand people, and had long encouraged youth to speak Lakota, saying:

> The language comes from the creator, so it doesn't belong to one of us. The language belongs to all of us. So my message to all of the young people—the young men, the young women, the boys, the girls—this is your language. When you learn it, you're going to be able to learn more about this beautiful thing called life, because that comes from Wakan Tanka. The opportunity to share your feelings, to share your thoughts, to express yourself comes with our language. And I ask you to take the courage. [speaking Lakota]. I believe that there will be a day that all of you will talk. [speaking Lakota]. Finally, in closing, I ask you to do this on behalf of all of us who are older than you. Take the courage to learn the language.[41]

Author and scientist Robin Wall Kimmerer, of the Potawatomi Nation, inspires me in a million ways with her insightful observations about our living Earth, language, and relationships. She has extensively chronicled the way nature appears in language, particularly contrasting English with Anishinaabe, the language her grandfather was forbidden from speaking at school. I was intrigued by her approach to nature pronouns.

Kimmerer points out that none of us from any culture would ever refer to our grandmother as "it." If someone else did so, we would probably be offended at the person's disrespect. Don't they know Grandma is a person, not a thing? Yet English speakers refer to Mother Earth as "it."[42]

In Kimmerer's culture and in other Indigenous cosmologies, trees, mountains, waterways, animals, and rocks are all persons as well as teachers who show us how to live in the world; they are not lifeless objects. She says, "Objectification of the natural world reinforces the notion that our species is somehow more deserving of the gifts of the world than the other 8.7 million species with whom we share the planet. Using 'it' absolves us of moral responsibility and opens the door to exploitation. When Sugar Maple is an 'it' we give ourselves permission to pick up the saw. 'It' means it doesn't matter."[43] In Anishinaabe, Sugar Maple is addressed as a living being, with the same words that are used for humans.

Kimmerer, after consulting with her Elders, suggested a new pronoun for the English language, one that is gender neutral, reflects a living world, and resists the linguistic imperialism that attempts to erase Indigenous languages and ways of thinking. The Elders stated that it was not the responsibility of their language to help the dominant society that pursued their genocide, but others acknowledged that "the reason we have held on to our traditional teachings is because one day, the whole world will need them."[44] Kimmerer also tells us that in the Anishinaabe language, words are added only after extremely careful thought and consultation to respect the sacredness of the language and to remember that it is an ancient holy gift to be used in service to others and Creation. With this in mind, Kimmerer explored further.

Kimmerer's colleague Stewart King put forward the Anishinaabe word *Bemaadiziiaaki,* which means "beings of the living earth," but Kimmerer recognized that English speakers needed simplicity and proposed the ending of this beautiful and life-affirming word—"ki"—to be introduced into the English language. While honoring the Indigenous language from which the word came, English speakers could use ki instead of he, she, or it to talk about our more-than-human family, such as Sugar Maple. We can now say, "Ki is giving us sap again."[45] Kimmerer enthusiastically conveys that the plural pronoun would, quite aptly, be "kin."

Internalizing and saying ki can help us to recognize the agency, intelligence, and aliveness of our more-than-human family. This realization can assist us with transforming our worldview, and ultimately how we behave in this world.

At this stage, the dominant culture has become stunted with narrow-minded frameworks that most often do not enhance the living world around us, including an errant language of predator commerce that serves and centers only our species. Just look at how we speak of "resources" instead of pronouncing the names of the actual living waters—the Mekong and the Amazon, the Mississippi and the Nile, the Caspian and the Mediterranean, the Volga and the Yangtze. We need to listen for our language in the land, in the water, in the sky, in the stars. It is only through this animate understanding of our integral connection to landscape that we can begin to revitalize the memory of Original Instructions, our responsibilities to the land. Consequently, this linguistic endeavor is crucial to the dominant culture transforming its worldview and healing its relationship with the land.

We need to explore and regenerate living-Earth languages with words that sprint off the page or our tongues like wild deer guiding us deep into our inner topography while simultaneously making us aware of vast grasslands, folded mountain crevices, forest thickets, and the ecologically mystical and unique terrain on which we are standing. We can explore words and language to help us become a beneficial companion species to each other and to our relatives in the more-than-human world.

Chapter 14:

Songlines Through the Landscape

I REMEMBER WALKING SWIFTLY to keep pace with Phil Prosser's long, confident gait. A tall, well-built man, and an Elder of the Noongar community, he was walking on his ancestral west Australian lands, where generations of Aboriginal Peoples have lived for more than sixty thousand years.

Phil Prosser was a significant leader, having served as president of the Aboriginal Veterans Affairs Association and chair of the Aboriginal Cultural Material Committee, and undertaken vital work to improve the lives of Aboriginal Peoples and establish and protect water and land rights after devastating European colonization of his homelands. But before I say more, it is important to know that Phil has since passed on to the other shores. When I was with him, I received his permission to share the stories that follow, and all that is conveyed here is in respect to Phil, his generosity, and his incredible lifelong work toward healing next generations.

When I traveled to Australia, I never dreamed I would have the honor of speaking with a Noongar Elder. Someone at the cultural center in Perth had suggested I meet with Phil regarding my art and ecology project (which was what brought me to the city), so with some hope and surely a bit of nervousness, I phoned him. To my joy, not only did Phil agree to speak with me, but he also told me he was excited to meet, that it meant a lot to him that someone should ask about the history of his people and the land.

At his request, Phil and I meet in Kings Park in Western Australia, located on Mount Eliza, which is the colonized name of Mooro Katta. The mountain is the highest peak in the region and overlooks the city of Perth, which sits on the original lands of the Noongar.[1] After offering gifts to Phil to express gratitude to him and his homelands, I wonder why he would want to meet at this location—although quite beautiful, it is one of the most touristic areas of the region. Yet, as we embark upon a long walk through the park, and Phil begins to orient me to the landscape, it becomes crystal clear: preceding European contact and colonial settlement, Mooro Katta was and remains a significant cultural and ceremonial site for the Whadjuk Noongar community, who in the past often hunted and camped in the region and drank sweet fresh water from the Kennedy Spring (Goonininup) located on the southern face of the mountain. Enduring beneath the tourist park attractions with their immaculately manicured lawns and gardens, and unwavering under the clamor and amnesiac sounds and structures of modernity, this is eternally Aboriginal land. The stories that emanate from this landscape remain, impervious to the assaults and vicissitudes of oppression, erasure, and subjugation.

Phil's eyes under his wide-brimmed hat capture my full attention, holding the symmetry of both hollows and light of a man who knows how to be resilient in the face of many harms. He is a survivor of the Stolen Generation, one of the thousands of Australian Aboriginal and Torres Strait Islander children, many of them of mixed race, seized from their Indigenous mothers between 1905 and the 1970s and taken to state- or church-run institutions to be violently inculcated with European culture. At least one in ten Indigenous Australian children were deprived of their identities, languages, families, and connection to specific lands, as the institutions often "lost" the records of children's birthplaces and their parents' names, as well as often changing their names.

Phil's mother was also a stolen child. After she died, Phil lived with his father and maternal grandmother, as Aboriginal lines are matriarchal. Phil's grandmother was a traditional Aboriginal woman, a healer with her upper body adorned with initiation marks. Police officers abducted Phil from her home in 1944 and took him to be schooled at a mission, and later a boarding school, where he faced severe abuse. He and other children were beaten so badly on the back with thick belts that they would roll around on the frozen grass to relieve the pain. He showed me his hand, covered in scars he earned by standing up for himself. This is part of the story of this land he wants me to know.

As we meander along one of the less populated trails, Phil relates the violations of his people and the many challenges they faced to create Kings Park, where he played an important role in designing the protected area and the landmarks. Along the way, I experience hearing an ongoing mytho-poetic verse, as Phil continuously offers names of significant sites, trees, animals, and plants and their cultural meanings. It dawns on me that we are traversing an ancient and modern animate cartography of stories, which Phil is generously imparting to me.

We make our way along a bushland trail flanked by wide swathes of *Banksia,* the stems topped by cylinder-like colorful bottle brushes. Phil motions his hands to the gently swaying grasses and says, "We are all woven together. This fact is the basis of everything we believe about the world—our relationship with Country. Every person, every tree, every rock was dreamed and sung into existence by our Creator Ancestors."

As the air swells with the song of a hidden bird, Phil begins to illuminate two intertwined topics I have long wished to learn about: the Dream Time and Songlines. Thousands and thousands of years ago, the Ancestors traversed the land, dreaming up rivers, mountains, humans, animals, and plants from the mountain ash tree to the tiny white rice flowers. As the Ancestors made their way over the emerging landscape, they sang songs of Creation, gifting the newly created humans their cosmology, Original Instructions, and Earth-based knowledge.

Phil gestures widely toward the gentle slope of Mooro Katta, our path upward lined by diverse pine species, and relays to me, "Everything in this landscape tells a story. It all has meaning for us. When we see a certain mountain, river, or rock formation, we remember things that happened there, both recently and long ago. We can trace the Dreaming Ancestors' movements in the shape of the land—these are called Songlines, because the Ancestors sang as they went, and we remember what they taught us every time we pass a waterway or valley marked by their bodies and songs. The land shows us the Ancestors' stories—it is a physical testimony of our culture."

I would later learn that the Indigenous Peoples of Australia possess the oldest uninterrupted civilization in the world, supported by the tradition of Songlines—the combination of physical landscape features and song that allows them to accurately recount their own history, knowledge of the flora and fauna, spirituality, and culture, and pass it on to others, generation after generation.[2] These Songlines contain volumes of Earth-based wisdom and detailed information about local plant and animal species, knowledge that helped Aboriginal Peoples to survive and thrive, find their way across the vast landscape, and care for Mother Earth.

Phil then reminds me that the land now called Australia carries many Indigenous groups, and they each speak their own languages and follow distinct traditions. "In the Dreaming Time is when the Ancestors traveled across the country singing into creation the world and creating the laws and ways of living for each of us," he says. "Because the Dreaming Time never ended—it is still with us, and we see creation happening around us every day. It is part of us."

We continue to walk, and Phil leads me to an overlook on the mountain from which we see a view of the azure Swan River, glistening under the bright sun. "We call this Derbal Yaragan," he says. "The river was created by two Rainbow Serpents, Wagyl. They are our Ancestors, and they formed many of the rivers in this region. They shaped the terrain with their bodies in the Dreamtime." I gaze down at the gently rippling water and imagine those two majestic rainbow-colored Wagyl gliding along the unformed land as Derbal Yaragan emerges in their wake.

I recall the words of poet Gary Snyder and his insightful descriptions of a trip he took to Australia and its storied landscape. In one of his essays, Snyder remarks:

> We were traveling by truck over dirt track west from Alice Springs in the company of a Pintupi elder named Jimmy Tjungurrayi. As we rolled along the dusty road, sitting back in the bed of a pickup, he began to speak very rapidly to me. He was talking about a mountain over there, telling me a story about some wallabies that came to that mountain in the dreamtime and got into some kind of mischief with some lizard girls. He had hardly finished that and he started in on another story about another hill over here and another story over there. I couldn't keep up. I realized after about half an hour of this that these were tales to be told while *walking,* and that I was experiencing a speeded-up version of what might be leisurely told over several days of foot travel. Mr. Tjungurrayi felt graciously compelled to share a body of lore with me by virtue of the simple fact that I was there.[3]

As we turn away from the river and continue down a trail lined by waving cycads, Phil mentions a beautiful painting by an Indigenous artist I had seen in Perth's cultural center, telling me that "our artwork illustrates the Dreamtime, and so do our dances and stories. You can see rock art in some places that depicts Creation stories, how that part of the land was created, and by which Dreaming Ancestor. We make art to continue to tell the stories and care for Country."

Phil leads me to a circle of stones, a place where women and men gathered. "The tribes met in council here and have done so for generations," he says. "Through our gatherings, we pass on our songs, knowledge, and stories. Some knowledge is not for everyone—it is only for certain people in the community. This is how we keep it safe from those who might change it. But some knowledge is open to anyone who cares to learn, as I am sharing with you now. People remember what they learn when the knowledge is in connection to a place. Not only do they remember what happened at a certain landmark when they see it, they also remember what they learned in that place, and the cultural practices there. It is the spirit of that place in a story."

I would later learn of the work of Lynne Kelly, known as the Memory Whisperer, who has researched memory and place. Our hippocampus remembers things better when associated with a particular location, especially one we are likely to see again, and when those memories are crystallized into song.[4] She explores in her research how Aboriginal Australians have been able to pass down oral encyclopedias of wisdom, with such high accuracy and consistency, for thousands of years via place and memory. Their stories encompass climactic shifts over time, even down to the changes in sea level millennia ago during the Ice Age. Every tree and flower, every bandicoot and gecko, has had the entirety of its physical appearance, behavior, nutritional needs, hunting and mating habits, and spiritual history recorded in song and place, and thus every being is deeply understood, respected, and appreciated as kin and fellow traveler through this life on Earth. I look around me and contemplate the awe-inspiring idea that the whole landscape of Mooro Katta was generated by song thousands of generations ago, and that it is also linked to a star map above in the navigational Songlines of the people here. Phil tells me, "We feel a connection to all of Creation because we tie our memories to the past, present, and future in the land. The Dream Time is always happening."

I contemplate how an umbilicus to the cosmos exists at the juncture of time, the landscapes that surround us, and cultural memory. To stand at this intersection is to grow as a person, inextricably connected to place and cosmos. The big story of life is in the land, and each of us is in that evolving story.

As I think about how the story is in the land, I return to *The Disobedience of the Daughter of the Sun* by Martín Prechtel, where we learn about the story and living presence of the mesmerizing lake called Mother Waters, known during ceremonies as Rumuxux Ruchiuleu, and about a Grandmother Volcano, where Sun Father and Grandmother Moon meet, in the pre-colonial Tz'utujil Maya Kingdom. This textured sentient landscape embodies the story of the Daughter of the Sun, and the local Indigenous Peoples honor the ecological and cultural relationship and cycles of their sacred lake, volcanoes, sun, and moon by ceremoniously retelling the story again and again. As Prechtel says, "To the Tz'utujil traditionalists the land is not some kind of detached matter that has a soul, but is a tangible living soul that has a heart. The soul of the tribal Earth feeds the people with what the wild and cultivated lands produce, which is known as the face of the Divine; and the people in turn are required to feed and maintain its heart. One of the sustaining foods of that heart is the telling of the story of The Daughter of the Sun and the rituals for which it forms a map of remembrance."[5] The story is embedded and remembered in the landscape, interwoven with Tz'utujil cultural and spiritual geographical identity.

Phil leads the way to the State War Memorial at the peak of Mooro Katta. We pause near the Court of Contemplation, ringed by bright flowers and stones bearing the names of battlefields on which Western Australians fell. "I joined the army when I was sixteen," he says. After his painful boarding school experience, twenty years in the military gave him a sense of freedom and self-respect. "I was tall and strong, good at football and cricket, and my physical abilities opened doors for me in the military that I would not have otherwise been able to access."

We proceed to the State War Memorial, built to commemorate those who died in the Boer War, World War I, World War II, the Korean War, and the Vietnam War. Phil conveys that it was important for the Aboriginal Peoples of Australia to receive public recognition for what they sacrificed for the country. Phil himself is a veteran of the Vietnam War. I visually take in the eighteen-meter obelisk overlooking Perth and the river as Phil relates the ways this memorial on Mooro Katta is now woven into the storied landscape of the Noongar—Dreamtime stories coexist simultaneously and are continuously evolving. This memorial is more than a structure of rock and placards etched with names and statements to honor veterans; it now is intertwined with the larger mythology of this hoary mountain and the Noongar Oral Tradition.

As we reach the edge of the park, a brown honeyeater, a small pollinator bird, flits across our path to a lemon-scented gum tree, where she breaks into a loud aria to greet the setting sun.

Before we say goodbye, Phil encourages me to see more of the Swan River. "The British came here and stole the land our people had lived on for thousands of years," he says. "They stole children from their mothers and forced us to go to their schools, where they beat us and tried to destroy our culture. But they could not make us forget who we are and the Dreaming. We see the Dreaming everywhere we look, and it is part of us, our identity. The land, the river, the Country, is alive because of the Dreaming. I hope you can take some time on the river."

On a warm night several days later, I visit Derbal Yaragan, the Swan River. I find the spot I scouted out earlier in an area recommended by Phil, a lush grassy area. Between two graceful peppermint trees, I sit, listening to the gentle ripples of Derbal Yaragan as she narrates her story. I look up and marvel at the majesty of stars that I cannot see from home, such as Alpha and Beta Centauri of the Southern Cross. Next to the Southern Cross stretches the Emu in the Sky, a constellation defined not by stars but by dark nebulae against the brightness of the Milky Way. This sky tells a different story than the one I am familiar with in North America, and I know that if I had binoculars, I would be able to see the Jewel Box, a beautiful compact galactic open star cluster with individual stars that appear to be different colors.

I contemplate Phil's stories and the Rainbow Serpent, the giver of life, who created the river long, long ago in the Dreamtime. According to Noongar Oral Tradition, the two Wagyl arose from the escarpment of Mount Eliza, a place called Ga-Ra-Katta, to fashion the Swan River and the Canning River, then rested at Goonininup, a portion of land at the foot of the mountain. The Wagyl deemed the Noongar people the guardians of the land, and Gooninınup became a sacred site for teaching, trading, and initiation. Perceiving Goonininup's significance, settlers reserved a small area for Indigenous use there in the 1830s. However, soon after that the Indigenous communities were legally banned from Perth by those who had usurped the land.

During the Dreamtime, the Wagyl continued east, molding waterways, lakes, and landforms, including sand dunes and the Darling Range, whose hills and valleys hold the shape of the Wagyl's body, now said to be represented by the Darling escarpment. Lakes sprang up in areas where the great Rainbow Serpent paused to rest, and the deity left droppings in the form of rocks. As the Wagyl scraped over the land, some of their scales dislodged and transformed into forests. Today, the Wagyl live deep in underground springs, dreaming still, as is shared in the telling of this story by Aboriginal Peoples.

In the quiet darkness by the river, with the busyness of the day over, I listen more deeply for the Wagyl under the ground, for the songs that mark the paths across the land, for long-lived Mooro Katta's stories that Phil has shared with me. Several words Phil spoke keep stirring within me about how the Dreamtime is renewed, over and again, and how Aboriginal Peoples continue in the tradition of their Dreaming Ancestors, singing the songs of the Country, of life, across the land. The Dreamtime exists in perpetuity, rejuvenating the land, the trees, the animals with the seasons and all of Creation.

I have had my eyes closed for some time in this inner state of listening when I feel compelled to come out of my reverie. I am quite startled—standing inches from my face in the moonlight is a black swan, staring back at me. Although Phil has told me stories about the black swans, called *Kooldjak, Gooldjak,* or *Maali* in the Noongar dialects, what I had not imagined is the beautiful bird's sheer size—I am sitting cross-legged on the grass, and the swan is looking down at me. Later, I would learn that black swans can grow up to four and a half feet tall, weigh up to twenty pounds, and have a six-foot wingspan.

Like the Rainbow Serpent, black swans inhabit Aboriginal Peoples' Dreamtime traditions, and they have lived alongside the local Noongar community for thousands of years, bugling and crooning on the lakes and rivers. They often migrate throughout the year from dry areas to rainy ones, where they sojourn to breed and raise their cygnets. These primarily obsidian-colored waterbirds display white flight feathers which, along with their bright red bills, give them a striking appearance. European colonizers saw this magnificent bird for the first time on the Derbal Yaragan, and accordingly called the waterway Swan River and named their settlement the Swan River Colony.

I remain completely still, not wishing to disturb the moment or the exquisite bird, and we continue to gaze at one another. In the presence of the swan, I feel the spirit of the land speaking to me, reminding me whose territory I am on. After some moments the great waterbird gently but briefly places her beak on top of my head before slowly turning and gliding into the river, sailing downstream and calling with a euphonic bugle song. In that moment, I feel somehow that I have become part of this land's story, even if only for a blink of an eye. Somehow, I am now inside the story Phil shared with me—as a humble and grateful visitor to the Noongar lands.

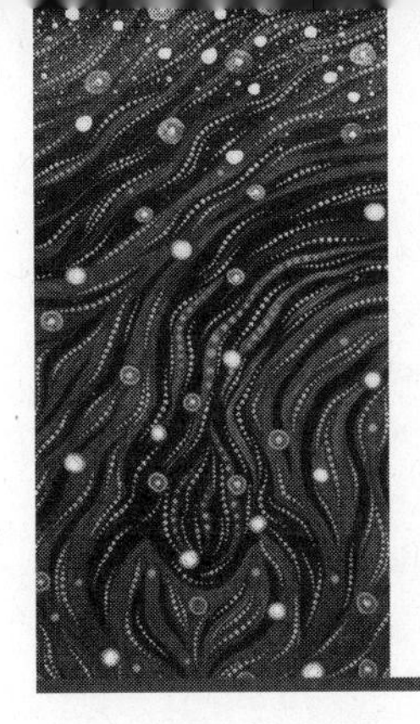

Chapter 15:

Building a Relationship with the Storied Land

IT IS IMPOSSIBLE TO TALK about the story of the land without discussing maps—political maps, topographic maps, thematic maps, economic maps, digital maps, and more—and how mapping has an extreme impact upon the way we perceive land. Mapping influences connection to place, and mapping conveys a particular narrative about the land. In light of this, it is essential to consider who is performing the mapping and naming of place, why they are mapping, and what the map tells us about the landscape via whose cultural or political lens. For instance, most maps designed and produced in Europe and North America are unsurprisingly drawn with the Northern Hemisphere on top, showing Global North dominance to the viewer.

Ancient maps were significantly more artistic than scientific, filled with fanciful illustrations that reminded users the map was not reality but a guide, infused with the mapmaker's point of view. In modern times, the scientific appearance of maps is meant to present an objective "reality" manufactured by people who are ostensibly unbiased authorities on geography.

Many of the borders between African, South and North American, and Asian countries were drawn by imperial powers, artificially dividing peoples who share the same language and culture and commingling those who do not. Maps chronicle colonialism, war, calls for sovereignty, land disputes, and occupations in an ever-changing geopolitical landscape, and they frequently omit the histories of original peoples. Just as history books are written by the conquerors, so cartographies are also in the hands of victors of colonization. When we look at a conventional map of the United States, we do not see the names of Indigenous Peoples and their Traditional Territories, and we generally do not see ancient place names reflected. Instead, we see an abundance of places named after primarily white men who colonized a people or region, or places named in the settler language.

When I zoom in on the digital map of Perth in Australia, I see Mount Eliza, not the name Mooro Katta, and I see the Swan River, not the name Derbal Yaragan. I cannot find Goonininup at all, only the Old Swan Brewery built upon Noongar sacred land. The biases preserved in map renderings of place continue into the digital age, manifesting on our phones, on our computers, and in our cars' GPS.

Part of shedding our subjugating mindset and conquering attitude toward the land involves decolonizing maps and retrieving Indigenous place names in the geographic information system. To challenge oppression and to disrupt, and then repair, our understanding of place, we need to reorient our cartographical practices to explore the full story of landscapes, specifically including parts that have been forcibly and systemically erased. Rescuing the beautiful words, names, and knowledge that come from the very song, sounds, and sentience of a place has everything to do with retrieving ourselves from a flattened, lifeless landscape, both of our inner worlds and the places we inhabit. This will also importantly uplift the original peoples of every land and remind us that most of us are guests in the land of First Nations Peoples and we need to respect their governance and knowledge.

Water scientist Kelsey Leonard noticed the invisibility of Indigenous Peoples and Indigenous land on maps in a 2014 project, when viewing the digital mapping interface of the Mid-Atlantic Ocean Data Portal. She and her colleagues then worked tirelessly to remedy the erasure. A citizen of the Shinnecock Nation, Dr. Leonard works for Indigenous water justice and brings forward solutions from Indigenous science to establish sustainable water and ocean governance. The map in question was one used throughout the United States by fishery management councils and federal and state agencies, and Dr. Leonard recognized the link between the absence of Indigenous communities on the map and their absence at the decision-making table: "There is no justice for Indigenous Peoples if we are not participating in the decision-making that affects our territories."[1]

Dr. Leonard advocates for maps designed by Indigenous communities, for Indigenous data sovereignty in the geographic information system, and for Indigenous languages to be included in mapping. This will allow for more cooperation between Indigenous communities, and between Indigenous and non-Indigenous communities, as we tackle issues common to all of us.

A group of A:shiwi (Zuni) artists in the American Southwest are also engaged with redesigning their own maps. The effort is led by the director of the A:shiwi A:wan Museum and Heritage Center, Jim Enote, who believes non-Indigenous people need to reevaluate their ideas of land. These new maps center Indigenous knowledge about land and challenge borders. By February 2018, the A:shiwi community had created thirty-two maps.[2]

The borders of the Zuni Indian Reservation were drawn by an executive order in 1877 and finalized in 1883, with no input from the A:shiwi People whose ancestral homeland had been constricted by the dictates of white men on the other end of the country. This was the fate of vast tracts of Indigenous land that had been seized and subsequently carved up by surveyors whose decisions were transformed into unchallengeable political "truths" through mapping.

I run my finger along the imposed straight imperial lines that mark the reservation borders in a New Mexico road atlas. This map embodies the story of white settler colonialism and eviscerates the actual history of the land and the peoples who had lived here for thousands of years, and continue to live here, a story that Jim Enote and the artists he works with have vowed to tell.

Today, the Zuni Reservation is home to twelve thousand Tribal members, though the sacred lands are much broader than the area represented by the fabricated lines on the map. As is shared in their tellings, the A:shiwi people's origin story begins underground, beneath four strata of the underworld, at what we now call the Grand Canyon, where they first emerged onto the Earth's surface. Within the Grand Canyon, the A:shiwi tended crops and thrived for a long time before venturing further, dividing themselves into three groups to explore the world beyond. One group, the medicine societies, traveled to the north. The second group traveled south. The third group, along with the medicine societies who had reunited with them, journeyed to the Middle Place, or Halona:wa, which we now call Zuni, New Mexico.[3]

It is this story of the land, marked by prayers of the journeying A:shiwi, that Jim Enote seeks to vivify. Along with a cultural advisory committee comprised of Elders, religious leaders, and Tribal members, he contemplated a series of "counter maps" that would apply Zuni names and illustrate a landscape from an Indigenous worldview, infused with prayer and culture.[4]

The application of Zuni names is key because colonization swallowed up so many Indigenous words and replaced them with English and Spanish place names, devoid of history or locally significant meaning. Unsurprisingly, many landmarks were named after colonizing missionaries, explorers, military personnel, and political figures, nearly all of them men. Philip Cash Cash, a linguistic anthropologist from the Nez Perce and Cayuse Peoples, emphasizes that saying and hearing these colonized place names over and over compounds the ancestral trauma experienced by Indigenous Peoples. Cash Cash adds that these names work to disappear the land's multi-realm history, making it more difficult to connect with the ancestral spirits and other numinous beings who live there—beings who carry ancient wisdom that can guide humanity in need today.[5]

The term "Zuni" is itself a designation by the US government for a community of people who refer to themselves as A:shiwi, or "human beings." The streets in and around the reservation originally began as trails the A:shiwi traveled to find food or water or visit sites of spiritual significance, but on maps they have names such as State Highway 53. As Jim Enote says, "Some of the trails are mentioned in songs, and those songs really say something about a destination: that trail goes to a place and the name of it is described as a destination, and how to get to that destination ... State Highway 53 doesn't tell you where you're going."[6]

Instead, the A:shiwi maps overflow with cultural knowledge, the accumulated intimacy that comes from generations of relationship-building and reciprocity with the land. These maps are not sterile grids devoid of place-based tradition, but ancient and modern narratives of the living land, full of history, intricate beauty, and prayer. They are story-filled, challenging dominant cultural worldviews about land, and about humanity's place and purpose among the rivers, valleys, plains, and forests of Earth.

Ronnie Cachini, a medicine man and rain priest who is fluent in the A:shiwi language, painted the map of the reservation, called Ho'n A:wan Dehwa:we, or "Our Land." His map is alive with color and includes blue lakes, snowy plateaus, villages, and large trees, many landmarks important to the community's ancestral history. Roads cross the map as thin lines, and a basket with prayer sticks appears in the center, symbolizing the prayers recited during the migration.

Cachini explains, "We're teaching these young families our migration history ... It is basically the creation of our people, creation of the world, our prayer. And so what we hope is that the map art would help the young people get back to where they came from."[7] To facilitate this, copies of the maps were produced for each A:shiwi family to hang in their homes, explore their history, and take pride in their culture.

Today, more maps are being made to acknowledge and honor Indigenous Peoples throughout North America, including an interactive digital map called Native Land.[8] You can enter your own address to discover whose land you are on, and you can view treaties. I like to check this map whenever I travel so at the very least, I can respect and pay tribute in some small way to the original inhabitants of each place I visit.

I first became conscious of cartographical discoveries close to home in my youth, when I was inspired to learn about wild food harvesting—foraging miner's lettuce, puffball mushrooms, spearmint, thimbleberries, salmonberries, rosehips, nasturtiums, and fiddlehead ferns in the Mendocino area.

In my mind's eye, each of the places where I gathered from the land began to coalesce into a seasonal map indicating the time of year I should visit. Special coordinates grew to include enchanted places that I loved to walk because they called to me, mystical spots along Big River or within ancient redwood groves where I could walk into another realm for a day and emerge with my inner world nourished and reassembled.

Living in relationship to place is not a static state—it is forever dynamic and requires continuous engagement and learning. Just like any human relationship, it is a practical as well as a sublime elegant entanglement. Even if we are not conscious of it, our very existence as beings of physical matter makes it impossible for us to live independently of a particular locality and the spirit and stories of that place, just as we as a species are inescapably intertwined with the welfare of the planet as a whole and its history.

With all of this in mind, in my youth I became obsessed with topographical maps of Northern California where I hiked, because I could see the sculptural shape of the terrain and imagine myself in that countryside, trekking up and down the mountains and chasing the rivers, yet those same maps lacked the actual stories and knowledge of the geographical strata, plants, animals, and sacred places in the landscape, such as described by Ronnie Cachini. At that age and enclosed in white privilege, I was unaware of colonization; all I knew was that something felt deeply amiss, that the land spoke to me in ways that no one seemed to be talking about. I read voraciously about Native Peoples of California, searching for stories of the land as my longing grew and the land whispered perennially.

This intense pull also carried over into cartographical and cultural pursuance of my ancestral heritage across the ocean, searching for old-time maps that may have stowed stories in the folds of ancient place-names or hidden in folktales, long-ago lore, and mythology. One such quest took me to the Rhine River in Germany during a time when I was traveling through Europe for work projects.

I was visiting with a German friend, Gabriele, in Bavaria. When out walking, we found ourselves in the low mists called the *nebel*. Shimmering silvery and mysterious, the mists conjure up the old tales of times gone past when Earth healers and bards freely walked the lands of ancient Europa. Stories seemed to filter through my cellular cognition while I drew my breath from these moist layers as an antique prayer.

Leaning over, I examined an iris flower with straight, sturdy leaves and an electric vibrant purple that shoots color into your belly. They are called sword lilies in German. In the days of old, when peoples of these lands listened to the natural winds voicing the silhouetted melodies of turning seasons, with no concept of separateness from the Earth, these sword lilies were understood to be relatives.

Gabriele told me with a tender frankness that she was only able to experience the mystical beauty of the land for the first time when she was a young woman and meandered with a friend's toddler through a forest near her home and saw absolute wonder in the child's eyes. For most of her life, she explained, the idea of sensing the trees, wind, sky, and stars as alive and connected to her core of being was unheard of—it was considered un-Christian and uncivilized to speak of the spiritual nature of the land.

Throughout Gabriele's childhood in the school system, like for many of us in mainstream society, mysticism was ostracized and there was only room for analytical, conventional science. Expressions of emotional or spiritual connections to nature were considered frivolous if not dangerous, as was curiosity about pre-Christian land practices. Jumping over fires during Beltane ceremonies, as her ancestors did, was co-opted by the church as the feast day of St. Walpurga, and gathering outdoors to celebrate nature was scorned if not for a church-related purpose.

Gabriele summed it up: "I felt strangled by Christianity." She longed to visit old trees, but they were almost always next to churches and owned by the church. She was interested in her Earth-honoring Celtic and Germanic tribal history, but this was never discussed in the university and was a topic disdained in part due to how the Nazis co-opted Celtic paganism in the service of white supremacism and nationalism. During World War II, this hijacking further poisoned the soil of relationship to ancestral Earth-based wisdom in Germany.

When I asked my friend about her history courses, she said the historical arc began with Neanderthals and progressed to the Roman Empire and Charlemagne, and then to medieval monks and cloisters, but nothing about Germanic tribes. She learned a great deal about church martyrs, particularly stories of women in heathen marriages who had converted to Christianity and were tortured and killed for their loyalty to the Church. These were presented as great female role models because they had denounced paganism, and many became saints. She recalls a morbid fascination with such women in history classes.

This severance from Earth knowledge, and from spiritual stories of the land, left her feeling hollow, as many of us have experienced. She conveyed to me that when she grew up there was always a sense among modern German people that their ancestors were primitive uneducated folk who belonged to pagan cults. It was proclaimed that the Church finally civilized her forebears, and rose to be the main house of spirituality and religion.

Gabriele told me that this conservative historical interpretation has greatly eased over the last twenty years, and she experiences that Germans are far more openly exploring diverse spiritualities today. Even though conventional German schools rarely if ever explore pre-Christian, Earth-centered history, an overall public awareness of an Earth-honoring past is actively translated now into the present and is a growing cultural phenomenon that is accepted, even if not a mainstream narrative. As a descendant of peoples connected to Germanic tribes through travel and trade, I felt a primordial tug to learn more.

It was from within this trail of thought that two German artist friends and I decided to visit the Rhine River to explore the old sagas of the land and the famous Loreley Rock in the Upper Middle Rhine Valley. To prepare, I collected folktales and maps of the area from books and colleagues.

Old European tales are multifaceted layered palimpsests within which traces of original stories remain, declaring their presence beneath strata of conquests and colonization, summoning us, the descendants who try to listen, to remember and regenerate them anew. To fully enter the marrow of a land-based poem, mythology, or story, I have found that it is often necessary to physically be in that place—to first make offerings to that land and then to humbly listen and look with all my senses open. There is a beautiful ecological context that we must become entangled with and lost in, with the hope that we might learn the story of that place, emerging from its very mountains, waterways, gorges, and valleys. We can also work to make some these connections through focused meditations, research, and communicating with people who live in the region if we cannot or choose not to travel. (Like many, I have greatly curtailed my travels.)

One early spring day, my companions and I eagerly drove the long ribbons of twisting highway to reach the Rhine. All along the way, when we stopped for coffee or lunch, people laughed when I asked about the Loreley. "Oh, the Loreley, she's not real. It's just a silly old tale."

The rocky Rhine River gorge had been carved and shaped by a million tons of water per year rushing through it over countless centuries. Five hundred years before the Common Era, Celtic peoples had settled the Loreley plateau and the Hühnerberg, nourished by the Rhine's abundant gifts of salmon. Up to one hundred years ago, the Rhine was the most significant watercourse in Europe to fish for salmon; around one million returned each year from the icy seas around Greenland to spawn in the Rhine tributaries until intense water pollution killed off many in the 1950s and newly constructed dams blocked their passage. Starting in the 1990s, however, measures have been taken to restore the salmon runs, including installation of channels and fishways around still existing dams. The salmon have now returned in great numbers and can swim all the way to Strasbourg, beyond which still lie impassable dams.

The storied Loreley, also spelled Lorelei and Lore Lay, is identified with a steep slate rock, 145 yards high, inside a curve on the eastern bank of the Rhine. Along the serpentine Rhine River, the bend inhabited by the Loreley is the narrowest and most perilous navigable section, at 175 yards in width. The origin of the name is disputed, but likely comes from the word *lurein,* or "murmuring," from an old regional German dialect, and a Celtic word for "rock"—*ley.* Loreley is the murmuring rock, and I hoped she would murmur her stories to us that day.

As was done to the river in the past century, writings about the Loreley have been polluted—the tales having been seized by patriarchal and Christian translations expressing derision against Mother Earth and against women, and the nineteenth-century version of the Loreley tale is no exception.

According to this legend, Lore Lay was a beautiful woman, a "temptress" of men. She threw herself into the Rhine in despair of a faithless lover and was transformed into a siren who lures boatmen to the rocks and their ultimate destruction. This version of the story largely comes to us through the work of the writers Heinrich Heine and Clemens Brentano. In Brentano's novel *Godwi,* he tells the story, in ballad form, of a woman whose irresistible beauty is attributed to her being a sorceress, who eventually threw herself from the rock that bears her name. Heine's poem presents her as a siren whose form and song enrapture boatmen, causing them to crash their boats in the treacherous current and rocky reefs.

Yet, stories of mystical occurrences at this dangerous bow of the river far predate Brentano and Heine. People living along the Rhine in the Middle Ages spoke of mountain spirits, nymphs, and dwarves whose mischief caused unsettling currents and reverberations from the rock. The dwarves also acted as oracles, as sailors and local peoples asked their questions at that spot and heard answers emanating from the rock.[9] But there was no talk of a temptress. Influences from Greek mythology—the sirens in Odysseus, the nymph Echo, a gaze that immobilizes—have certainly influenced the nineteenth-century writers.

The more recent stories I read about the Loreley were contorted and changed, but fortunately, the roots of the tale still sometimes managed to stay alive in the cultural mycorrhizal network and survive. Each word of an old folktale is a puzzle to unravel, a thread to follow. What would we find at the Rhine, if anything? Who and what was the Loreley?

As we neared the river, we examined our maps and decided to take a ferry directly to the famed Loreley stone, which juts into the Rhine. After driving our car onto the ferry, we stepped out on the open deck to embrace the majestic river. Wide and rushing, she brought tears to my eyes. There we were, floating upon the ancient waters in a modern motorboat, traveling where perhaps some of my forebears had traveled. The Rhine's viridescent hair was braided into countless strands of water that plaited past, present, and future reflections.

Roadways, railroads, factories, restaurants, tourist sites, shipping docks, and castles on high cliffs enveloped the riverbanks, yet the river's natural beauty still held my heart. The ship's captain announced that we had arrived at the Loreley stone. We parked at the base of a huge slate cliff and began the trek up the path, passing lush trees and rugged faces that appeared in the primeval stone.

We hiked for hours up and down hills, unsure but listening and searching until I pulled out a copy of a map that I had found in an old history book. It pointed the way to a nearby archeological site on the outskirts of Koblenz. Perhaps this site would provide a clue to the story of the Loreley, and the people who once lived here and spoke the names of the land. We returned to the car and drove to the site, only to find it locked behind the steel gates of a military base. I only learned of the significance of the archeological site —the Goloring—many years later.

Created between 1200 and 800 BCE during the Urnfield culture of the Bronze Age, when many Central Europeans practiced a living-Earth cosmology, the Goloring is an ancient earthwork shaped into a ringed trough surrounded by an embankment, part of which is covered with trees. This structure, 208 yards in diameter, qualifies in the archeological world as a henge and is similar in proportion to Stonehenge. It is unique in Germany, one of the most important Celtic artifacts remaining in Central Europe, and it continued to hold legal, social, cultural, and religious significance for local peoples through to the Iron Age.[10] Dr. Josef Röder, who researched the structure extensively, believes people gathered at the Goloring from all over the Maifeld and the Neuwied Basins.[11]

The sacred calendar henge was a place where Celtic peoples met and seasonal solstices were celebrated within a holy site that mirrored their worldview of a living universe. Reflecting the cycle of seasons, procession of the stars, and Earth-based wisdom, many ancient ceremonial structures like Goloring were round. Many ritual paths at sacred sites, such as the

Goloring henge, follow the passage of the sun across the horizon, with participants in ceremonies there bearing lights in a further parallel to the sun and the solar deities.[12] Vestiges of seasonal rites continue today in folk festivals in Europe, including May Day celebrations where people dance in a spiraling circle with ribbons around a maypole—which is traditionally a tree, calling remembrance and renewal of the Tree of Life or World Tree.

When I visited the site of the Goloring, I sensed that behind the military blockade lay something that yearned to be seen—and had a tale to tell. As we stood at the gate, we wondered what befell this ancient site that was off-limits to us (I have since learned that guided tours are now offered). And although at the time, I had not known of the ceremonial henge, somehow being in the presence of this sacred site drew us closer into the hidden landscape.

Evening had arrived in purple and peach hues, and we looked up at the sky, wondering what to do next given the late hour. We decided to return to the Loreley stone and turned toward the car just as a man passed on his bicycle and asked what we were doing there, three women out on a fairly untraveled dirt road at dusk. When I told him of our quest to learn about the Loreley, he chuckled and said, "Well, there is nothing there, sorry to say. Nothing at all." Again, another voice of dismissal came our way.

Night surrounded us as we drove, yet a bright, almost full moon soon rose over the eastern edge of the plateaus. As I contemplated the moonlight glinting on the Rhine River, inspiration pulled at me. I felt that we needed to visit the *opposite* side of the great river, directly across from the Loreley stone. We crossed at the nearest bridge, parked the car, and stood facing the Loreley cliff, the Rhine River directly in front of us, flowing and turning in her never-ending dance toward the sea. In the moonlight, we could perfectly see the elegant silhouette of the Loreley outcropping.

In that moment, I remembered that each and every tale of the Loreley mentioned an echo, and the old sagas had even described the Loreley song as an echo from the Rhine. I had assumed this referred to the rushing sounds of the river passing through the narrowest rocky curve of the waterway, but now I sensed that there was more. I told my friends, "I believe the Loreley is in the land herself, that the echo *is* the Loreley within the landscape."

We silently walked to the water's edge and studied the ancient stone on the far shore. Unsure of what would happen, we each took a deep breath and lifted our voices in unison, singing out across the river "Loreley, Loreley!"

To our amazement and great joy, we heard the powerful Loreley. She called back to us from across the river, answering with two melodious reverberations, singing her own name. Loud and clear, high above the commotion and uproar of trains and cars, we heard our voices blended with hers, echoing in petroglyphic harmonics, resonating from the slate cliff wall. We laughed and cried together as we realized that the Loreley was alive in the beguiling silver-black singing stone. Her spirit and song is in the land, in the naturally chiseled shape of the landscape. Massive and strong beyond the reach of modernity's life-crushing enterprises was the primordial echo forever etched into the steep riverbank that had also sung to our ancestors.

Today, as we witness and experience ever more sacred landscapes paved over, clear-cut, or mined by extractive industries, and flooded or burned through climate disruption, we need to listen even more attentively to the story of the land that still emerges and lives on, to let it nourish and guide us through the perilous and unpredictable times that are here and coming.

When I learn the true histories of the many peoples of our world and the vibrant stories of the land, when I act upon calls for justice, rights, liberation and Earth protection, I see the possibilities of a different future than what is offered by the brutality of the extractive, colonial, patriarchal paradigm. All these stories, different ways of knowing, actions we can take, and the magnificence of Nature, help us to remember our true place and responsibilities in the greater universe. This remembering is imperative, and one of the most highly sophisticated and strategic actions we can take in this time of the Anthropocene.

The Earth is eloquently speaking about time, memory, reciprocity, equity, and a storied landscape. As well, the Earth is speaking in the language of catastrophic infernos, droughts, and deluges that reflect destructive human activity. Can we hear Mother Earth's clarion call to galvanize a much-needed transformation of the dominant cultural worldview in this very moment? By deeply listening and acting with humility and courage, we have the opportunity to open a portal to alive, equitable, majestic, and reciprocal worldviews—a metamorphosis deeply needed to summon and build Earth-centered communities for all, now and in the times ahead.

Reader's Guide

PLEASE BE WELCOME TO VISIT https://ospreyoriellelake.earth/ where you can have access to an in-depth reader's guide and other resources to further explore themes presented in *The Story is in Our Bones.*

Acknowledgments

I MEANDER MY WAY ALONG one of the banks of Big River, which runs on the south end of my hometown of Mendocino in Northern California. The river here meets the Pacific Ocean in a stunning estuary that the Me-tum'mah, or Mitom Pomo, call Buldam, which means blow-hole. Buldam refers to the blowholes that erupt around the bay from sea caves that are secreted along the shore. When ocean waves hurl and slam into these caves, huge masses of water shoot upward from openings in the ground, as if a great whale is clearing its airway in a booming water song to the sky.

Before colonization, the Mitom kept a seasonal site, and later a permanent village, near the mouth of Big River where I now walk—something I always want to remember and respect. Big River exists, like all rivers, because of a myriad of contributions from creeks, rivulets, streams, brooks, and tributaries all flowing together, gaining volume and momentum as the river is pulled from its mountain source to the sea.

In this same way, my book would not exist without the vibrant, loving, and brilliant contributions of beloved family, friends, colleagues, and mentors whose steady waters fed the river in me that formed my manuscript.

Many mentors, scholars, and authors have influenced my thinking. Casey Camp-Horinek of the Ponca Nation has been a dear friend and mentor for over a decade, and I thank her for her extraordinary wisdom, powerful leadership, and call to protect Nature that reaches globally. I wish to thank the remarkable Martín Prechtel whose profound knowledge and acumen of speech have deeply touched me at every level of being and are remaking worlds and healing generations of people at a time of great crisis. I thank dear friend and scholar Riane Eisler for her enduring and crucial work on dominator and partnership models of society, which has continued to inform my work for decades. Brian Swimme continues to tell the Universe Story with enormous dedication and grace, and I deeply appreciate his friendship and commitment to this transformative narrative.

It is an incredible honor to work in the global climate justice movement with an extraordinary community who are not only my colleagues but often dear friends and extended family. Through this movement, as well as working with systems thinkers and artists in a variety of fields, I have had the great fortune to meet truly brilliant and courageous leaders, and have had my ideas and understandings shaped and refined via countless conversations and inputs to my work. Thank you to: Eriel Tchekwie Deranger, Jacqui Patterson, Leila Salazar-Lopez, Shannon Biggs, Atossa Soltani, Maude Barlow, Cormac Cullinan, Melina Laboucan-Massimo, Michelle Cook, Waste Win Young, Tara Houska, Ruth Nyambura, Nati Greene, Pennie Opal Plant, Kandi White, Linda Sheehan, Wanda Kashudoha Culp, Nina Simons, Kenny Ausubel, Sarah Drew, Dal LaMagna, Monique Verdin, Pandora Thomas, Jessi Parfait, Neema Namadamu, Tzeporah Berman, Sally Ranney, Chip Commins, Jim Garrison, Tracey Osborne, Clare Dubois, Jeremy Lent, Lisa Ferguson, Liane Schalatek, Randy Hayes, Emily Arasim, Katherine Quaid, Ashley Guardado, Jahanarna Mangus, Li Stargazer, Blythe Reis, Linda Adams, Bridget Burns, Tamara Toles O'Laughlin, Carmen Capriles, Corrina Gould, Pearl Praise Gottschalk, Patricia Gualinga, Nina Gualinga, Helena Gualinga, Ena Santi, Blanca Chancoso, Penny Livingston, Joanna Macy, Christi Belcourt, David Solnit, Nuna Teal, Justin Winters, Clayton Thomas-Muller, Natalie Isaacs, Farhana Yamin, Sônia Bone Guajajara, Thilmeeza Hussain, Hannah Greep, Sohrob Nabatian, Ananda Lee Tan, Alana Conner, and Steve Lillo. I want to express especially big buckets of gratitude to colleagues and friends who generously read and commented on chapters in the manuscript: Frances Roberts-Gregory, Clara Vondrich, Gopal Dayaneni, Pat McCabe (Woman Stands Shining), Lori Goshert, Harriet Shugarman, Jeff Conant, Robin Milam, Victor Menottti, and Trish Weber. All final analysis is my responsibility alone. Please forgive if I have missed anyone's name here, you all matter very much to me, as well as the movement and vision we share.

A mountain of thanks goes to Amanda Lewis for her superb editorial work, keen insights, and humor all along the way. Huge praise goes to Mary Reynolds Thompson for her incredible encouragement for this project, acuity in reading the manuscript, and excellent editorial work. Many thanks to Murray Reiss for the fine-tuned copyedit.

I am filled with immense gratitude for Rob West for believing in this book at the onset and welcoming me to the amazing family of New Society Publishers who are living and practicing their social and ecological values. It is a pleasure and honor to be one of their authors.

There are hardly words to express my appreciation for Wyolah Garden and Gabriele Schwibach who are long-time friends who have been by my side for the entire journey of the book. Their dedication to the organization I founded, the Women's Earth and Climate Action Network, which is central to the content and context of this book, is something for which I am forever grateful.

I am very thankful for the everlasting support of Annie Whiteside, my beloved sister. Very simply put, she is the best sister ever, one of the kindest, most caring, and loving people I know.

There are several significant contributors to my book river journey that are no longer in this realm but need to be thanked as ancestors who have influenced me: Audre Lorde, Marija Gimbutas, Paula Gunn Allan, Ladonna Brave Bull Allard, and Thomas Berry. And I thank my own ancestors and all they lived through that allowed me to arrive in this life.

In closing, I am in awe of and in deep appreciation of all the climate justice leaders and people working for transformative change at this critical juncture in the human venture. We need to keep on together and never give up on our beautiful planet and all generations to come. Thank you, each and every one, for the incredible loving and fierce work you do every day, seen and unseen, that makes all the difference in the world.

Credits

The portion of Chapter 1 about the mythology of fire's origin has drawn upon the chapter "Around the Fire" from my previous book: *Uprisings for the Earth* (Ashland, Oregon: White Cloud Press, 2010).

Parts of Chapter 4 have been adapted from an article I wrote on the subject: "The Voices of Amazon Women and a Visionary Declaration to Protect Indigenous Lands," *Common Dreams,* November 18, 2018.

I wrote more about the Escazú Agreement in the following article, and parts of a section in Chapter 4 are adapted from it: "Ratifying the Escazú Agreement Will Support Women Land Defenders and Protect Nature," *Ms.*, November 9, 2020.

Several sections of the opening paragraphs to Chapter 5 have been adapted from an article I wrote on the subject: "Women Rising for the Earth," *Earth Island Journal,* Autumn 2018.

The story in Chapter 6 about North Dakota is adapted from the following article: Emily Arasim and Osprey Orielle Lake, "Women on the Front Lines Fighting Fracking in the Bakken Oil Shale Formations," *EcoWatch,* March 12, 2016.

Parts of Chapter 7 are adapted from articles I wrote on the subject: "Rights of Nature and an Earth Community Economy," *Tikkun,* January 28, 2013; "Declaring the Rights of Nature Protects the Future for Us All," *The Environment,* April 2022, 12–14.

Part of Chapter 8's opening paragraph has been adapted from an article I co-authored: Wanda Kashudoha Loescher Culp and Osprey Orielle Lake, "America's 'Climate Forest' Is Under Attack. Biden Can Protect It.," *Fix,* January 14, 2021.

Endnotes

Chapter 1: Worldviews Are a Portal

1. Daniel Politi, "California's Camp Fire Is Fully Contained After Killing at Least 85 People," *Slate,* November 25, 2018.
2. Alejandra Borunda, "The Science Connecting Wildfires to Climate Change," *National Geographic,* September 17, 2020.
3. Women and Population Division, Sustainable Development Department, "Women and Sustainable Food Security," Food and Agriculture Organization of the United Nations, accessed August 8, 2022.
4. Zhike Lv and Chao Deng, "Does Women's Political Empowerment Matter for Improving the Environment? A Heterogeneous Dynamic Panel Analysis," *Sustainable Development* 27, no. 4 (July/August 2019): 603–612.
5. "Why Women," WECAN International, accessed July 1, 2021.
6. Thalif Deen, "Women Spend 40 Billion Hours Collecting Water," *Inter Press Service,* August 31, 2012.
7. Osprey Orielle Lake and Katherine Quaid, "Ratifying the Escazú Agreement Will Support Women Land Defenders and Protect Nature," *Ms.,* November 9, 2020.
8. Nafeez Ahmed, "Capitalism Will Ruin the Earth by 2050, Scientists Say," *Vice,* October 21, 2020. This article uses the year 2050 as a reference, but I believe we must address these issues much sooner.
9. You can read more about Pandora Thomas's work on her website: https://www.pandorathomas.com/about.
10. *Global Warming of 1.5 °C,* IPCC, accessed July 1, 2021, www.ipcc.ch/sr15/.
11. IPCC Sixth Assessment Report, *Climate Change 2022: Impacts, Adaptation and Vulnerability,* IPCC, February 27, 2022; Mitchell Beer, "'Buried' Science Shows Fast Carbon Cuts Can Stabilize Temperatures in 3–4 Years," *The Energy Mix,* February 18, 2022.
12. The topic of omphalos stones draws from a chapter in my previous book: "Of Redwoods and Whales, Jewel Baskets and Roots," *Uprisings for the Earth* (Ashland, Oregon: White Cloud Press, 2010).

13. For a detailed explanation of racialized trauma in the body: Resmaa Menakem, *My Grandmother's Hands: Racialized Trauma and the Pathway to Mending Our Hearts and Bodies* (Las Vegas: Central Recovery Press, 2017).
14. John Thackara, "When Value Arises from Relationships, Not from Things," *John Thackara* (blog), March 8, 2018.
15. Oral story-telling by Sky Road Webb on Saturday May 25, 2019, Point Reyes Station, CA 94956.
16. Peter Byrne, "Tamál Húye: Coast Miwoks Fight for Recognition of Point Reyes' Indigenous History," *Pacific Sun*, May 5, 2021.
17. Isabelle Gerretsen, "Fight Fires with Indigenous Knowledge, Researchers Say," *Thomson Reuters Foundation News*, August 13, 2018.
18. A portion of this section has drawn upon the chapter "Around the Fire" from my previous book: *Uprisings for the Earth* (Ashland, Oregon: White Cloud Press, 2010).
19. Gaston Bachelard, *The Psychoanalysis of Fire* (Boston: Beacon Press, 1964), 7.
20. James G. Frazer, *Myths of the Origin of Fire* (1930; repr., London: Routledge, 2019).

Chapter 2: The Story is in Our Bones: Origin Stories to Remake Our World

1. "Las hijas del maiz," *Saramanta*, accessed July 17, 2021.
2. Linda A. Newson, *Life and Death in Early Colonial Ecuador* (Norman: University of Oklahoma Press, 1995), 164.
3. Oral comments at the International Rights of Nature Global Summit, 2014.
4. Ibid.
5. Drew Dellinger, "The Most Radical Thing a Person Can Think About" (Recording from the 2019 Religion & Ecology Summit at CIIS, San Francisco, CA, March 15, 2019).
6. Neil deGrasse Tyson and Donald Goldsmith, *Origins: Fourteen Billion Years of Cosmic Evolution* (New York: W.W. Norton & Company, 2004), 25.
7. F. David Peat, "Blackfoot Physics and European Minds," *Future* 29 (1997): 563–573.
8. Susan Bridle, "Comprehensive Compassion: An Interview with Brian Swimme," *What Is Enlightenment?*, no. 19 (Spring–Summer 2001).
9. Massimo Bonasorte, "The Dogon's Extraordinary Knowledge of the Cosmos and the Cult of Nommo," *Ancient Origins*, May 17, 2018.
10. Chris Deziel, "Animals That Share Human DNA Sequences," *Sciencing*, July 20, 2018.
11. Gen. 1:26 KJV.
12. Catherine Nixey, *The Darkening Age: The Christian Destruction of the Classical World* (London: Macmillan, 2017).

13. Gleb Raygorodetsky, "Indigenous Peoples Defend Earth's Biodiversity—But They're in Danger," *National Geographic,* November 16, 2018.
14. Melissa K. Nelson, "Introduction: Lighting the Sun of Our Future—How These Teachings Can Provide Illumination," in *Original Instructions: Indigenous Teachings for a Sustainable Future,* ed. Melissa K. Nelson (Rochester, VT: Bear & Company, 2008), 5.
15. Ibid., 7.
16. Rebecca Adamson, "First Nations Survival and the Future of the Earth," in *Original Instructions: Indigenous Teachings for a Sustainable Future,* ed. Melissa K. Nelson (Rochester, VT: Bear & Company, 2008), 33–34.
17. Darold Treffert, "Genetic Memory: How We Know Things We Never Learned," *Scientific American,* January 28, 2015.
18. Vidal Jaquehua, "The Story of the Origin of the People of the Andes," *Adios Adventure Travel,* accessed January 22, 2023.

Chapter 3: Ancient Trees and Ancestral Warnings

1. Kate Goldbaum, "What Is the Oldest Tree In the World?," *Live Science,* August 23, 2016.
2. Suzanne Simard, *Finding the Mother Tree: Discovering the Wisdom of the Forest* (New York: Alfred A. Knopf, 2021), 5.
3. Mother Tree Project, https://mothertreeproject.org/.
4. Heidi Toth, "Indigenous Knowledge Offers New Approach to Help Forests Adapt to New Conditions," Phys.org, August 16, 2019.
5. Anita Sethi, "Robert Macfarlane on the Enchanted Childhood Words Technology Has Disrupted," *Financial Review,* October 5, 2017.
6. MJL, "What is the Tree of Life (Etz Chaim)?" *My Jewish Learning,* accessed July 25, 2021.
7. Colm Moriarty, "Sacred Trees in Early Ireland," *Irish Archaeology,* August 20, 2013.
8. George W. Robinson (trans.), *The Life of Saint Boniface by Willibald* (Cambridge, MA: Harvard University Press, 1916), 63.
9. Mike Shanahan, "What Happened When Christian Missionaries Met Kenya's Sacred Fig Trees," Under the Banyan, April 11, 2018.
10. Dennis Sumrak, "Navajos Will Never Forget the 1864 Scorched-Earth Campaign," Historynet (originally published in the October 2012 issue of *Wild West*).
11. Hannah Ritchie, "The World Has Lost One-third of Its Forest, but an End of Deforestation Is Possible," *Our World in Data,* February 9, 2021.
12. Julie C. Aleman, "Deforestation in Tropical Africa Is Not as Bad as Previously Thought," *The Conversation,* April 10, 2018.

13. Carolyn Cowan, "Forest Loss in Mountains of Southeast Asia Accelerates at 'Shocking' Pace," *Mongabay,* June 28, 2021.
14. Priscilla Jordão, "The Amazon: Vital for Our Planet," *DW Akademie,* August 28, 2019.
15. Fiona Harvey, "Amazon Near Tipping Point of Switching from Rainforest to Savannah – Study," *The Guardian,* October 5, 2020.
16. Jeff Tietz, "The Fate of Trees: How Climate Change May Alter Forests Worldwide," *Rolling Stone,* March 12, 2015.
17. Ibid.
18. Covering Climate Now, "The Corona Connection," *The Nation,* March 18, 2020.
19. Ibid. For further reading on the link between deforestation and diseases such as COVID-19, read the aforementioned article as well as the following: Sonia Shah, "Think Exotic Animals Are to Blame for the Coronavirus? Think Again," *The Nation,* February 18, 2020; and John Vidal, "Destroyed Habitat Creates the Perfect Conditions for Coronavirus to Emerge," *Scientific American,* March 18, 2020.
20. John C. Cannon, "Forests and Forest Communities Critical to Climate Change Solutions," *Mongabay,* August 8, 2019.
21. Lisa Song and Paula Moura, "An Even More Inconvenient Truth: Why Carbon Credits for Forest Preservation May Be Worse Than Nothing," *ProPublica,* May 22, 2109; Alia Al Ghussain, "The Biggest Problem with Carbon Offsetting Is That It Doesn't Really Work," *Greenpeace,* May 26, 2020.
22. Nafeez Ahmed, "World Bank and UN Carbon Offset Scheme 'Complicit' in Genocidal Land Grabs – NGOs," *The Guardian,* July 3, 2014; Friends of the Earth International, *The Great REDD Gamble,* October 2014.
23. Tesni Clare, "This Is Not a Forest," *The Ecologist,* February 21, 2020.
24. Black Elk, John G. Neihardt, and Raymond J. DeMallie, *Black Elk Speaks: Being the Life Story of a Holy Man of the Oglala Sioux* (Albany, NY: SUNY Press, 2008), 218.
25. Kirsten Weir, "Nurtured by Nature," *Monitor on Psychology* 51, no. 3 (April/May 2020): 50.
26. Miguel Sabino Díaz et al., "Assembly and Council Trees in the Valley of Carranza (Bizkaia)," *Labayru Fundazioa,* May 18, 2018.
27. Nkwichi, "Under the Mango Tree," *Nkwichi Blog,* accessed July 26, 2021.
28. R. Bruce Allison, *Every Root an Anchor: Wisconsin's Famous and Historic Trees* (Madison: Wisconsin Historical Society Press, 2005), 6.
29. Bruce E. Johansen, "Dating the Iroquois Confederacy," *Akwesasne* Notes 1, no. 3 and 4 (Fall 1995): 62–63.

Chapter 4: A Visionary Declaration from the Amazon

1. Parts of this chapter have been adapted from an article I wrote: Osprey Orielle Lake, "The Voices of Amazon Women and a Visionary Declaration to Protect Indigenous Lands," *Common Dreams,* November 18, 2018.
2. Ibid.
3. "Kawsak Sacha, Sarayaku, Living Forest," *Kawsak Sacha,* accessed May 28, 2022.
4. Asyl Undeland, "Indigenous Land Rights -- A Critical Pillar of Climate Action," *Development and a Changing Climate* (blog), *World Bank,* November 19, 2021; Damian Carrington, "Indigenous Peoples by Far The Best Guardians of Forests – UN report," *The Guardian,* March 25, 2021.
5. Author's conversation with Helena Gualinga, July 2018, Quito, Ecuador.
6. José Gualinga and Pueblo Originario Kichwa de Sarayaku, *Kawsak Sacha – Living Forest,* Amazon Watch, accessed August 13, 2021.
7. Ibid.
8. Osprey Orielle Lake, "The Voices of Amazon Women and a Visionary Declaration to Protect Indigenous Lands."
9. Kelsey Flitter, "Ecuador's Supreme Court Makes Historic Ruling Recognizing Indigenous Right to Consent Over Oil and Mining Projects," *Amazon Frontlines,* February 2022; Kimberley Brown, "Ecuador's Top Court Rules for Stronger Land Rights for Indigenous Communities," *Mongabay,* February 9, 2022.
10. Osprey Orielle Lake, "The Voices of Amazon Women and a Visionary Declaration to Protect Indigenous Lands."
11. Ibid.
12. "Antonio Donato Nobre: The Magic of the Amazon: A River that Flows Invisibly All Around Us," TED (YouTube channel), uploaded September 19, 2014.
13. Bettina Ehrhardt, *Amazonia's Flying Rivers: No Forest No Water* (Munich: bce films & more, 2018).
14. Chelsea Harvey, "Amazon Rain Forest Nears Dangerous 'Tipping Point'," *Scientific American,* March 8, 2022.
15. You can follow their activities through their Facebook page: https://www.facebook.com/MujeresAmazonicasDefensoras.
16. Osprey Orielle Lake, "The Voices of Amazon Women and a Visionary Declaration to Protect Indigenous Lands."
17. "International Women's Day Delegation to Reject New Oil Contracts - Stand with Indigenous Women of the Ecuadorian Amazon Announced by WECAN," WECAN International, February 17, 2016.

18. Women's Earth & Climate Action Network, "Women of the Amazon Defend Their Homeland Against New Oil Contract on International Women's Day," *EcoWatch*, March 7, 2016.
19. "Over Five-Hundred Indigenous Women of the Amazon and Allies March for Climate Justice, Indigenous Rights on International Women's Day," *WECAN International*, March 9, 2016.
20. Ibid.
21. Ibid.
22. Ibid.
23. Freddy Cuevas and Peter Orsi, "Court Files Show Bid to Tar Slain Honduran activist Caceres," *The Seattle Times*, May 3, 2016.
24. Solomon Sogbandi, "Sole Witness to Berta Cáceres Murder: 'It Was Clear She Was Going to Get Killed'," *Huffington Post*, December 6, 2017.
25. Anna van Ojik, "Agua Zarca: Indigenous Fight Against Dam Costs Lives," Both ENDS, accessed May 30, 2022.
26. "Meet Berta Cáceres," The Goldman Environmental Prize.
27. "Hija de Berta Cáceres: Ella era una luchadora firme," *La Prensa*, March 5, 2016.
28. "Former Dam Executive Found Guilty in the Killing Of Berta Cáceres," *Mongabay*, July 6, 2021.
29. In June of 2022, Castillo was sentenced to a jail term of twenty-two years and six months: "Honduras Sentences Former Executive for Murder of Environmental Activist," *Reuters*, June 20, 2022.
30. Nina Lakhani, "Berta Cáceres Assassination: Ex-head of Dam Company Found Guilty," *The Guardian*, July 5, 2021.
31. Benjamin Wachenje, "Defending Tomorrow," *Global Witness*, July 29, 2020.
32. Chris Madden, "Five years on from Berta Cáceres' Assassination, Impunity Reigns in Honduras – Updated," *Global Witness*, March 2, 2021.
33. Alex Pashley, "UN Envoy Warns of Environmental Activist Murder 'Epidemic'," *The Guardian*, March 18, 2016.
34. Osprey Orielle Lake, "The Voices of Amazon Women and a Visionary Declaration to Protect Indigenous Lands."
35. WECAN, "A Red Flag on International Women's Day: Women Land and Rights Defenders Facing Violence," *Medium*, March 7, 2018.
36. Though this phrase was new to me and to some of my colleagues, it was adapted from the works of Greek poet Dinos Christianopoulos (born in 1931), who was ostracized from his society, possibly because he was gay. The Indigenous Zapatista often use the phrase, translated into Spanish, in their fight for sovereignty in Mexico.
37. I wrote more about the Escazú Agreement in the following article, and parts of this section are adapted from it: Osprey Orielle Lake, "Ratifying the Escazú

Agreement Will Support Women Land Defenders and Protect Nature," *Ms.*, November 9, 2020.

38. "Women Speak" database on biodiversity and forest protection: https://womenspeak.wecaninternational.org/category/biodiversity-protection/
39. Emily Arasim and Osprey Orielle Lake, "Women on the Front Lines Fighting Fracking in the Bakken Oil Shale Formations," *Indigenous Women Rising,* March 12, 2016. This message is often rendered as graffiti, in Spanish, throughout Latin America, as in the artwork of Mujeres Creando, an anarcho-feminist collective in Bolivia. It is often chanted in protests and marches.
40. Osprey Orielle Lake, "The Voices of Amazon Women and a Visionary Declaration to Protect Indigenous Lands."

Chapter 5: She Rises

1. John P. Schmal, "Oaxaca: A Land of Amazing Diversity," *Indigenous Mexico,* September 19, 2019.
2. Ibid.
3. Several sections of the opening paragraphs to this chapter have been adapted from an article I wrote on the subject: Osprey Orielle Lake, "Women Rising for the Earth," *Earth Island Journal,* Autumn 2018.
4. Joseph Lee, "Dakota Access Pipeline Operator Loses Legal Battle," *Grist,* February 22, 2022.
5. "DIVEST INVEST PROTECT," accessed June 8, 2022. "Indigenous Women's Divestment Delegations," *WECAN International,* accessed June 8, 2022. Kim Maida, "Divest, Invest, Protect: Indigenous Women Lead Divestment Campaign," *Cultural Survival,* March 2, 2018; Michelle L. Cook, "Striking at the Heart of Capital: International Financial Institutions and Indigenous Peoples' Human Rights," in *Standing with Standing Rock: Voices from the #NoDAPL Movement,* eds. Nick Estes and Jaskiran Dhillon (Minneapolis: University of Minnesota Press, 2019), 103–157.
6. Author's recorded interview with Phyllis Young, 2016.
7. "Why Women?," *WECAN International,* accessed June 8, 2022.
8. For more information, please see the following resources: *Gender and the Climate Change Agenda* (Women's Environmental Network, 2010); UNFPA, State of World Population 2009.
9. Summer Blaze Aubrey, "Violence Against the Earth Begets Violence Against Women: An Analysis of the Correlation Between Large Extraction Projects and Missing and Murdered Indigenous Women, and the Laws That Permit the Phenomenon Through an International Human Rights Lens," *Arizona Journal of Environmental Law & Policy* 10, no. 1 (Fall 2019); Brandi Morin, "Pipelines, Man Camps and Murdered Indigenous Women in Canada," *Al Jazeera,* May 5, 2020.

10. "Women Speak: Stories, Case Studies and Solutions from the Frontlines of Climate Change," *WECAN International,* accessed June 8, 2022.
11. Osprey Orielle Lake, "Women Rising for the Earth."
12. Sam Eaton, "These Indian Women Said They Could Protect Their Local Forests Better Than the Men In Their Village. The Men Agreed," *The World,* March 29, 2016.
13. Osprey Orielle Lake, "Women Rising for the Earth."
14. "Why Women?," *WECAN International.*
15. Diane Harriford and Becky Thompson, "'Say It Loud, I'm Black and I'm Proud': Organizing Since Katrina," *Fast Capitalism* 4, no. 1 (2008).
16. "Why Women?," *WECAN International.*
17. "Stand with Indigenous Peoples of Brazil, the Amazon and the Climate," Women's Earth and Climate Action Network (WECAN) International (YouTube channel), uploaded April 29, 2019.
18. WECAN International, *Gendered and Racial Impacts of the Fossil Fuel Industry in North America and Complicit Financial Institutions: A Call to Action for the Health of Our Communities and Nature in the Climate Crisis* (2021), 2.
19. Jessica Merino, "Women Speak: Ruth Nyambura Insists on a Feminist Political Ecology," *Ms.,* November 15, 2017.
20. Nemonte Nenquimo, "A Message of Indigenous Resistance and Inspiration from the Amazon," *Common Dreams,* November 29, 2018.
21. Rachel Riederer, "An Uncommon Victory for an Indigenous Tribe in the Amazon," *The New Yorker,* May 15, 2019.
22. Jennifer Hassan, "Greta Thunberg Says World Leaders' Talk on Climate Change Is 'Blah Blah Blah'," *Washington Post,* September 29, 2021.
23. Dr. Vandana Shiva, "Empowering Women," *BBC World Trust,* June 10, 2004.
24. Author's conversation with Joanna Macy.
25. Author's recorded interview with Phyllis Young, 2016.
26. "Women in Peacekeeping," *United Nations Peacekeeping,* accessed June 8, 2022.
27. Marija Gimbutas, *The Living Goddesses* (Berkeley: University of California Press, 1999), 8.
28. Marija Gimbutas, *The Language of the Goddess: Unearthing the Hidden Symbols of Western Civilization* (San Francisco: Harper and Row, 1989), xviii–xix.
29. Marija Gimbutas, *The Goddesses and Gods of Old Europe.*
30. Kenneth Bauer, "New Homes, New Lives: The Social and Economic Effects of Resettlement on Tibetan Nomads (Yushu Prefecture, Qinghai Province, PRC)," *Nomadic Peoples* 19, no. 2 (2015), 209–220.
31. Marija Gimbutas, "Old Europe in the Fifth Millennium B.C.," in *The Indo-Europeans in the Fourth and Third Millennia,* ed. Edgar C. Polomé (Ann Arbor: Karoma Publishers, Inc., 1982), 1–2.

32. *Ancient Greece: An Illustrated History* (New York: Marshall Cavendish, 2011), 25.

Chapter 6: Tracing and Healing the Assault on Women

1. Carolyn Merchant, "The Scientific Revolution and *The Death of Nature*," *Isis* 97, no. 3 (September 2006), 526.
2. Silvia Federici, *Caliban and the Witch: Women, the Body and Primitive Accumulation* (Brooklyn: Autonomedia, 2004), 48.
3. Eph. 5:23 KJV.
4. Barbara Ehrenreich and Deirdre English, *Witches, Midwives, and Nurses: A History of Women Healers* (New York: The Feminist Press at CUNY, 2010), 35.
5. Silvia Federici, *Witches, Witch-Hunting, and Women* (Oakland, CA: PM Press, 2018), 12.
6. Silvia Federici, *Witches, Witch-Hunting, and Women*, 21.
7. John Demos, *The Enemy Within: A Short History of Witch-Hunting* (New York: Penguin Group, 2008), 3.
8. Ehrenreich and English, *Witches, Midwives, and Nurses: A History of Women Healers*, 56.
9. Ibid., 28.
10. Megan Johnson, "The Healthcare Future Is Female," athenahealth, February 14, 2018. While these trends are encouraging, it is important to note that many women left their jobs during the Covid-19 pandemic to care for children. Even in 2020 and 2021, women were the ones normally expected to leave their careers to perform necessary care work (as well as the emotional labor of the home, which is another big topic). Women's contributions to professional journals also sharply declined during the pandemic, while men's submissions went up. It is too soon to speculate on the long-term effects of these developments.
11. This phenomenon is often called "bikini medicine," meaning that male doctors assume that female bodies are just like male bodies aside from the parts covered by a two-piece swimsuit.
12. CDC Newsroom, "Racial and Ethnic Disparities Continue in Pregnancy-Related Deaths," Centers for Disease Control and Prevention, September 5, 2019.
13. Dyan Elliott, *Proving Woman: Female Spirituality and Inquisitional Culture in the Later Middle Ages* (Princeton, NJ: Princeton University Press, 2004), 297–299.
14. Ehrenreich and English, *Witches, Midwives, and Nurses: A History of Women Healers*, 36.
15. Tish Thawer, *The Witches of BlackBrook* (Taylor, MO: Amber Leaf Publishing, 2015).
16. Seema Yasmin, "Witch Hunts Today: Abuse of Women, Superstition and Murder Collide in India," *Scientific American*, January 11, 2018; Charlotte Müller,

"Witch hunts, Not Just a Thing of the Past," *Deutsche Welle,* August 10, 2020; Laure Gautherin, "Where Witch Hunts Are Not a Metaphor — And Women Are Still Getting Killed," *Worldcrunch,* February 16, 2020.
17. The Silver Spoons Collective, "I AM WITCH-Tales from the Roundhouse Exhibition," *Crowdfunder,* accessed January 13, 2022.
18. Jessica Green, "Can't Get Ahead At Work? Don't Trust Other Women? Blame the Witch Hunts! Psychotherapist Claims Women Hold Back Due to Inherited 'Self-destructive' Traits Like 'A Fear of Being Heard' That Ancestors Needed to Survive," *Daily Mail,* December 10, 2021.
19. Harriet Sherwood, "300 Years on, Will Thousands of Women Burned as Witches Finally Get Justice?," *The Guardian,* September 13, 2020.
20. Ben Panko, "Last Person Executed as a Witch in Europe Gets a Museum," *Smithsonian,* August 29, 2017.
21. Matt Lavietes and Elliott Ramos, "Nearly 240 Anti-LGBTQ Bills Filed in 2022 So Far, Most of Them Targeting Trans People," *NBC News,* March 20, 2022; Jaclyn Diaz, "Florida's Governor Signs Controversial Law Opponents Dubbed 'Don't Say Gay'," *NPR,* March 28, 2022.
22. The Trevor Project, 2021 National Survey on LGBTQ Youth Mental Health (West Hollywood, California: The Trevor Project, 2021); The Trevor Project, 2022 National Survey on LGBTQ Youth Mental Health (West Hollywood, California: The Trevor Project, accessed June 12, 2022).
23. Duane Brayboy, "Two Spirits, One Heart, Five Genders," *Indian Country Today,* September 7, 2017.
24. Kelsey Borresen, "Here's What It Means to Be 'Two-Spirit,' According to Native People," *Huffington Post,* September 8, 2022.
25. Duane Brayboy, "Two Spirits, One Heart, Five Genders."
26. Annalisa Merelli, "UN Study Finds Almost 90% of Men And Women Are Biased Against Women," *Quartz,* March 5, 2020.
27. "Facts and Figures: Ending Violence Against Women," *UN Women,* November 2019.
28. "#MeToo Is at a Crossroads in America. Around the World, It's Just Beginning," *The Washington Post,* May 8, 2020; Epstein Becker Green, "The Global Impact of #MeToo Movement," Lexology, February 19, 2021.

Chapter 7: Listening to Black and Indigenous Women and Debunking the Myth of Whiteness

1. Marilyn La Jeunesse, "The 19th Amendment Only Really Helped White Women," *Teen Vogue,* August 16, 2019.
2. Ibid.
3. "Medical Exploitation of Black Women," *Equal Justice Initiative,* August 29, 2019.

4. Kimberlé Crenshaw, "Demarginalizing the Intersection of Race and Sex: A Black Feminist Critique of Antidiscrimination Doctrine, Feminist Theory and Antiracist Politics," *University of Chicago Legal Forum* 1989, No. 1 (1989).
5. Combahee River Collective, "The Combahee River Collective Statement" (Zillah Eisenstein, 1978).
6. Rachel Elizabeth Cargle, "When Feminism Is White Supremacy in Heels," *Harper's Bazaar*, August 16, 2018. There are many other authors who have explained how white feminism failed BIPOC women; here are just a few: Rafia Zakaria, *Against White Feminism: Notes on Disruption* (New York: W.W. Norton & Co., 2021); Mikki Kendall, *Hood Feminism: Notes from the Women That a Movement Forgot* (New York: Viking, 2020); Kimberlé Crenshaw, "Demarginalizing the Intersection of Race and Sex: A Black Feminist Critique of Antidiscrimination Doctrine, Feminist Theory and Antiracist Politics."
7. Constance Grady, "The Waves of Feminism, and Why People Keep Fighting Over Them, Explained," *Vox*, July 20, 2018.
8. Audre Lorde, "Learning from the 60s," in *Sister Outsider: Essays and Speeches* (Berkeley: Crossing Press, 2007), 183.
9. Audre Lorde, "The Uses of Anger: Women Responding to Racism," *Black Past*, August 12, 2012.
10. Audre Lorde, "An Open Letter to Mary Daly," in *Sister Outsider: Essays and Speeches.*
11. Ibid.
12. Audre Lorde, *The Black Unicorn* (Toronto: George T. MC Lead Ltd., 1978), 6.
13. Paula Gunn Allen, *The Sacred Hoop: Recovering the Feminine in American Indian Traditions* (Boston: Beacon Press, 1986), 267.
14. Ibid.
15. Maile Arvin, Eve Tuck, and Angie Morrill, "Decolonizing Feminism: Challenging Connections between Settler Colonialism and Heteropatriarchy," *Feminist Formations* 25, no. 1 (2013), 9.
16. Kathleen Martens, "Cowesses Identifies 300 of 751 Unmarked Graves," *APTN National News*, October 8, 2021.
17. "U.S. Report Identifies Burial Sites Linked to Boarding Schools for Native Americans," *NPR*, May 11, 2022.
18. Author's conversation with Louise Wakerakats:se Herne, June 2020.
19. Roxanne Dunbar-Ortiz, "Columbus Day Helped Italians Become 'White', Roxanne Dunbar-Ortiz Explains," *Teen Vogue*, October 8, 2021.
20. Noel Ignatiev, *How the Irish Became White* (New York: Routledge, 1995).
21. Roxanne Dunbar-Ortiz, "Columbus Day Helped Italians Become 'White', Roxanne Dunbar-Ortiz Explains," *Teen Vogue*, October 8, 2021.

22. Roxanne Dunbar-Ortiz, *Not "a Nation of Immigrants": Settler Colonialism, White Supremacy, and a History of Erasure and Exclusion* (Boston: Beacon Press, 2021), 157.
23. Brigit Katz, "New Orleans Apologizes for 1891 Lynching of Italian-Americans," *Smithsonian*, April 15, 2019.
24. Louis Nevaer, "The Lynching That Gave Us Columbus Day," *Medium*, June 18, 2020.
25. "About," *Black Lives Matter*, accessed June 8, 2022, https://blacklivesmatter.com/about/.
26. Elana Schor, "'Camp Auschwitz' Shirt Among Anti-Semitic Signs Raising Alarms in Capitol Riot," *Tampa Bay Times*, January 13, 2021; Monica Hesse, "Capitol Rioters Searched for Nancy Pelosi in a Way That Should Make Every Woman's Skin Crawl," *Washington Post*, February 10, 2021.
27. Author David Abram coined the phrase "more-than-human" In 1996.

Chapter 8: Worldviews of Our Ancestral Lineages

1. David W. Anthony, *The Horse, the Wheel, and Language: How Bronze-Age Riders from the Eurasian Steppes Shaped the Modern World* (Princeton: Princeton University Press, 2007); Valeria Kovtun, "Cucuteni-Trypillia: Eastern Europe's Lost Civilization," *BBC Travel*, August 6, 2021.
2. Joan Marler, "Women and Men in Catalhoyuk," The Association for a Free University of Women, March 2004.
3. Esther Jacobson, *The Deer Goddess of Ancient Siberia: A Study in the Ecology of Belief* (Leiden: Brill, 1992), 1.
4. Judith Shaw, "The Reindeer Goddess," *Feminism and Religion*, December 18, 2016.
5. Ibid.
6. Betty De Shong Meador, *Inanna, Lady of Largest Heart: Poems of the Sumerian High Priestess* (Austin: University of Texas Press, 2000), 171.
7. Diane Wolkstein and Samuel Noah Kramer, *Inanna, Queen of Heaven and Earth: Her Stories and Hymns from Sumer* (New York: Harper Perennial, 1983).
8. Joshua J. Mark, "Inanna," *Ancient History Encyclopedia*, October 15, 2010.
9. Jeffrey L. Cooley, "Inana and Šukaletuda: A Sumerian Astral Myth," *Kaskal* 5, 2008, 161–172.
10. Elizabeth Ofosuah Johnson, "Nana Buluku, the Revered Goddess and Supreme Deity of West Africa and the Caribbean," *Face2Face Africa*, July 31, 2018.
11. Teresa N. Washington, *Our Mothers, Our Powers, Our Texts: Manifestations of Aje in Africana Literature* (Ọya's Tornado, 2015), 63–64.
12. Monica Sjöö and Barbara Mor, *The Great Cosmic Mother: Rediscovering the Religion of the Earth* (San Francisco: HarperOne, 1987), 165–167.
13. Aaron J. Atsma, "Demeter," *Theoi Project*, accessed February 20, 2023.

14. Mark Cartwright, "Demeter," *Ancient History Encyclopedia,* November 12, 2019.
15. Evangelia Hatzitsinidou, "Goddess Demeter and the Wrath of Nature," *Greek-gods.info,* April 4, 2021.
16. Alan R. Sandstrom, "The Tonantsi Cult of the Eastern Nahua," in *Mother Worship: Theme and Variations,* ed. James J. Preston (Chapel Hill: UNC Press, 1983), 25–50.
17. Griselda Alvarez Sesma, "A short history of Tonantzin, Our Lady of Guadalupe," *Indian Country News,* May 18, 2009.
18. "Quan yin," *Encyclopedia.com,* accessed February 20, 2023.
19. Christina Feldman, "She Who Hears the Cries of the World," *Lion's Roar,* May 12, 2020.
20. Albert G. Mackey, *A Lexicon of Freemasonry* (London: Richard Griffin and Company, 1860), 156.
21. Thomas Berry, *The Great Work: Our Way into the Future* (New York: Bell Tower, 1999), 69–70.

Chapter 9: Offering and Tending to the Land

1. Sibyl Diver, Mehana Vaughan, Merrill Baker-Médard, and Heather Lukacs, "Recognizing 'Reciprocal Relations' to Restore Community Access to Land and Water," *International Journal of the Commons* 13, No. 1 (2019).
2. Ibid.
3. Ibid.
4. Ibid.
5. Alan Friedlander, Eric Brown, and Mark Monaco, "Coupling Ecology and GIS to Evaluate Efficacy of Marine Protected Areas in Hawaii," *Ecological Applications* 17, no. 3 (2007): 715–30.
6. "Community-Based Subsistence Fishing Areas - VOS 8-1," Voice of the Sea TV (YouTube channel), uploaded February 25, 2021.
7. Tracy Barnett, "Lyla June on the Forest as Farm," *The Esperanza Project,* November 15, 2019.
8. Lyla June has since completed her dissertation, called *Architects of Abundance: Indigenous Food Systems, Indigenous Land Management, and the Excavation of Hidden History.* Find her TED talk here: "3000-year-old Solutions to Modern Problems | Lyla June | TEDxKC," TEDx Talks (YouTube channel), uploaded September 29, 2022.
9. Barnett, "Lyla June on the Forest as Farm."
10. Ibid.
11. Donna L. Feir et al., "The Slaughter of the Bison and Reversal of Fortunes on the Great Plains" (working paper, National Bureau of Economic Research, Inc., Cambridge, MA, August 2022).

12. Dr. Gleb Raygorodetsky, "Reimagining the Post-Pandemic 'Normal': Learning from Indigenous Peoples about Reconciling Culture and Nature," *Cultural Survival*, May 16, 2020.
13. Ibid.
14. Matthew D. Moran, "Bison Are Back, and That Benefits Many Other Species on the Great Plains," *The Conversation*, January 18, 2019.
15. "NDN Collective Celebrates Buffalo Release with Rosebud Sioux Tribe's Wolakota Regenerative Buffalo Range & Wildlife Sanctuary," *NDN Collective*, October 22, 2021.
16. Alison Henry, "Rosebud Sioux Tribe Will Create the Largest Native-owned and Managed Bison Herd in North America," *World Wildlife Fund*, May 7, 2020.
17. Lucy Purdy, "The US Tribe Bringing Buffalo Back from the Brink," *Positive News*, April 14, 2022.
18. Robinson Meyer, "The Amazon Rainforest Was Profoundly Changed by Ancient Humans," *The Atlantic*, March 2, 2017.
19. Umberto Lombardo, José Iriarte, Lautaro Hilbert, Javier Ruiz-Pérez, José M. Capriles, and Heinz Veit, "Early Holocene Crop Cultivation and Landscape Modification in Amazonia," *Nature* 581, 2020, 190–193.
20. Robinson Meyer, "The Amazon Rainforest Was Profoundly Changed by Ancient Humans."
21. Suzana Camargo, "In a Drier Amazon, Indigenous People Recalibrate Their Relationship with Fire," *Mongabay*, August 6, 2021.
22. Rufo Quintavalle, "Politics and People Fuelling Amazon Rainforest Fires," *Canada's National Observer*, September 17, 2019; Meg Kelly and Sarah Cahlan, "The Brazilian Amazon Is Still Burning. Who Is Responsible?," *The Washington Post*, October 7, 2019; Zoe Sullivan, "The Real Reason the Amazon Is on Fire," *TIME*, August 26, 2019.
23. Carolina Schneider Comandulli, "Designing for Life," *Scientific American* 326, no. 5 (May 2022): 70–83.
24. Ibid.
25. Max Paschall, "The Lost Forest Gardens of Europe," *Resilience*, October 8, 2020.
26. Ibid.
27. Wil Roebroeks et al., "Landscape Modification by Last Interglacial Neanderthals," *Science Advances* 7, no. 51 (December 15, 2021); Stéphane Peyrégne et al., "Nuclear DNA from Two Early Neandertals Reveals 80,000 Years of Genetic Continuity in Europe," *Science Advances* 5, no. 6 (June 26, 2019).
28. Alic William Gray, "Origins of Agriculture," *Britannica*, July 29, 2022.
29. "Our Beliefs," *SAAFON*, accessed June 23, 2022.
30. John D. Liu, *Earth's Hope: Learning How to Communicate the Lessons of the Loess Plateau to Heal the Earth* (2007), 2.

31. John D. Liu, *Hope in a Changing Climate* (Fairfax, VA: Environmental Education Media Project, 2009).
32. Alexandra Groome, "Meet John D. Liu, the Indiana Jones of Landscape Restoration," *Regeneration International,* March 7, 2016.
33. "Cultural Landscape of Honghe Hani Rice Terraces," UNESCO, accessed August 28, 2022; Gary Jones, "The Marvel of China's Multi-generational Rice Terraces," *BBC,* October 25, 2021.
34. La Via Campesina, *Seeds in Peasant and Indigenous Culture, for Food Sovereignty of the Peoples* (Bagnolet, France, May 2022), 4.
35. Pratibha Parmar, *Alice Walker: Beauty in Truth* (Kali Films, 2013).
36. For more information on this program, please visit WECAN's site: https://www.wecaninternational.org/women-for-forests.
37. Marine Gauthier and Riccardo Pravettoni, "Clashing Over Conservation: Saving Congo's Forest and Its Pygmies," *The Guardian,* August 30, 2016.
38. "UPDATE: WECAN Women for Forests in the Democratic Republic of Congo," *WECAN Newsletter,* November 9, 2020.
39. Jessica Wanless, "DR Congo: The World's Most Neglected Displacement Crisis," *Al Jazeera,* June 1, 2022.
40. Marine Gauthier, "New Legislation to Protect the Rights of the Indigenous Pygmy Peoples in the DRC," *International Union for Conservation of Nature,* August 5, 2022.
41. Emily Arasim, "Planting Hope in the DR Congo: Women Lead First Ever Reforestation Effort in Itombwe," *WECAN International,* October 26, 2015.
42. Linnea Tanner, "Celtic Sovereignty Goddess," *Apollo's Raven,* June 11, 2018.
43. Ibid.
44. "Hieros Gamos," *Encyclopedia.com,* accessed February 13, 2022.
45. Tanner, "Celtic Sovereignty Goddess."
46. Jessie L. Weston, *From Ritual to Romance* (Cambridge: Cambridge University Press, 1920); Alfred Nutt, *Studies on the Legend of the Holy Grail: With Especial Reference to the Hypothesis of Its Celtic Origin* (London: D. Nutt, 1888).

Chapter 10: Composting the Cultural Toxins of Colonization and Capitalism

1. C.G. Jung, *Alchemical Studies,* vol. 13, Collected Works, par. 335.
2. "Treatment Methods for Contaminated Soil," *Enva,* accessed February 15, 2022.
3. Onur Ulas Ince, *Colonial Capitalism and the Dilemmas of Liberalism* (Oxford University Press, 2018), 4.
4. Ellen Meiksins Wood, "The Agrarian Origins of Capitalism," *Monthly Review,* July 1, 1998.

5. Ibid.
6. Jason Hickel, "Degrowth: A Call for Radical Abundance," *Jason Hickel* (blog), October 27, 2018.
7. Ibid.
8. John Bellamy Foster and Brett Clark, "The Paradox of Wealth: Capitalism and Ecological Destruction," *Monthly Review*, November 1, 2009; Hickel, "Degrowth: A Call for Radical Abundance."
9. Shannon Hall, "Exxon Knew About Climate Change Almost 40 Years Ago," *Scientific American*, October 26, 2015.
10. "Statement of Dr. James Hansen, Director, NASA Goddard Institute for Space Studies," *Greenhouse Effect and Global Climate Change: Hearings Before the Committee on Energy and Natural Resources, United States Senate, One Hundredth Congress*, first session (Washington: US Senate Committee on Energy and Natural Resources, 1988), 39–41.
11. Naomi Klein, *This Changes Everything: Capitalism vs. The Climate* (New York: Simon & Schuster, 2014), 19.
12. Ibid., 19.
13. Ibid., 18.
14. Ibid.,178.
15. "Press Release: Indigenous Peoples' Caucus Opening Statement," International Indigenous Peoples' Forum on Climate Change, November 3, 2021.
16. Yessenia Funes, "Yes, Colonialism Caused Climate Change, IPCC Reports," *Atmos*, April 4, 2022.
17. Natural Resources Defense Council, *Beavers: Nature's Wetland Ecosystem Engineers* (NRDC Wildlife Coexistence Series, November 2017).
18. Nafeez Ahmed, "Capitalism Will Ruin the Earth by 2050, Scientists Say," *Vice*, October 21, 2020. This article uses the year 2050 as a reference, but I believe we must address these issues much sooner.
19. William J. Ripple et al., "World Scientists' Warning of a Climate Emergency," *BioScience* 70, no. 1 (January 2020): 8–12.
20. "Just Transition: A Framework for Change," *Climate Justice Alliance*, accessed June 26, 2022.
21. "What Is an Ecovillage?," *Global Ecovillage Network*, accessed August 23, 2022.
22. Jason Hickel, "Degrowth: A Call for Radical Abundance," *Jason Hickel* (blog), October 27, 2018.
23. Francesco Sarracino and Małgorzata Mikucka, "Consume More, Work Longer, and Be Unhappy: Possible Social Roots of Economic Crisis?," *Applied Research in Quality of Life* 14 (2019): 59–84.
24. Hickel, "Degrowth: A Call for Radical Abundance."

25. Read more about Doughnut Economics and view the diagram here: Kate Raworth, *A Safe and Just Space for Humanity: Can We Live Within the Doughnut?* (Oxford: Oxfam International, 2012).
26. Clive Thompson, "Is Sucking Carbon Out of the Air the Solution to Our Climate Crisis?," *Mother Jones*, November + December 2021.
27. Thomas Day et al., *Corporate Climate Responsibility Monitor 2022* (New Climate Institute, February 2022), 45; Patrick Galey, "'Net-zero' Promises from Major Corporations Fall Short, Climate Groups Say," *NBC News*, February 7, 2022.
28. "World on Track for 2.4°C Warming Despite 2030 Pledges: Analysis," *Al Jazeera*, November 9, 2021.
29. Rafael Calel et al., *Do Carbon Offsets Offset Carbon?* (Centre for Climate Change Economics and Policy Working Paper 398/Grantham Research Institute on Climate Change and the Environment Working Paper 371. London: London School of Economics and Political Science, November 2021); Frédéric Mousseau, *Evicted for Carbon Credits: Norway, Sweden, and Finland Displace Ugandan Farmers for Carbon Trading* (Oakland, CA: The Oakland Institute, 2019).
30. Marina Stavroula Melanidis and Shannon Hagerman, "Competing Narratives of Nature-based Solutions: Leveraging the Power of Nature or Dangerous Distraction?," *Environmental Science & Policy* 132 (June 2022): 273–281.
31. Ibid.
32. Grayson Badgley, "Systematic Over-crediting in California's Forest Carbon Offsets Program," *Global Change Biology* 28, no. 4 (February 2022): 1433–1445; Lisa Song, "An Even More Inconvenient Truth," *ProPublica*, May 22, 2019.
33. Climate False Solutions, *Hoodwinked in the Hothouse: Resist False Solutions to Climate Change*, 3rd ed. (2021).
34. Patrick Greenfield, "Revealed: More Than 90% of Rainforest Carbon Offsets by Biggest Certifier are Worthless, Analysis Shows," *The Guardian*, January 18, 2023.
35. Desirée Yépez, "Time's Most Influential Indigenous Activist: 'Capitalism Destroys The Whole Planet,'" *Huffington Post*, October 16, 2020.
36. Klein, *This Changes Everything*, 156; Ahmed, "Capitalism Will Ruin the Earth by 2050, Scientists Say."
37. Adam Aron, "Mark Jacobson on the Clean Energy Transition," *Resilience*, May 24, 2021.
38. Elsa Dominish, Nick Florin, and Rachael Wakefield-Rann, *Reducing New Mining for Electric Vehicle Battery Metals: Responsible Sourcing Through Demand Reduction Strategies and Recycling* (Sydney: Institute for Sustainable Futures, University of Technology, for Earthworks, April 2021); Anita Parbhakar-Fox et al., "Tapping Mineral Wealth in Mining Waste Could Offset Damage from New Green Economy Mines," *The Conversation*, May 30, 2022.

39. Mark Z. Jacobson et al., "100% Clean and Renewable Wind, Water, and Sunlight All-Sector Energy Roadmaps for 139 Countries of the World," *Joule* 1, no. 1 (September 6, 2017); "Our 100% Renewable Energy Vision," *The Solutions Project,* accessed March 1, 2022.
40. Michelle T. Dvorak et al., "Estimating the timing of geophysical commitment to 1.5 and 2.0 °C of global warming," *Nature Climate Change* 12, no. 6 (2022): 547–552; Mark Hertsgaard, Saleemul Huq, and Michael E. Mann, "How a little-discussed revision of climate science could help avert doom," *The Washington Post,* February 23, 2022; Mitchell Beer, "'Buried' Science Shows Fast Carbon Cuts Can Stabilize Temperatures in 3-4 Years," The Energy Mix, February 18, 2022.
41. *Banking on Climate Chaos: Fossil Fuel Finance Report 2023* (San Francisco, April 2023).
42. "Governments' Fossil Fuel Production Plans Dangerously Out of Sync with Paris Limits," United Nations Framework Convention on Climate Change, October 20, 2021.
43. Robert Higgs, "Eisenhower and the Military-Industrial Complex," *The Independent Institute,* March 21, 2011.
44. Neta C. Crawford, *Pentagon Fuel Use, Climate Change, and the Costs of War,* Cost of War Series (Brown University, Watson Institute, November 13, 2019); Katya Forsyth and Frederick Kerr, "The Toxic Relationship Between Oil and the Military," Global Center for Climate Justice, March 22, 2022.
45. "The Call for a Phase-out of Fossil Fuels Has a Powerful, Global History. This Is How We Got Here.," The Fossil Fuel Non-Proliferation Treaty, accessed July 16, 2022.
46. Tzeporah Berman and Nathan Taft, "Global Oil Companies Have Committed to 'Net Zero' Emissions. It's a Sham," *The Guardian,* March 3, 2021.
47. Dallas Goldtooth and Alberto Saldamando, *Indigenous Resistance Against Carbon* (Washington, DC: Oil Change International, 2021).
48. Documented by author, COP26, Glasgow, Scotland, 2021.
49. Kaitlin Smith, "More Than Monsters: The Deeper Significance of Wendigo Stories," *Facing Today,* November 30, 2021.
50. Jack D. Forbes, *Columbus and Other Cannibals* (New York: Seven Stories Press, 2011), xvi.
51. Yamuna Hrodvitnir, "Wendigo: The Monster Born of Starvation and Greed," *Medium,* January 25, 2020.
52. Vincent Schilling, "8 Myths and Atrocities About Christopher Columbus and Columbus Day," *Indian Country Today,* October 15, 2020; History.com editors, "Why Columbus Day Courts Controversy," HISTORY, October 9, 2020.
53. Forbes, *Columbus and Other Cannibals,* 22.

54. Ibid. 24.
55. Stephanie Van Hook, "Winona LaDuke on Earth-based Economics in 'the Time of the Seventh Fire'," Nonviolence Radio, August 31, 2020.
56. Jack D. Forbes, "Columbus and Other Cannibals: History and Prophecy for the 2000's," UC Davis Department of Native American Studies, July 2008.
57. Tyson Yunkaporta, *Sand Talk: How Indigenous Thinking Can Save the World* (New York: HarperOne, 2020), 28.
58. Ibid., 27–29. Additionally, Restorative Justice can be included as another path that seeks to bring about a holistic approach to wrong-doings: Dennis Sullivan and Larry Tifft, eds., *Handbook of Restorative Justice: A Global Perspective* (London: Routledge, 2018).
59. This story is adapted from the following article: Emily Arasim and Osprey Orielle Lake, "Women on the Front Lines Fighting Fracking in the Bakken Oil Shale Formations," *EcoWatch*, March 12, 2016.
60. While oil production in the Bakken has not yet recovered at the time of this writing, they unfortunately plan to expand in late 2022: Haley Chinander, "Bakken Oil Production Lagging Despite Higher Prices," Federal Reserve Bank of Minneapolis, October 18, 2021; Brandon Evans, Richard Frey, and Felix Clevenger, "Expansion Project Looks to Boost Bakken Production, Squeeze Out Canadian Exports," *S&P Global*, January 31, 2021.
61. Brandi Morin, "Pipelines, Man Camps and Murdered Indigenous Women in Canada," *Al Jazeera*, May 5, 2020; First Peoples Worldwide, "Violence from Extractive Industry 'Man Camps' Endangers Indigenous Women and Children," University of Colorado Boulder, January 29, 2020.
62. NCAI Policy Research Center, *Research Policy Update: State of the Data on Violence Against American Indian Women and Girls*, National Congress of American Indians, October 2021.
63. Siham Zniber, "Tribal and Environmental Groups Sue Trump Administration over Methane Emissions Rollbacks," *Earthjustice*, September 15, 2020.
64. Miranda Fox, "Earthjustice Applauds Methane Pollution Proposal," *Earthjustice*, November 2, 2021.
65. Isaac Stone Simonelli, Maya Leachman, and Andrew Onodera, "Gaslit: How One Native American Tribe Is Battling for Control Over Flaring," *Indianz.Com*, February 28, 2022.
66. Ursula K. Le Guin, "Video: Ursula's Acceptance Speech: Medal for Distinguished Contribution to American Letters," Ursula K. Le Guin, November 19, 2014.
67. Mark Fisher, *Capitalist Realism: Is There No Alternative?* (Winchester: Zero Books, 2010), 2.
68. Fisher, *Capitalist Realism*, 2.

69. "About Us," *Rethinking Economics,* accessed March 3, 2022.
70. Jason Hickel and Martin Kirk, "Don't Be Scared About the End of Capitalism—Be Excited to Build What Comes Next," *Fast Company,* September 11, 2017.

Chapter 11: Reciprocal Relationships with People and Land

1. Osprey Orielle Lake, "Rights of Nature and an Earth Community Economy," *Tikkun,* January 28, 2013.
2. For more information on gift economies: Genevieve Vaughan, *The Maternal Roots of the Gift Economy* (Toronto: Inanna Publications & Education Inc., 2019); Anitra Nelson and Frans Timmerman, eds., *Life Without Money: Building Fair and Sustainable Economies* (London: Pluto Press, 2011); Christopher A. Gregory, *Gifts and Commodities* (London: Academic Press, 1982); Lewis Hyde, *The Gift: Creativity and the Artist in the Modern World,* 25th Anniversary ed. (New York: Vintage Books, 2007).
3. Robin Wall Kimmerer, "The Serviceberry: An Economy of Abundance," *Emergence Magazine,* December 10, 2020.
4. Robin Wall Kimmerer, *Braiding Sweetgrass* (Minneapolis: Milkweed Editions, 2013), 180.
5. Max Fisher and Emma Bubola, "As Coronavirus Deepens Inequality, Inequality Worsens Its Spread," *The New York Times,* March 16, 2020; Joseph Stiglitz, "Conquering the Great Divide," *Finance & Development,* Fall 2020; Ian Goldin and Robert Muggah, "COVID-19 Is Increasing Multiple Kinds of Inequality. Here's What We Can Do About It," *World Economic Forum,* October 9, 2020.
6. "Inequality and the Care Economy," *Inequality.org,* accessed February 24, 2021.
7. Genevieve Vaughan, "Theory and Practice of the Gift Economy," *Gift Economy,* July 7, 2011.
8. For more information about the care economy: Riane Eisler, *The Real Wealth of Nations: Creating a Caring Economics* (Oakland, CA: Berrett-Koehler Publishers, Inc., 2007); The Care Collective et al., *The Care Manifesto: The Politics of Interdependence* (New York: Verso, 2020).
9. Yewande Omotoso and Abigail Aguilar, "Care at the Core," *Greenpeace,* March 8, 2022.
10. US Congress, House, *Recognizing the Duty of the Federal Government to Create a Green New Deal,* HR 109, 116th Congress, 1st session, introduced in House February 7, 2019.
11. *A Feminist Agenda for a Green New Deal: Principles and Values,* February 20, 2019.
12. United Frontline Table, *A People's Orientation to a Regenerative Economy,* June 2020, 12. You can read the whole document there.

13. Martín Prechtel, *Long Life, Honey in the Heart* (Berkeley: North Atlantic Books, 1999), 345.
14. Ibid., 346.
15. Ibid., 347.
16. Ibid., 349.
17. "Civic Lab Online: The Land Back Movement," Spokane County Library District, November 1, 2021.
18. "Manifesto," *LANDBACK*, accessed March 1, 2021.
19. Ibid.
20. Ruth Hopkins, "What Is the Land Back Movement? A Call for Native Sovereignty and Reclamation," *Teen Vogue*, October 12, 2021.
21. Marianne Brooker, "Scientists Support Indigenous Land Rights," *The Ecologist*, January 30, 2020; Hopkins, "What Is the Land Back Movement? A Call for Native Sovereignty and Reclamation."
22. Harmeet Kaur, "Indigenous People Across the US Want Their Land Back—and the Movement Is Gaining Momentum," *CNN*, updated November 26, 2020.
23. Pennelys Droz, "The Healing Work of Returning Stolen Lands," *Yes!*, November 15, 2021.
24. Jenna Kunze, "50 Acres of Ancestral Homeland Repatriated to the Wiyot Tribe," *Native News Online*, August 24, 2022.
25. Alex Fox, "Indigenous Peoples in British Columbia Tended 'Forest Gardens'," *Smithsonian*, April 29, 2021.
26. Jim Robbins, "How Returning Lands to Native Tribes Is Helping Protect Nature," *Yale Environment 360*, June 3, 2021.
27. Brooke Migdon, "Lumber Company Returns Waterfront Property to Native American Tribe in Washington State at No Cost," *The Hill*, December 25, 2021.
28. Brian Melley, "California Redwood Forest to be Returned to Tribal Group," *LA Times*, January 25, 2022.
29. Claire Elise Thompson, "Returning the Land," *Grist*, November 25, 2020.
30. Jullan Brave NoiseCat, "'It's About Taking Back What's Ours': Native Women Reclaim Land, Plot By Plot," *The Huffington Post*, March 22, 2018.
31. Guananí Gómez-Van Cortright, "How Indigenous People Got Some Land Back in Oakland," *Bay Nature*, December 13, 2022.
32. "Lisjan (Ohlone) History & Territory," *Sogorea Te' Land Trust*, accessed March 8, 2022.
33. Ayana Young, "Corrina Gould on Settler Responsibility and Reciprocity," *For the Wild*, November 11, 2020.
34. Ibid.
35. "Dr. Cutcha Risling Baldy – Indigenous Voices for Decolonized Futures," *Bioneers*, December 6, 2020.

36. Commission to Study and Develop Reparation Proposals for African Americans Act, H.R. 40, 117th Congress. (2021).
37. Thai Jones, "Slavery Reparations Seem Impossible. In Many Places, They're Already Happening.," *Washington Post,* January 31, 2020; Rodney Brooks, "How Reparations Would Work Today," *Quartz,* October 6, 2020.
38. Summer Sewell, "There Were Nearly a Million Black Farmers in 1920. Why Have They Disappeared?" *The Guardian,* April 29, 2019.
39. "Today in Labor History: Black Farmer-union Leader Murdered by Sheriff's Posse," *People's World,* July 15, 2013.
40. John Francis Ficara and Juan Williams, "Black Farmers in America," *NPR,* February 22, 2005.
41. Leah Penniman, "The Untold History of Black Farming," *Omega,* accessed March 9, 2022.
42. Ibid.
43. Chelsea Steinauer-Scudder, "The Seeds of Ancestors," *Emergence Magazine,* October 7, 2019.
44. "Reparations," *Soul Fire Farm,* accessed March 1, 2021.
45. Matto Mildenberger, "The Tragedy of the *Tragedy of the Commons," Scientific American,* April 23, 2019.
46. "Elinor Ostrom: Facts," *The Nobel Prize,* accessed March 9, 2022.
47. M. Cox, G. Arnold, and S. Villamayor Tomás, "A Review of Design Principles for Community-based Natural Resource Management," *Ecology and Society* 15, no. 4 (2010): 38.
48. Maude Barlow, "Water as a Commons: Only Fundamental Change Can Save Us," *The Wealth of the Commons,* accessed March 4, 2021.
49. Tamara Pearson, "The Scam That Water Billionaires Run on Poor Countries," *New Age,* March 23, 2022; Maria Cristina Rulli, Antonio Saviori, and Paolo D'Odorico, "Global Land and Water Grabbing," *PNAS* 110, no. 3 (January 2, 2013): 892–897.
50. Alexandra Shimo, "While Nestlé Extracts Millions of Litres from Their Land, Residents Have No Drinking Water," *The Guardian,* October 4, 2018.
51. Sam Leavitt, "ESG Spotlight: Water Woes Plague Beverage Titans Coke and Pepsi," *Amenity Analytics,* March 15, 2022.
52. Adam Davidson-Harden, *Local Control and Management of Our Water Commons: Stories of Rising to the Challenge* (Ottawa: The Council of Canadians, 2008), 25.
53. Ibid., 29.
54. Suhas Wasave, "A Glance of Fishermen's Cooperative Societies of Various Countries Around the Globe," Journal of Extension Systems 31, no. 1 (June 2015).

55. Jon Mathieu, "Prologue: Why Switzerland," in *Balancing the Commons in Switzerland: Institutional Transformations and Sustainable Innovations*, ed. Tobias Haller (Abingdon: Routledge, 2021).
56. David Sneath, "State Policy and Pasture Degradation in Inner Asia," *Science Magazine* 281, no. 5380 (1998): 1147–1148.
57. Carly Ryan, "Atlanta Creates the Nation's Largest Free Food Forest with Hopes of Addressing Food Insecurity," *CNN*, March 2, 2021; *Urban Food Forest at Browns Mill* (Facebook page); https://www.facebook.com/urbanfood forestat brownsmill/.
58. Ebony Joy, "Tariq El Nahl: Herbal Collective," *Collective Journal*, February 10, 2022.
59. "School of the Commons," *Urban Sustainability Exchange*, accessed March 10, 2022.
60. David Bollier, *The Commoner's Catalog for Changemaking: Tools for the Transition Ahead* (Great Barrington, Massachusetts: Schumacher Center for a New Economics, 2021).
61. Ibid., 4.
62. Ibid., 5.
63. "Okla Hina Ikhish Holo People of the Sacred Medicine Trail," *WECAN International*, accessed March 11, 2022.
64. "Women for Food Sovereignty and Food Security," *WECAN International*, accessed March 11, 2022.
65. "Growing Indigenous Food Sovereignty: Okla Hina Ikhish Holo Network in Bvlbancha & the Gulf South," Women's Earth and Climate Action Network (WECAN) International (YouTube channel), uploaded August 13, 2021.
66. Oral comments at the International Rights of Nature Global Summit, 2014.
67. Juan Francisco Salazar, "Buen Vivir: South America's Rethinking of the Future We Want," *The Conversation*, July 23, 2015.
68. Zuhal Yesilyurt Gunduz, "Another World Is Possible: The Post-COVID-19 World – Buen Vivir," *Vox Lacea*, August 19, 2020.
69. Oliver Balch, "Buen Vivir: The Social Philosophy Inspiring Movements in South America," *The Guardian*, February 4, 2013.
70. Natasha Chassagne, "A Reset for Unprecedented Times," *YES!*, November 3, 2020.
71. Natasha Chassagne, *Buen Vivir as an Alternative to Sustainable Development: Lessons from Ecuador* (Abingdon: Routledge, 2021), 83.
72. Chassagne, "A Reset for Unprecedented Times."
73. Helena Norberg-Hodge, "The Multiple Benefits of Economic Localization," *Catalysta*, September 20.2014.
74. *Transition Network*, accessed July 17, 2022, https://transitionnetwork.org/.

75. Ben Raskin, *The Community Gardening Handbook: The Guide to Organizing, Planting, and Caring for a Community Garden* (Brighton, UK: Leaping Hare Press, 2017); Monica M. White, *Freedom Farmers: Agricultural Resistance and the Black Freedom Movement* (Chapel Hill: University of North Carolina Press, 2018); Leah Penniman, *Farming While Black: Soul Fire Farm's Practical Guide to Liberation on the Land* (White River Junction, VT: Chelsea Green Publishing, 2018).
76. Robert Graves, *The White Goddess: A Historical Grammar of Poetic Myth* (New York: Farrar, Straus and Giroux, 1972), 248–249.
77. Olga Stanton, "Pysanka – a Magical Egg of Ukrainians," from *MagPie's Corner - East Slavic Rituals, Witchcraft And Culture* (Facebook page), March 10, 2022.

Chapter 12: Rights of Nature: A Systemic Solution

1. Dallas Goldtooth, "Indigenous Peoples Take Lead at D12 Day of Action in Paris – Official Response to COP21 Agreement," *The Ruckus Society,* December 14, 2015.
2. "Indigenous Women of the Americas Defenders of Mother Earth Treaty Compact 2015," *Indigenous Women Rising,* September 27, 2015.
3. Indigenous Environmental Network, "VIDEO: Historic Kayak Action Demands Indigenous Peoples Rights in Paris Climate Accord," *Global Climate Convergence,* December 2015.
4. Ibid.
5. Parts of this chapter are adapted from an article I wrote on the subject: Osprey Orielle Lake, "Rights of Nature and an Earth Community Economy," *Tikkun,* January 28, 2013.
6. Rachel Carson, *Silent Spring* (New York: Houghton Mifflin Harcourt, 1962), 99.
7. Emily Arasim and Osprey Orielle Lake, "Indigenous Women of the Americas Protecting Mother Earth: Struggles and Climate Change Solutions," *WECAN International,* June 16, 2015.
8. Ibid.
9. Atossa Soltani and Kevin Koenig, "U'wa Overcome Oxy: How a Small Ecuadorian Indigenous Group and Global Solidarity Movement Defeated an Oil Giant, and the Struggles Ahead," *Multinational Monitor* 25, no. 1–2 (January–February 2004).
10. Parts of this chapter are adapted from an article I wrote on the subject: Osprey Orielle Lake, "Declaring the Rights of Nature Protects the Future for Us All," *The Environment,* April 2022, 12–14.
11. Rights of Nature Sweden, "The Sami Parliament Endorses the Universal Declaration of the Rights of Mother Earth," *Intercontinental Cry,* June 26, 2018.

12. Christopher Stone, "Should Trees Have Standing?—Toward Legal Rights for Natural Objects," *Southern California Law Review* 45 (1972): 455.
13. Roderick Frazier Nash, *The Rights of Nature: A History of Environmental Ethics* (Madison: University of Wisconsin Press, 1989).
14. Thomas Berry, *The Great Work: Our Way into the Future* (New York: Bell Tower, 1999).
15. "Dashboard," *Global Forest Watch,* accessed July 1, 2022; "UN Report: Nature's Dangerous Decline 'Unprecedented'; Species Extinction Rates 'Accelerating,'" *United Nation: Sustainable Development,* May 6, 2019; Rebecca Lindsey and Luann Dahlman, "Climate Change: Global Temperature," Climate.gov, March 15, 2021; Stefan Lovgren, "Rivers and Lakes are the Most Degraded Ecosystems in the World. Can We Save Them?" *National Geographic,* March 1, 2021.
16. You can read the full constitution here. Articles 71–74 deal specifically with Rights of Nature: Ecuador Constitution 2008 (revised 2021), accessed December 20, 2022, https://www.constituteproject.org/constitution/Ecuador_2021?lang=en.
17. "Road Widening at Vilcabamba River and Recognition of Rights of Nature, Ecuador," *Environmental Justice Atlas,* January 30, 2018.
18. Osprey Orielle Lake, "Declaring the Rights of Nature Protects the Future for Us All," 14.
19. "In Wake of Toxic Dumping, Tamaqua Borough Passes Rights of Nature Ordinance, USA," *Environmental Justice Atlas,* March 25, 2019.
20. "Ponca Tribe of Oklahoma Makes History Declaring Rights of Rivers," *Indian Country Today,* July 18, 2022.
21. Alexandria Herr, "The Line 3 Pipeline Protests Are About Much More Than Climate Change," *Grist,* June 9, 2021.
22. Alex Brown, "Cities, Tribes Try a New Environmental Approach: Give Nature Rights," *Pew Trusts,* October 30, 2019.
23. Latoya Abulu, "Oil Pipeline on Native Lands Ramps up as Canada Honors Its Indigenous People," Stop Line 3, "September 30, 2021.
24. Osprey Orielle Lake and Katherine Quaid, "Indigenous Women Lead the Movement to Stop Line 3 Pipeline: 'This Is Everything We Have'," *Ms.,* May 24, 2021.
25. "Line 3 Pipeline Opponents File Suit on Behalf of Wild Rice," *AP News,* August 5, 2021.
26. Camilla Breen, "Why Nature Needs Legal Rights: The Manoomin Case," *Minnesota Women's Press,* July 28, 2022.
27. Alex Brown, "Cities, Tribes Try a New Environmental Approach: Give Nature Rights."

28. You can read the UDRME at this site: https://pwccc.wordpress.com/programa/.
29. Phil Lane, Jr., "The Fulfillment of the Prophecy of the Reunion of the Condor and Eagle Is Accelerating," *Four Worlds International Institute,* June 12, 2014.
30. Cormac Cullinan, "A Tribunal For Earth: Why It Matters," International Rights of Nature Tribunal, June 1, 2012.
31. "Opening Ceremony: Sarayaku, Goldtooth, 4 Directions, Camp Horinek," Global Alliance for the Rights of Nature – GARN (YouTube channel), uploaded March 18, 2017.
32. "Pablo Solon: Fundacion Solon, Global South," Global Alliance for the Rights of Nature – GARN (YouTube channel), uploaded March 18, 2017.
33. Ibid.
34. "Maude Barlow, Council of Canadians, and Michal Kravcik, Slovakia," Global Alliance for the Rights of Nature – GARN (YouTube channel), uploaded March 14, 2017.
35. Ibid.
36. Ibid.
37. Fahad Saeed, Carl-Friedrich Schleussner, and William Hare, "Why Geoengineering Is Not a Solution to the Climate Problem," *Climate Analytics,* February 12, 2018; Mark G. Lawrence et al., "Evaluating Climate Geoengineering Proposals in the Context of the Paris Agreement Temperature Goals," *Nature Communications* 9, no. 3734 (September 13, 2018).
38. You can read transcripts of Bassey's and my statements, and see the videos, at the following site: https://www.garn.org/crimes-against-nature-paris-statement/.
39. "Dr Vandana Shiva, Case Presenter: Agro-food Industry/GMOs," Global Alliance for the Rights of Nature – GARN (YouTube channel), uploaded July 19, 2016.
40. Ibid.
41. "GMO Witness: Andre Leu (IFOAM)," Global Alliance for the Rights of Nature – GARN (YouTube channel), uploaded July 19, 2016.
42. "Belen Paez: Pachamama Foundation," Global Alliance for the Rights of Nature – GARN (YouTube channel), uploaded March 11, 2017.
43. "Ruth Nyambura: African Biodiversity Network – Nigeria," Global Alliance for the Rights of Nature – GARN (YouTube channel), uploaded January 8, 2017.
44. "Judge Statement Attosa Soltani," Global Alliance for the Rights of Nature – GARN (YouTube channel), uploaded July 1, 2016.
45. "Fracking Witness: Shannon Biggs," Global Alliance for the Rights of Nature – GARN (YouTube channel), uploaded July 21, 2016.
46. Ibid.
47. "Fracking Witness: Kandi Mosset," Global Alliance for the Rights of Nature – GARN (YouTube channel), uploaded July 21, 2016.

48. *Hoodwinked in the Hothouse: Resist False Solutions to Climate Change*, 3rd ed. (2021).
49. Linda Sheehan, "Prosecutor Linda Sheehan Closing Statement," Global Alliance for the Rights of Nature, December 5, 2015.
50. The role of civil society leaders is critical to best outcomes for climate negotiations, and I have centered their voices throughout the book. Here I am referencing key women leaders who were instrumental in steering governments to secure the Paris Climate Agreement and creating its architecture: Laurence Tubiana, Christiana Figueres, Jennifer Morgan, Farhana Yamin, Rachel Kyte, and Achala Abeyansinghe.
51. Helen Moewaka Barnes and Tim McCreanor, "Colonisation, Hauora and Whenua in Aotearoa," *Journal of the Royal Society of New Zealand* 49, no. 1 (2019): 19–33.
52. The following book discusses the Treaty in detail: Malcolm Mulholland and Veronica Tawhai, *Weeping Waters: The Treaty of Waitangi and Constitutional Change* (Wellington: Huia Publishers, 2017).
53. Kennedy Warne, "A Voice for Nature," *National Geographic*, 2018.
54. Waitangi Tribunal, *The Whanganui River Report* (WAI 167, 1999), 106.
55. "Te Awa Tupua (Whanganui River Claims Settlement) Bill - Third Reading - Part 13," inthehouseNZ (YouTube channel), uploaded March 14, 2017.
56. Eleanor Ainge Roy, "New Zealand River Granted Same Legal Rights as Human Being," *The Guardian*, March 16, 2017.
57. Linda Tuhiwai Smith, *Decolonizing Methodologies: Research and Indigenous Peoples*, 2nd ed. (London: Zed Books, 2012), 4.
58. Shannon Biggs, "Rivers, Rights and Revolution: Learning from the Māori," Movement Rights, February 13, 2017.
59. Kennedy Warne, "A Voice for Nature."
60. Carolyn Merchant, *The Death of Nature* (San Francisco: Harper & Row, 1980), 3.

Chapter 13: Worldviews Conjured by Words

1. Read more about this concept here: Thomas F. Thornton, "*Tleikw Aaní*, the "Berried" Landscape: The Structure of Tlingit Edible Fruit Resources at Glacier Bay, Alaska," *Journal of Ethnobiology* 19, no. 1 (Summer 1999): 37, https://ethnobiology.org/sites/default/files/pdfs/JoE/19-1/Thornton.pdf.
2. George Steiner, *After Babel: Aspects of Language and Translation* (Oxford University Press, 1975), xiv.
3. Jeff Jetton, "The Secret of Noam: A Chomsky Interview," *Brightest Young Things*, March 9, 2011.
4. Martín Prechtel, *The Disobedience of the Daughter of the Sun: A Mayan Tale of Ecstasy, Time, and Finding One's True Form* (Berkeley: North Atlantic Books, 2005), 110.

5. Ibid., 110.
6. Ibid., 112–113.
7. Mary Isabelle Young, *Pimatisiwin: Walking in a Good Way* (Manitoba: The Prolific Group, 2005), 47–48.
8. Toshiki Osada, "Mundari," in *The Munda Languages,* ed. Gregory D.S. Anderson (Abingdon, UK: Routledge, 2008), 107, 121.
9. Many of the Munda have converted to Hinduism or Christianity over time, often syncretizing their Earth-centered practices with those religions.
10. Barry Lopez and Debra Gwartney, *Home Ground: A Guide to the American Landscape* (San Antonio: Trinity University Press, 2006), 87, 393.
11. Francesca Castagnetti, "Lost in Translation: Speaking the Language of the Land," *The Ethnobotanical Assembly,* Issue 3, Summer 2019.
12. Alison Flood, "Oxford Junior Dictionary's Replacement of 'Natural' Words with 21st-century Terms Sparks Outcry," *The Guardian,* January 13, 2015.
13. Robert Macfarlane, "The Word-hoard: Robert Macfarlane on Rewilding Our Language of Landscape," *The Guardian,* February 27, 2015.
14. Ibid.
15. Manchán Magan, "A Magical Vision is Hidden in the Irish Language – We Need to Rediscover It," *The Irish Times,* July 14, 2018.
16. Manchán Magan, *Thirty-Two Words for Field: Lost Words of the Irish Landscape* (Dublin: Gill Books, 2020), 32.
17. Ibid., 33.
18. Don Lincoln, "Quantum Foam," *Fermilab Today,* February 1, 2013.
19. "'Song of Amergin': A Cinematic Poem Short Film in Ireland Directed by Arnie Hensman (2019)," *Cinematic Poems,* August 17, 2019. There, you may view a video in which the poem is recited in the original Irish.
20. Robert Graves, *The White Goddess: A Historical Grammar of Poetic Myth* (London: Faber & Faber, 1948), 12.
21. Laura Kiniry, "What We Can Learn from the Sámi Tradition of Reading Snow," *Condé Nast Traveler,* December 2, 2020.
22. Jan Åge Riseth et al., "Sámi Traditional Ecological Knowledge as a Guide to Science: Snow, Ice and Reindeer Pasture Facing Climate Change," *Polar Record* 47, no. 3 (July 2011): 202–217.
23. David Stringer, "When Grasshopper Means Lightning," *Langscape* 5, no. 1 (January 2016): 18.
24. Black Elk, Raymond J. DeMallie, and John G. Neihardt, *The Sixth Grandfather: Black Elk's Teachings Given to John G. Neihardt* (Lincoln, Nebraska: University of Nebraska Press, 1984), 291–292.
25. Namiko Abe, "What Were the Old Names for the Months in Japanese?," *ThoughtCo,* February 28, 2020.

26. Manchán Magan, *Thirty-Two Words for Field: Lost Words of the Irish Landscape,* 2.
27. Leanne Hinton, *Flutes of Fire: Essays on California Indian Languages* (Berkeley: Heyday Books, 1994), 50.
28. Ibid., 51.
29. Ibid., 58.
30. Ibid., 59.
31. "Ogham," *Forestry Focus,* accessed December 25, 2022.
32. Damian McManus, *A Guide to Ogam* (Maynooth: An Sagart, 1991), §3.15.
33. "Wade Davis: Cultures at the Far Edge of the World," TED (YouTube channel), uploaded January 12, 2007.
34. "Indigenous Languages Hold the Key to Understanding Who We Really Are," Survival International, accessed April 25, 2022.
35. Lenzy Krehbiel-Burton, "Interior Department Releases First Report on Indian Boarding School Impact," *Tulsa World,* May 12, 2022.
36. Ibid.
37. Leanne Hinton, *Flutes of Fire: Essays on California Indian Languages,* 181.
38. Sara Kehaulani Goo, "The Hawaiian Language Nearly Died. A Radio Show Sparked Its Revival," *NPR,* June 22, 2019.
39. Amy Goodman, "As Pandemic Rips Through Indian Country, Indigenous Communities Work to Save Elders & Languages," *Democracy Now!,* January 22, 2021.
40. Ibid.
41. Amy Goodman, "As Pandemic Rips Through Indian Country, Indigenous Communities Work to Save Elders & Languages"; *Wakan Tanka* is often translated into English as "Great Spirit."
42. Robin Wall Kimmerer, "Living Beings as Our Kith and Kin: We Need a New Pronoun for Nature," *The Ecologist,* April 25, 2015.
43. Ibid.
44. Ibid.
45. Ibid.

Chapter 14: Songlines Through the Landscape

1. Not to be confused with another Mount Eliza located near Melbourne, a city that also has a Kings Park.
2. Briana Diopenes, "The AIATSIS Songlines Project: Preserving Aboriginal Cultural Songs," *Integrate Sustainability,* June 19, 2020; Hannah Devlin, "Indigenous Australians Most Ancient Civilisation on Earth, DNA Study Confirms," *The Guardian,* September 21, 2016.
3. Gary Snyder, *The Practice of the Wild: Essays* (Albany, CA: North Point Press, 1990), 82.

4. Duane W. Hamacher, "The Memory Code: How Oral Cultures Memorise So Much Information," *The Conversation,* September 26, 2016.
5. Martin Prechtel, *The Disobedience of the Daughter of the Sun,* 87.

Chapter 15: Building a Relationship with the Storied Land

1. Kelsey Leonard, "Putting Indigenous Place-Names and Languages Back on Maps," *Esri,* Winter 2021.
2. Chelsea Steinauer-Scudder, "Counter Mapping," *Emergence Magazine,* February 8, 2018.
3. Ibid.
4. Ibid.
5. Brian Oaster, "How Place Names Impact the Way We See Landscape," *High Country News,* May 1, 2022.
6. Chelsea Steinauer-Scudder, "Counter Mapping."
7. Ibid.
8. The interactive map called *Native Land* can be found at the following website: https://native-land.ca/.
9. Fritz Achilles et al., *Loreley: Signs and Legends* (Bingen: Druckwerkstätte Leindecker Gmbh, 1998), 1.
10. "The Goloring," *University of California San Diego,* December 2, 2007, Archive. December 12, 2012.
11. Josef Röder, *Der Goloring* (Düsseldorf: Bonner Jahrbuch, 1948), 97.
12. Ibid., 117.

Index

About the Author

OSPREY ORIELLE LAKE is founder and executive director of the Women's Earth and Climate Action Network (WECAN), and sits on the executive committee of the Global Alliance for the Rights of Nature, and on the steering committee for the Fossil Fuel Non-Proliferation Treaty. She writes for *The Guardian, Common Dreams, The Ecologist,* and others, and is author of *Uprisings for the Earth.* She lives in the San Francisco Bay Area on Coast Miwok lands.

Also Available by Osprey Orielle Lake

Uprisings for the Earth: Reconnecting Culture with Nature

ABOUT NEW SOCIETY PUBLISHERS

New Society Publishers is an activist, solutions-oriented publisher focused on publishing books to build a more just and sustainable future. Our books offer tips, tools, and insights from leading experts in a wide range of areas.

We're proud to hold to the highest environmental and social standards of any publisher in North America. When you buy New Society books, you are part of the solution!

- This book is printed on **100% post-consumer recycled paper,** processed chlorine-free, with low-VOC vegetable-based inks (since 2002).
- Our corporate structure is an innovative employee shareholder agreement, so we're one-third employee-owned (since 2015)
- We've created a Statement of Ethics (2021). The intent of this Statement is to act as a framework to guide our actions and facilitate feedback for continuous improvement of our work
- We're carbon-neutral (since 2006)
- We're certified as a B Corporation (since 2016)
- We're Signatories to the UN's Sustainable Development Goals (SDG) Publishers Compact (2020–2030, the Decade of Action)

At New Society Publishers, we care deeply about *what* we publish — but also about *how* we do business.

To download our full catalog, sign up for our quarterly newsletter, and learn more about New Society Publishers, please visit newsociety.com

ENVIRONMENTAL BENEFITS STATEMENT

New Society Publishers saved the following resources by printing the pages of this book on chlorine free paper made with 100% post-consumer waste.

TREES	WATER	ENERGY	SOLID WASTE	GREENHOUSE GASES
81 FULLY GROWN	6,500 GALLONS	34 MILLION BTUs	280 POUNDS	35,200 POUNDS

Environmental impact estimates were made using the Environmental Paper Network Paper Calculator 4.0. For more information visit www.papercalculator.org